Design and Organization of Computing Structures

James H. Herzog
Oregon State University

Franklin, Beedle & Associates • 8536 SW St. Helens Drive, Suite D • Wilsonville OR • 503.682.7668

Dedication

To my parents Earl (1912–1992) and Helen, who made it possible; to my students, Aaron through Yuepeng, who made it challenging; and to my wife Beverly, who made it happen.

Publisher & Editor	Jim Leisy (jimleisy@teleport.com)
Production Manager	Bill DeRouchey
Manuscript Editor	Tom Sumner
Production Editor	Jeff Tade
Cover Design	Steve Kline
Manufacturing	Malloy Lithographing, Inc.

Produced and manufactured in the United States of America.

Library of Congress Cataloging-in-Publication Data

Herzog, James H.
    Design and organization of computing structures / James H. Herzog.
        p.   cm.
    Includes index.
    ISBN 0-938661-97-3
    1. Computer architecture.   2. Electronic digital computers--Design
and construction.   I. Title.
QA76.9.A73H481995
004.2'2--dc20   95-21652
                                                                CIP

Preface

Motivation

This book addresses the dual challenge of presenting the core concepts of digital computer architecture along with the design strategies for implementing them. These complementary bodies of knowledge comprise a *science* because all the concepts and methods are backed by solid mathematical and engineering design principles. They are an *art* because creativity and inspiration produce "a ha!" experiences in the practitioners who make major contributions to real-world applications. Thomas Edison attributed the success of his work to "one percent inspiration and ninety-nine percent perspiration." Inspiration cannot be learned from a book, but by using good design strategies, the perspiration portion can be more productive.

My goal in writing this book is to bridge the conceptual gap between classical bit-level logic systems and more complex digital systems such as computers. A solid platform for future creative activity is formed by understanding the organizational structure of a digital computer and having a design methodology for implementing the structure. This book also prepares students for more quantitative design approaches encompassing real-world economic issues, design optimization, and performance analysis.

In the 1980s we learned that marvelous new devices could be created using relatively modest microprocessors. In the 1990s we are discovering that more customized hardware platforms may be required to support new and innovative processing algorithms. These systems are more complex than those typically designed by truth table and state diagram descriptions. They are the ubiquitous digital devices that provide the intelligence and control functions for consumer products, data acquisition, flexible manufacturing systems, medical electronics, and aerospace systems. In their simplest form they are the multitude of devices that move, store, and modify digital information for useful purposes; in their advanced form they are new hardware structures for implementing digital computers.

The impetus of this new thinking is the emergence of sophisticated CAD (computer-aided design) tools and the automation of semiconductor fabrication facilities that can produce very-large-scale integration (VLSI) devices at low cost. The modern engineer and computer scientist must know the appropriate design techniques and principles to use these tools intelligently and effectively. A major motivation in creating this book is my belief that engineers and computer scientists require a methodology, other than ad hoc techniques, to transform algorithmic descriptions of devices and processes into reality.

Approach

This book stresses a structured approach for designing computing devices. Processing algorithms are first converted to sequential processing operations, called microoperations. System design consists of implementing a structure of information-processing components, called the datapath, for performing the microoperations. A second structure, the controller, generates control signals to guide the datapath to perform the microoperations in the proper sequence. This approach is initially used to implement several uncomplicated data-processing algorithms. When the process is extended to include a program memory, digital computer structures are created.

The computer structures in the early portion of this book are for educational insight rather than high performance. The instruction set selection and the process of instruction execution is much more easily grasped in a relatively simple environment. The interrelationship of the architecture and design methodology is also less complex in such systems. Later chapters expand on variations possible in the processor unit, the memory organization, input/output, and the control unit. Modifications in instruction processing, such as RISC, are also examined.

Several actual computer architectures are introduced in the text. In addition to complementing the earlier discussion, this material can introduce the instruction set selection available with modern computers. Programming of such computers, using assembly language, is encouraged to form a tighter bond between hardware and software of the devices.

Course Level

This book is targeted at sophomore or junior undergraduate students in computer engineering, electrical engineering, and computer science who have completed a course in logic design. Appendix A can provide a starting point for students without this background. The text supports a one-semester course with a title such as "Computer Structure and Organization." The goal of such a course is to introduce the student to the basic structure of digital computers, their operation, instruction sets, and design. Such a course might also include some programming exercises in assembly language on a personal computer or single board microprocessor. Introductory material and instruction sets for four typical platforms—the Intel 8086, Motorola 6800, the Intel 8051 microcontroller, and the AMD 29000 RISC processor—are included. The material is also suitable for self-study and as refresher material for practicing engineers.

At Oregon State University, this course is taken by all students in our Electrical Engineering, Computer Engineering, and Computer Science programs. We regard the material as an essential prerequisite for our courses in VLSI design, microprocessors, and digital communication and networks. This course is also used to prepare focussed students for a second-level architecture course based on more quantitative approaches. A typical text for such a second-level course would be *Computer Architecture: A Quantitative Approach*, by John Hennessy and David Patterson.

Organization of the Book

Chapter 1, "Components and Structures," reviews components required for implementing register, memory, processing, interconnection, and control activities. It presents the basic premise that the design of digital systems should be based on a functional decomposition of control and information-processing activities.

Chapter 2, "The Processing Algorithm," introduces the microoperation as a processing primitive for digital structures. Sequences of microoperations are expressed in a register transfer language. The microoperation diagram is introduced as an alternative way for expressing the sequential algorithms associated with digital processing.

Design approaches for processing digital information are considered in Chapter 3, "Hardware Structures for Data Processing." Pipelined, ripple (cascade), and parallel processors are compared, evaluated, and contrasted. Iterative algorithms are presented. Bit-slice processors are examined as an approach to modular processing devices.

The structure and design of the control unit, both hardwired and microprogrammed, is covered in Chapter 4, "The Controller." These devices, when combined with the processor structures of Chapter 3, create working computing devices.

Data representation and arithmetic, as it applies to computing devices, is considered in Chapter 5, "Data Representation and Arithmetic." Addition, subtraction, multiplication, and division are covered. Techniques for speeding up arithmetic, such as the use of array devices, are presented.

In Chapter 6, "The Digital Computer," the treatment of digital systems is expanded to include programmable systems and digital computers. BART (Boolean Algorithmic Register Trainer) is used to study the instruction set and implementation of a simple instructional minicomputer.

RISC processors and their associated pipelined instruction processing are considered in Chapter 7, "Instruction Set Processing." This modern approach to computer design offers an alternative structure to the dominant complex instruction set computers (CISC) of the 1980s.

Chapter 8, "System Software and Representative Computer Architectures," is concerned with the system software for a digital computer and the operation of an assembler. This chapter expands on the methods of implementing instruction sets by examining the structure, instruction set, and instruction formats of four representative computers. With additional material, programming exercises on one or more of the machines can be performed.

Memory devices and memory structures such as cache memory and virtual memory are studied in Chapter 9, "Memory Structure and Management." Several techniques are presented for managing systems containing a hierarchy of memory elements.

Chapter 10, "System Input/Output and Interfacing," covers techniques for exchanging information between asynchronous and semiautonomous devices, including handshake signals. These techniques are applied to the input/output system of a digital computer. Interrupt structures and direct memory access are also considered. This chapter concludes with material on standardized buses such as Multibus and the IEEE 488 instrumentation bus.

Modern high-performance parallel computing structures are introduced in Chapter 11, "Parallel and High-Performance Computing." This chapter also includes material on pipelined, dataflow, and superscalar architectures.

Chapter 12, "Digital Communication and Networking," provides introductory material on digital communications and computer network structures. It provides an overview of data encoding, error control, and communication protocols. The chapter concludes with a discussion of token ring, token bus, and Ethernet, the most common structures for local area networks. The increased interrelationship of computing and communication structures suggests that students should be exposed to material such as this early in their curriculum.

Appendix A, "Foundations in Digital Logic," provides a starting point for students without previous experience in logic design. For better prepared students, it can be used as review material prior to beginning the main material for the course.

Appendix B, "Design For Test," introduces the major concepts related to the integration of testing techniques into the design process. The relatively new IEEE 1149.1 testing standards represent an important advance in this area.

The core material is in Chapters 1 through 10. This material, and selected portions of the remaining chapters, is appropriate for a one-semester, three-hour course. If time constraints are severe (for example a quarter system), the material will probably need to be shortened. Portions of Chapters 7, 8, and 10 may be eliminated without destroying the major thrust of the text.

Acknowledgments

I would like to thank the following manuscript reviewers whose valuable contributions helped to refine the presentation of this material:

Haim Barad	*Intel Corporation*
Jerry Daniels	*Brown University*
Thomas Lang	*University of California, Irvine*
Wayne Lu	*University of Portland*

Internet and World Wide Web (WWW) Support

I have established a Web page as a supplement to this text and as a service to a wide range of professionals involved in the design, fabrication, and application of computers. Web services are available through most university computing centers and several commercial services and provide a convenient, interactive method to search and access text and images for a wide range of topics. The Web page associated with this text can be reached at:

```
http://www.ece.orst.edu/~herzog/docs.html
```

This page provides a variety of services. It initiates the exploration of a wide variety of topics related to the design and use of digital computers. Current topics include information on innovative computer architectures, new components, design projects at various universities, and sources of historical information and images. I sincerely welcome your comments and suggestions. I will try to respond quickly to questions and concerns. A section of the page will be devoted to corrections, clarifications, and reader feedback on portions of the book.

I intend to revise this page on a regular basis in response to new material, suggestions, and information provided by browsers. In this way, the page will evolve to provide timely and interesting information to the computer community.

I may also be reached by email at:

```
herzog@ece.orst.edu
```

Contents

1
Components and Structures

The Abacus

We humans invented the Abacus
So leaders could easily keep track of us
It's worked for three thousand years
Without transistors or gears
And that's why today there's no lack of us

Prologue

The abacus, a mechanical computational device developed in China over 3000 years ago, is representative of the earliest attempts to use digital techniques for computation. The abacus and its operator combine data, processing hardware, and a processing algorithm to produce a functioning computational system.

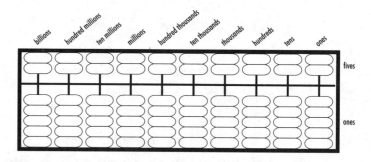

Figure 1.1   The Chinese Abacus

As shown in Figure 1.1, the abacus consists of a mechanical frame with multiple columns of sliding beads. Each column is separated into two portions by a crossbar. Each bead below the crossbar has a value of one; each above the crossbar has a value of five. The first column is the ones column, the second is the tens column, and so on. The numeric value depends on the position of the beads on the mechanical frame. Data (beads) are manipulated according to a processing algorithm to perform addition, subtraction, multiplication, and division.

Addition is performed by shifting beads toward the crossbar for each decimal position. Accumulations of 10 in a column are cleared and carried to the next column. A skilled operator can perform addition, subtraction, multiplication, division, square roots, and even cube roots. The abacus, despite its roots in antiquity, is still used in many retail stores in Asia.

1.1   Introduction

Data processing systems involve three interrelated components: **data** to be processed, **hardware** to manipulate the data, and a **processing algorithm** to control and sequence the actions. In the past fifty years, processing hardware has dramatically advanced from mechanical devices through primitive electronic components to today's sophisticated use of high-density integrated logic circuitry.

Integrating data, hardware, and algorithms effectively and efficiently to meet performance objectives is known as **design** of computing structures. Design is a highly creative activity. It is the process of developing a set of specifications and then creating an actual device to satisfy these specifications. Design starts with an abstract idea and a set of criteria and ends with a working device.

Good design creates a smoothly functioning system from a disordered assortment of components. Good designs are well documented, allowing for modifications. A good design can be tested for performance and is always influenced by equipment limitations, economic considerations, and time constraints.

Design objectives have changed dramatically in the past several decades. Previous generations of designers created microprocessors, calculators, and high-performance digital computers. The next generation will create new processing devices for music synthesis, speech recognition, computer vision, distributed computation, digital control and monitoring systems, artificial intelligence, character recognition, robotics, parallel computation, virtual reality, computer games, medical diagnosis, and digital communication networks.

Effective design of digital structures requires the designer to work with data, hardware, and a processing algorithm. There is a common misconception that creating complex digital systems involves tremendous intellectual insight. Good designs don't just happen: they are the result of carefully planned activity. The goal of this book is to provide a pathway for the one percent of inspired ideas to be smoothly transformed into reality.

In this book you will study techniques for designing computing structures. You will start with hardware components such as registers, arithmetic units, and memory arrays. Such components are readily available as components and are included in component libraries associated with the design of very large scale integrated circuit (VLSI) components. The systems of interest are more complex than those designed using the truth table and state diagram approaches developed in logic design. They require greater hardware flexibility and higher performance than microprocessors allow. These systems range from unique computer architectures to the ubiquitous clever digital devices needed to implement medical instrumentation, signal processing, robots, intelligent appliances, and gadgets to make our lives more interesting.

This chapter introduces the hardware components used in computing structures. These components are more complex than gate-level devices and less complex than microprocessors. Such components provide the ability to store, move, and process multiple bits of related information—the word. Additional material related to logic devices and logic design can be found in Appendix A. The chapter concludes with a discussion of the advantages of decomposing a complex system into a **datapath** and a **controller**. The datapath incorporates the data processing actions; the controller provides a sequence of control signals enabling a processing algorithm to be implemented.

1.2   Data and Control Signals

Computing structures must deal with two fundamentally different types of information, data and **control signals**. Data words are fixed-width information units that represent numeric and text information. They are generated by computation or produced by measurement devices and external sources. System components modify, move, and store them as a unit.

Control signals are variable width information units asserted to produce actions at system component **control points**. For example, a **control signal** can cause a register to load a data word. The word "asserted" is used to avoid the confusion associated with such implementation details as positive and negative logic. When control signals are asserted, their associated action is performed by a processing element. The sequential interplay of data words and control signals produces a device capable of performing a processing algorithm on data.

While control signals can originate from many sources, a useful design strategy concentrates their generation and organization in a separate structure called a controller. Figure 1.2 shows a structure in which the controller provides control signals to registers and an arithmetic/logic unit (ALU). The control signals specify the device actions and sequence the computation algorithm. This concept will be further developed later in this chapter.

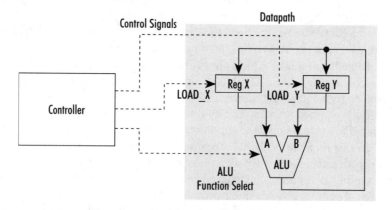

Figure 1.2   Assertion of Control Signals at Control Points

1.3   Logic Gates

This and the following sections examine some characteristics of devices that are useful building blocks for the design of digital systems. Most of the emphasis will be on using higher order components, usually fabricated as MSI (Medium Scale Integration) logic components, or available in parts libraries associated with VLSI (Very Large Scale Integration) design environments.

Most digital designs make limited use of standard logic gates, such as AND, OR, NOT, NAND, and NOR. These gates are sometimes called "glue" because they interconnect other devices into a working system. In modern design techniques, interconnecting logic is often provided by **programmable logic devices** (PLDs) such as programmable logic arrays (PLAs), array logic, and read-only memories (ROMs). If necessary, Appendix A may be used as a review reference for logic components and logic design.

Tri-state buffers use a control input in addition to the data input. The device output is either the input logic level (control input asserted) or a high impedance (control input not asserted). These devices are useful when multiple logic sources must be connected to a single connection point. For example, several data sources may share a data bus. Only one data source is placed in an active mode; the remaining units are in tri-state. If two devices actively attempt to simultaneously drive a shared logic bus, correct performance is unlikely.

1.4 Arithmetic and Logic Devices

From the perspective of the designer, components can be classified into several broad classes of devices. Arithmetic and logic devices, as shown in Figure 1.3, accept n-bit data words from one or more sources. The words are combined to produce an m-bit output data word. These devices have no storage capability. Examples include an adder to add two binary words; a complement device to perform a bit-by-bit complement on a single input word; and standard logic circuits involving AND, OR, and NOT gates. Some devices, such as an arithmetic and logic unit (ALU) are often given a distinctive shaped symbol. ALU devices may have a control input to specify the data conversion operation to be performed.

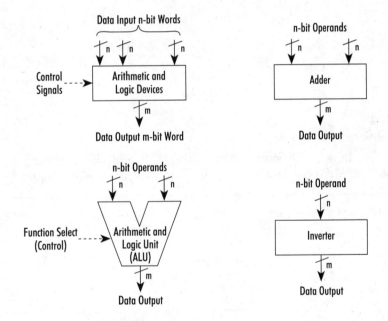

Figure 1.3   Arithmetic and Logic Devices

The following table shows the set of logic operations and arithmetic operations performed by a 74LS382 4-bit ALU. The three control inputs $S_2 S_1 S_0$ can specify eight different operations on the two input words: three arithmetic functions, three logic functions, and constants 0000 and 1111.

Function Select			Function
S_2	S_1	S_0	
0	0	0	$F = 0000$
0	0	1	$F = B + \overline{A} + C_{in}$ (subtract)
0	1	0	$F = A + \overline{B} + C_{in}$ (subtract)
0	1	1	$F = A + B + C_{in}$ (add)
1	0	0	$F = A \text{ XOR } B$
1	0	1	$F = A \text{ OR } B$
1	1	0	$F = A \text{ AND } B$
1	1	1	$F = 1111$

Table 1.1

In addition to the operand inputs, the ALU accepts an input carry, C_{in}, and generates a carry out, C_{out}. ALUs may be cascaded to accommodate longer word lengths.

1.5   Decision Elements

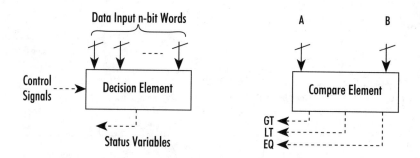

Figure 1.4   Decision Elements

Decision elements, as shown in Figure 1.4, accept *n*-bit data words as inputs and produce one or more binary decision bits as status outputs. A compare device, for example, compares two binary inputs, A and B, to determine whether they are have a greater-than, less-than, or equal relationship. Status signals generated by decision elements provide information needed to modify the sequencing of control signals.

1.6   Data Routing Elements

Data routing elements, shown in Figure 1.5, route data to and from multiple locations. A multiplexer, for example, receives multiple input bits. One input bit is selected and routed to the multiplexer output. A demultiplexer routes a single data input bit to a selecting output pin. A control signal specifies actions of both devices. When data words, rather than bits, are routed, multiple devices are used—one for each bit in a data word. When clear by context a diagram such as shown in Figure 1.5B indicates the routing of complete data words.

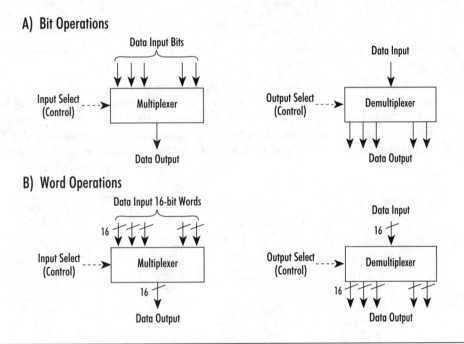

Figure 1.5   Data Routing Components

A **crossbar switch**, as shown in Figure 1.6, provides a wide variety of simultaneous data paths between system components. The crossbar switch contains a set of input rows and output columns, with each element of a row or column providing an interconnection link for one data bit. Each place where a row and column intersect is a possible connection point for the row and column links. The switch mechanism can be replicated to provide switching capability for n-bit data words.

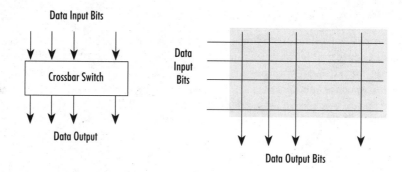

Figure 1.6   The Crossbar Switch

The internal connection of the crossbar switch is controlled by an AND gate which detects when a row and column are simultaneously selected for connection. A tri-state buffer provides the actual connection. Figure 1.7 shows a connection structure for two inputs and two outputs. Crossbar switches allow any input data link to be connected to any or all output data links. No more than one input link can be simultaneously connected to the same output data link.

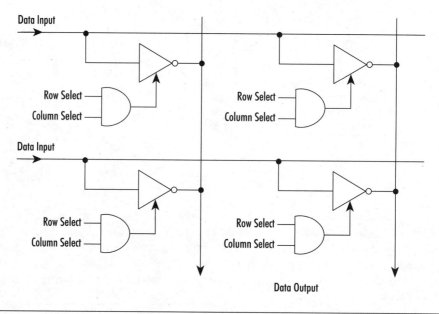

Figure 1.7   Internal Connection Pattern for a Crossbar Switch

1.7   Data Registers

Data storage devices are capable of storing one or more data words. A register is an array of clocked, synchronous flip-flops (shown in Figure 1.8A) capable of accepting and storing multiple bits of information. The grouping of flip-flops is given a single symbol as in Figure 1.8B. Since all registers include a clock input, its notation is usually omitted.

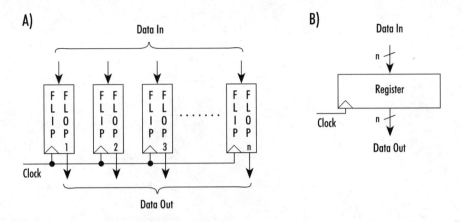

Figure 1.8   Data Storage Devices

In Figure 1.9 a multiplexer has been combined with a register to add capabilities. When control signal LOAD is asserted the multiplexer provides new data to be stored in the register. When LOAD is not asserted, the clock causes the current register data to be reloaded into the register.

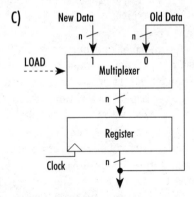

Figure 1.9   A Register with Load Control

1.8   Register-to-Register Data Transfer

Registers are the primary storage components in computing structures. They can be referenced by either a single letter designation, such as A or B, or by a suggestive mnemonic such as NEW_DATA or SUM. The width of the data word may be shown with a "/" on the data lines, or omitted if it is clear that all operations are on words. In Figure 1.10 registers NEW_DATA and SUM provide words of unspecified width to the adder. Register SUM accepts the results of the addition process. In register SUM, the load feature has been incorporated in the register. Clock inputs are not shown.

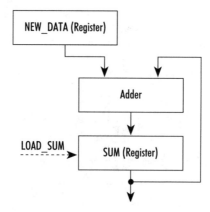

Figure 1.10   Use of a Data Register in Computing Structures

To express the processing action, a register transfer statement is used. For the action indicated:

SUM←SUM + NEW_DATA

This statement describes a synchronous data transfer. On the next active transition of the system clock, register SUM receives the sum of the current contents of register SUM and the value in register NEW_DATA.

The ability to transfer information from one register to another is a key element in the design of digital systems. Such data transfers must be synchronized with the system clock. To reliably implement a register transfer operation, several conditions must be met:

- **Data Register.** Register transfer operations require clocked data registers. All transfer operations are fully synchronized with the active transition of the system clock.
- **Control Signals.** Control signals specifying the register operation must be present and stable an adequate length of time in advance of the active clock transition.
- **Data.** Data to be stored in registers must be present and stable at the input of the register for an adequate length of time, known as the **set-up time**, before the active transition of the clock. The presence of arithmetic and logic devices introduces propagation delays that must be considered in the design.
- **Clock.** The system clock must be chosen at a rate to accommodate all processing delays, propagation delays, and register set-up times. A clock rate which is too high will result in unreliable data transfers. It is also important that all clocked elements in the design receive their clock signal simultaneously.

Timing of Data Transfers

The system clock initiates data transfers between registers. The timing of events is very important. All data processing actions are performed in the time frame of a **clock cycle**. The sequence of operations for a register-to-register data transfer is shown in Figure 1.11.

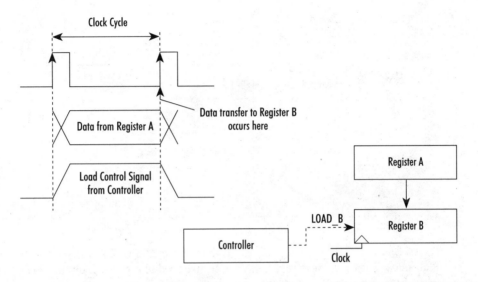

Figure 1.11   Sequence of Data Transfer Operations within a Clock Cycle

Every clocked device responds to the active transition of the system clock. This is usually, but not necessarily, the low-to-high clock transition. In Figure 1.11, a "↑" represents the active transition. A clock cycle is the period between the active transition of two clock pulses. The cycle begins *after* the active transition of a clock pulse of the previous cycle. This transition updates the contents of all registers. The registers retain their contents, unchanged, until the clock cycle ends with the next active transition of the system clock.

During the clock cycle, data is routed from register A, the source register, to register B, the destination register. The LOAD_B command from the controller is asserted at register B. At the end of the clock cycle, the active transition of the system clock causes the new value to replace the old value in register B. It is important to note that the contents of register B do not actually change until the active transition of the system clock at the end of the clock cycle.

Example 1.1:   *Transfer of Information between Registers*

Figure 1.12 shows the timing of an idealized data transfer from register A to register B. Data from register A must be stable at the input of register B for a minimum period of time, the set-up time, before the clock pulse. Register B must be instructed, by means of the LOAD_B control signal, that the following clock pulse is to perform a load operation. The clock pulse causes the data from register A to be accepted and stored in register B. All events associated with a register transfer event are initiated directly or indirectly by the system clock. A register transfer event requires one clock cycle.

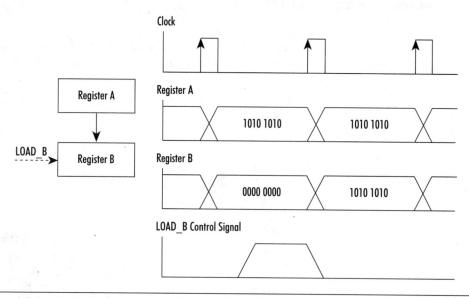

Figure 1.12   Data Exchange in a Register Transfer Operation

In Example 1.1, register A initially has the contents 10101010_2; register B has contents 00000000_2. The contents of register A are transferred to register B by the next clock pulse. The old contents of register B, 00000000_2, are destroyed. Since register A has no control signal, it is assumed to retain its current contents of 10101010_2 after the clock pulse.

Figure 1.13 illustrates another data transfer situation; register A and register B simultaneously exchange their data contents.

A←B,
B←A;

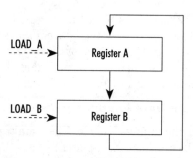

Figure 1.13   Data Exchange between Two Registers

Since this looks a bit like a juggling act, there is some natural concern about the reliability of such a transfer. Indeed, there are potential problems if the data transfer is not performed by clocked synchronous registers. In the first, data from register A might be accepted by register B and then immediately transferred back to register A. This process could continue and cause the contents of registers A and B to oscillate. The final value in the registers would be unpredictable. This could be a problem if a latch is used as a storage element, but it is not possible if clocked synchronous registers are used. These devices allow only one transition of register contents per clock cycle. The second problem, known as **clock skew**, can occur if different registers receive their clock signals at slightly different times. This might happen, for example, if the clock signal is delayed by passage through several logic gates. Clock skew delays the action of some components and may cause some registers to accept data before others, which in turn can cause undesired actions. Additional information about clock skew can be found in Appendix A.

1.9   Control of Register Operations

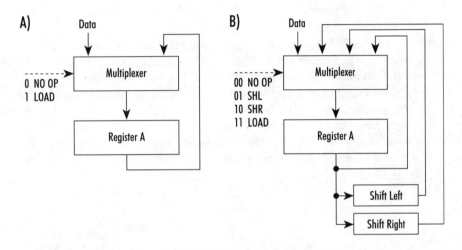

Figure 1.14   Using a Multiplexer to Implement Register Operations

Figure 1.14A shows a register which can either load input data or do nothing. A multiplexer selects either the external input data when LOAD is asserted or the current register contents when LOAD is not asserted. This latter event is a commonly referred to as "No Operation" or NOP. In every clock cycle the register continuously updates its contents with the selected data source.

Control Signal	Register Transfer	Description
LOAD = 0	A←A;	Current data reloaded.
LOAD = 1	A←DATA;	New data loaded.

Table 1.2

Additional features, such as shift right (SHR) and shift left (SHL), are included in the register shown in Figure 1.14B. The two encoded control inputs select one of the possible data sources for the register. Table 1.3 shows the possible operations.

Control Signal	Register Transfer	Comment
00	A←A;	Current data reloaded (NO OP).
01	A←SHL(DATA);	Data shifted left before loading.
10	A←SHR(DATA);	Data shifted right before loading.
11	A←DATA;	New data loaded.

Table 1.3

The 74LS323 8-bit universal shift/storage register has similar characteristics to those shown in Figure 1.14B. Available in a 20-pin integrated circuit, it has provisions for 8-bit storage, output control, load control, synchronous clear, shift left, and shift right.

When a single access port is used for both input and output, a combination of a register and a tri-state buffer is used. This is shown in Figure 1.15A. The device requires two control signals—one for a load operation and one to control the tri-state buffer. Table 1.4 enumerates the possible data transfer operations. This type of register works well on a shared data bus. In Figure 1.15B, the tri-state buffer has been incorporated within the symbol for the register.

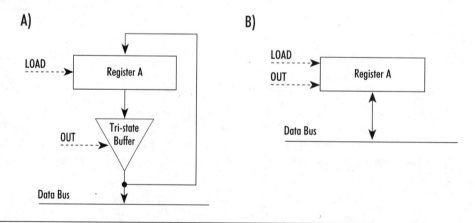

Figure 1.15   A Register with a Single Input/Output Port

Control Signals	Register Transfers	Description
LOAD = 0 (not asserted) OUT = 0 (not asserted)	A←A;	Reload current value.
LOAD = 1 OUT = 0	A←DATA;	Load external data.
LOAD = 0 OUT = 1	A←A, BUS←A;	Register contents placed on the BUS.
LOAD = 1 OUT = 1	A←A, BUS←A;	Register contents placed on the BUS. Load value from BUS (which is A).

Table 1.4

Example 1.2:   Analysis of a Small Computation Device

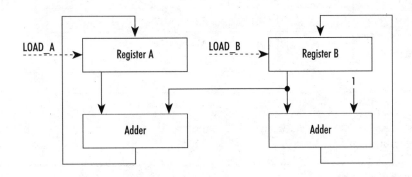

Figure 1.16   Datapath for Example System

A hardware description for a small computing structure is shown in Figure 1.16. All registers, arithmetic units, and decision elements are 8 bits. Initially, register A and register B both have a value of 00000001. The system is provided with the following sequence of control signals in five successive clock cycles.

Clock Cycle	Control Signal
t_1	LOAD_A
t_2	LOAD_B
t_3	LOAD_A, LOAD_B
t_4	LOAD_A
t_5	LOAD_A, LOAD_B

Table 1.5

Determine the contents of register A and register B for each clock cycle.

Solution:

The following table summarizes the activity in each clock cycle. Each clock cycle begins with the assertion of a control signal. The clock cycle ends with the active transition of the clock, which changes the contents of the registers. All register contents are expressed in decimal.

Clock Cycle	Control Signal	Register Transfers	Register Contents (Before Clock)		Register Contents (After Clock)	
			A	B	A	B
t_1	LOAD_A	A←A + B	1	1	2	1
t_2	LOAD_B	B←B + 1	2	1	2	2
t_3	LOAD_A, LOAD_B	A←A + B, B←B + 1	2	2	4	3
t_4	LOAD_A	A←A + B	4	3	7	3
t_5	LOAD_A, LOAD_B	A←A + B, B←B + 1	7	3	10	4

Table 1.6

Note that in clock cycles t_3 and t_5 the register transfer operations are performed simultaneously.

1.10   Register File

A modest size grouping of registers may be included in a **register file**, as shown in Figure 1.17. ABUS or BBUS provides access to data from the registers. Two addresses, A_ADDRESS and B_ADDRESS, select the register contents to be placed on ABUS and BBUS. A third address, C_ADDRESS, selects the destination register for data provided on CBUS. In Figure 1.17, a register file has been combined with an arithmetic and logic unit (ALU) to provide a structure capable of executing computational algorithms.

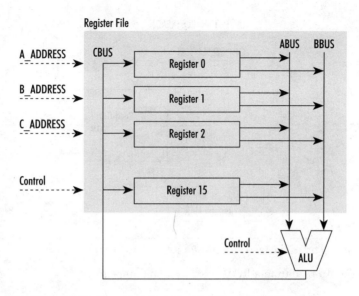

Figure 1.17   A Register File

1.11   RAM and ROM Memory Devices

Random access memory (RAM) and read-only memory (ROM) both access information based on an address. An address of n bits is capable of selecting any one of the 2^n locations in the memory.

System memory devices are an important part of a digital computing structure. RAM stores data, tables of values, and system programs. ROM is used for tables of data constants, as a replacement for logic devices, and as an important component in microprogrammed controllers. It is important to have efficient structures for storing and acquiring information. This section considers several different types of memory and the control structures needed to make them useful.

Read-only Memory (ROM)

ROM is usually implemented using a memory address register (MAR) to provide an n-bit address and a memory data register (MDR) to store retrieved information, as shown in Figure 1.18. Control lines include a CHIP SELECT (CS) and READ. In some ROM memory, READ is replaced by an equivalent signal, OUTPUT ENABLE (OE).

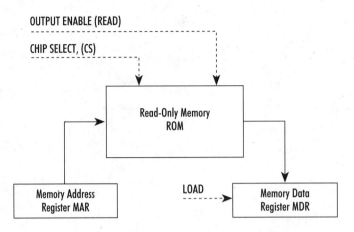

Figure 1.18   Register Organization Used for Memory Access

A data read from a ROM memory would appear as:

Register Transfer (Memory Read)	Comment
MDR←ROM(MAR);	MDR receives contents of ROM with location specified by the MAR.

Table 1.7

All output functions of a ROM are inactive unless the CHIP SELECT (CS) has been asserted. In some circuit implementations, the ROM remains in a low power consumption mode when CS is not asserted. The READ signal controls the tri-state output of the ROM. During a memory read operation, both CHIP SELECT and OUTPUT ENABLE must be asserted to cause data to be placed on the ROM output lines. Because of their tri-state outputs, ROMs may be placed directly on system buses.

In Figure 1.19, several ROMs, each with n address bits, are used to cover a larger address space requiring p address bits. The p address bits from the address register can specify 2^p memory locations. Each of the ROMs can provide 2^n locations. Therefore:

$$\text{number of ROMs required} = m = \frac{2^p}{2^n} = 2^{p-n}$$

The memory address register uses p data bits to specify an address. The $p-n$ higher order bits of the address are decoded to select the ROM containing the required information. The remaining n bits are directed to the address lines of each ROM. Control signal READ is sent to all memory devices. Only the selected chip responds by placing data on the bus for storage in the memory data register. Non-selected chips provide a high impedance through tri-state gates.

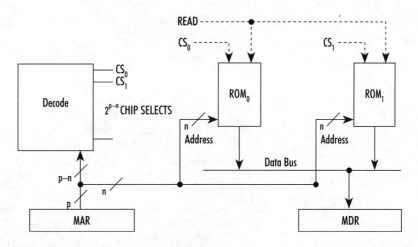

Figure 1.19   Use of Multiple Memory Devices

Random Access Memory (RAM)

For storage of large quantities of information, RAM is used as a replacement for data registers. Logically, a RAM is a collection of registers for which access is limited by an output multiplexer and input demultiplexer. Information is accessed or stored by specifying a RAM address. Historically, many different physical phenomena, ranging from delay lines to magnetic cores, have been used for digital memory. In most current systems, electronic memories dominate other memory forms.

Two forms of RAM—static random access memory (SRAM) and dynamic random access memory (DRAM)—are in widespread use. SRAM uses a flip-flop as a data storage mechanism and has a small set of control signals. DRAM uses the charge storage of a capacitor for memory. Since a charged capacitor will discharge in a relatively short time, a DRAM requires the added complexity associated with memory refreshing. Unless otherwise noted, all RAM references will refer to SRAM.

Figure 1.20 shows an example system implementation of RAM storage. The memory address register and CHIP SELECT circuitry function the same as they do with a ROM. The timing for the read cycle for a RAM (or ROM) is shown in Figure 1.21. When the ADDRESS, CHIP SELECT, and the OUTPUT ENABLE lines have been asserted, the device provides data to its output terminals after a delay called the access time. Data from the RAM must be valid for a period greater than the set-up time for the memory data register. In most actual RAM devices, asserted signals correspond to a low voltage.

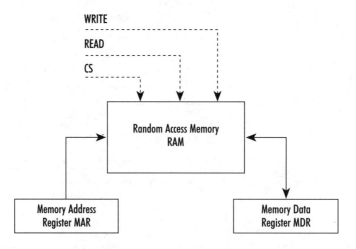

Figure 1.20   Organization of a RAM Memory for Data Storage and Access

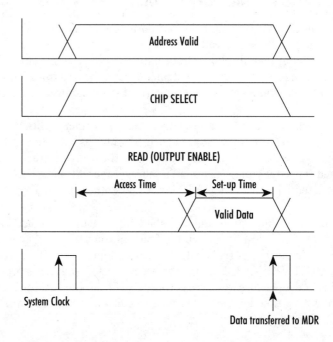

Figure 1.21   Timing for a RAM (or ROM) Memory Read

If the combined access time and set-up time exceeds one clock period, provisions must be made to extend the read cycle for additional clock cycles to allow the memory data register to reliably capture the data.

Most solid state memory devices are fabricated for applications with computers and microprocessors. The input and output data lines are shared using a common access pin. This makes them well suited for bus systems. Tri-state buffers disconnect memory devices from the bus by providing a high impedance on their data lines when not selected.

A write, which transfers data from the memory data register to a location specified by the memory address register, uses the WRITE control signal. In many implementations, the device is in a read mode when WRITE is not asserted. A data write would appear as follows:

Register Transfer (Memory Write)	Comment
RAM(MAR)←MDR;	Location MAR of the RAM memory receives the contents of the MDR.

Table 1.8

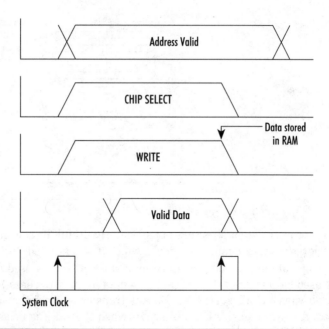

Figure 1.22   Timing for a RAM Memory Write

The timing diagram for a single-cycle write operation is shown in Figure 1.22. During the clock cycle, address bits and data bits are provided to the RAM along with control signals. The exact timing sequence of address, data, and control signals is unimportant as long as all have their correct value prior to the end of the clock cycle. There is no clock input to a RAM. Instead, the trailing edge of the WRITE signal triggers the storage of information in the memory cell. It is important that the duration of the WRITE signal be long enough to allow an adequate set-up time for the RAM to capture the data. In some cases the combination of long set-up time for the RAM and delays in providing stable data may require more than one clock period to complete a write operation. Since the trailing edge of WRITE indicates the end of the write operation, it is important that the WRITE signal not have any false transitions after the CHIP SELECT has been asserted. Such transitions might cause unwanted data to be erroneously copied into a memory cell.

1.12   Last-In-First-Out Data Storage—The Stack

RAM stores and retrieves information based on an address. Some applications require a device that store and retrieve information based on the sequential order in which the data is generated. Possibilities include last-in-first-out (LIFO) storage, a **stack**, and first-in-first-out (FIFO) storage, a queue.

Stack organization is similar to a stack of plates with a spring-loaded mechanism. Additional plates may be added only to the top of the pile. Only the last plate placed on the pile may be removed. This uncovers another plate, which is then on top of the pile.

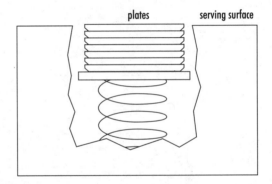

Figure 1.23   Stack Storage of Plates

A stack information structure is particularly useful for calculators and computers. It stores and retrieves information in the order in which it is generated.

Two operations may be performed on a stack. PUSH is used for the storage of information; POP is used for its retrieval. In Figure 1.24 data in the stack data register (SDR) is transferred to the stack when PUSH is asserted. Data is transferred from the stack to the SDR when POP is asserted. A status signal, FULL, indicates when the stack is incapable of performing additional pushes. EMPTY indicates the stack is incapable of performing POPs.

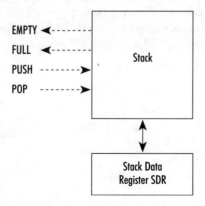

Figure 1.24   A Stack Data Storage Structure

A Stack Constructed from Registers

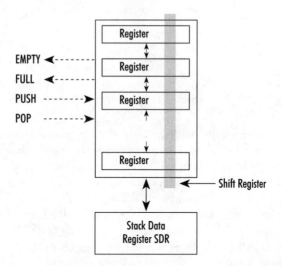

Figure 1.25   A Stack Implemented with Bidirectional Shift Registers

A register stack may be constructed from a group of bidirectional shift registers, as shown in Figure 1.25. Each shift register holds one bit from each of the registers in the stack. The PUSH command is equivalent to a SHIFT_UP of the contents in the shift register; the lowest elements in the shift registers receive their input from the stack data register. Overflow information from the uppermost location is lost.

The POP command is equivalent to a SHIFT_DOWN operation of the contents of the shift registers. During POP, garbage data is inserted in the highest location.

The FULL and EMPTY conditions can be detected with an up/down counter. After initializing to a value of 0, the counter increments with every PUSH and decrements with every POP. If the value exceeds a threshold determined by the storage capacity, logic provides a FULL status. If the counter value is 0, logic provides an EMPTY status.

Example 1.3:   A Calculator Based on a Stack Structure

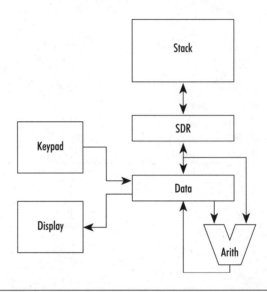

Figure 1.26   A Datapath Structure for a Calculator

Stacks make dealing with number storage simple since no explicit addresses are required. This makes them useful for data management in small calculators. An example architecture for a calculator is shown in Figure 1.26. A register, DATA, is used as an input register for keypad data. When data is entered into DATA from the keyboard, the current contents are pushed to the SDR. DATA also provides information for the display. Two types of operation are allowed, ENTER and arithmetic operations. Only the ADD operation will be considered.

In addition to its input and display requirements, DATA is an extension of the stack and can participate in PUSH and POP operations. Registers DATA and SDR provide the data inputs for an arithmetic circuit.

Operation	Register Transfers	Comment
ENTER	SDR←DATA, STACK←SDR;	The PUSH operation.
+	DATA←DATA + SDR, SDR←STACK;	A POP operation.

Table 1.9

ENTER pushes information to the stack. ADD combines two operands, SDR and DATA, to produce a single operand placed in DATA. This allows room in SDR for an element popped off the stack.

Using a calculator with this type of memory requires that the operands be entered before entering the specified operation. This is referred to as **Reverse Polish Notation** or RPN in honor of the Polish mathematician, Jan Lukasiewicz, who developed it.

For example, the addition of four integers A, B, C, and D could be performed as shown in Table 1.10:

Integer	Operation	Comment
A	ENTER	Store the first number on the stack.
B	ENTER	Store the second number on the stack.
C	ENTER	Store the third number on the stack.
D	+	C + D is in the DATA register.
	+	B + C + D is in the DATA register.
	+	A + B + C + D is in the DATA register.

Table 1.10

Table 1.11 shows an alternative method that could also be used:

Integer	Operation	Comment
A	ENTER	Stores the first number on the stack.
B	+	A + B is in the DATA register.
C	ENTER	C is placed on the stack.
D	+	C + D is in the DATA register.
	+	A + B + C + D is in the DATA register.

Table 1.11

Example 1.4:   A Stack Constructed with RAM

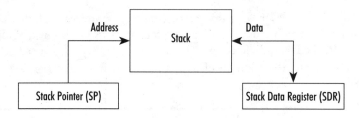

Figure 1.27   Implementation of a Stack with RAM

Stacks constructed with registers are limited to a relatively shallow depth. A stack may be constructed using RAM and the organization shown in Figure 1.27. The RAM control signals have been omitted. A **stack pointer** (SP) serves as an address register for the RAM. A stack data register (SDR) contains the data to be placed on the stack. It is also used as a destination register for data removed (popped) from the stack. The SP points to the first available RAM location for storing data. This is usually called the **top of the stack**. The origin of the stack, or bottom, is a fixed location. In the clock cycle identified as PUSH, two register transfer operations are performed simultaneously.

Operation	Register Transfers	Comment
PUSH	STACK(SP)←SDR, SP←SP + 1;	Store the data in RAM. Increment the stack pointer.

Table 1.12

In the PUSH operation, data is transferred to memory and the SP is incremented. Both operations may be performed simultaneously.

The POP operation transfers information from the memory to the memory data register. Since the SP is incremented after a PUSH to point to the first available RAM location, it must be decremented prior to retrieving data. The following must be performed in sequential clock cycles identified as POP1 and POP2.

Operation	Register Transfers	Comment
POP1	SP←SP − 1;	Decrement the stack pointer.
POP2	SDR←STACK(SP);	Retrieve the data.

Table 1.13

The PUSH and POP form a pair which store information on the stack and then retrieve it in a last-in-first-out manner. PUSHes and POPs can be performed in any order as long as some care is used to assure that a POP is not performed when the stack is empty or a PUSH is not performed when the stack is filled to its capacity.

Some stacks are implemented using the convention that the stack pointer points to the last filled memory location rather than the first empty location. If this is the case, the algorithm for the PUSH and POP changes slightly. In the PUSH, the SP must be incremented before a data transfer from the memory data register. In the POP, the SP is decremented simultaneously with the data transfer from the RAM.

It is also possible to implement a stack to fill the available address space from the top down rather than from the bottom up. This involves decrementing the stack register for a PUSH and incrementing it for a POP.

1.13   System Interconnection Structure

Within an information processing system, data must be routed between multiple components. The source and destination elements are a set of data registers as shown in Figure 1.28. There are multiple ways of accomplishing the routing, as shown in Figure 1.29.

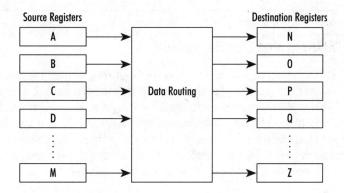

Figure 1.28   Data Routing within the Datapath Components

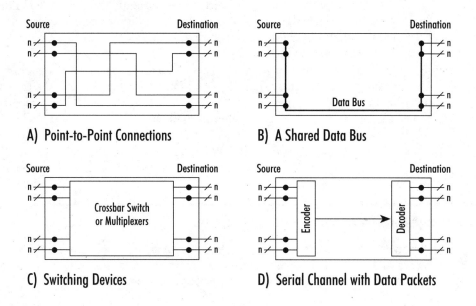

Figure 1.29   Interconnection Structures

Point-to-point Connections

Point-to-point interconnection is the least complicated method of connecting a system; no control signals are required for the interconnection structure. Dedicated n-bit communication paths establish register interconnections. Allocating a path to all source/destination pairs assures that there will always be a communication path between the source and destination. Since data paths are not shared, this method may lead to dedicating a large percentage of the device area to interconnection paths. This is a definite disadvantage in producing high-performance integrated circuits where area allocations are of prime importance.

Bus Structures

A bus achieves economy by providing a single n-bit communication path which links multiple source and destination registers. Tri-state buffers on all of the registers control access to the bus. Since the path must be shared, the availability of the communication path may influence the efficiency of the processing algorithm. To move multiple pieces of information, it may be necessary to send them sequentially and provide additional temporary storage registers at the destination. Multiple buses may be used to increase concurrent information flow. When using buses, the buses themselves are passive; all control signals for providing or accepting bus information are sent to the source and destination devices.

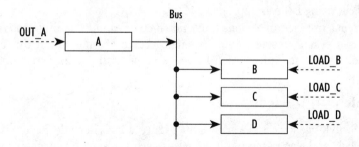

Figure 1.30   Routing a Source Register to Multiple Destinations

In Figure 1.30 a single source register, A, provides information to several possible destinations. To implement a data transfer to one register, for example B, only the OUT_A and LOAD_B control points would be asserted. In Figure 1.31 the bus provides both an input and an output to each of the registers. Each uses an internal tri-state buffer to isolate the input and output operations.

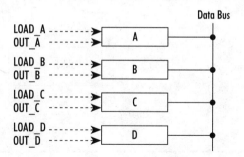

Figure 1.31   A Bus Interconnection Structure

Transferring information on a bus requires two control signals: one for the source and the other for the destination. The timing for a bus data transfer is identical to the timing required for a register-to-register transfer using a dedicated data path. The source puts the information on the bus near the start of the clock period. The destination uses the next clock pulse to capture the information from the bus. All inputs to the register must be stable before the active clock transition. This means that all information paths and all mode controls must be established before the active clock transition.

Two possible ways of indicating a transfer of the contents of register A to register B are as follows:

B←BUS, BUS←A;

           or

B←BUS←A;

Both convey the intent of using the system bus in the information transfer. When bus use is implicit, using B←A is appropriate.

A bus imposes an obvious limitation on performing simultaneous register transfers. Since the bus is a system resource, only a single source of information can use it in any clock period. Although only one information source is possible, there may be many simultaneous destination registers.

Switching Devices

Switching devices, as shown in Figure 1.29C, include logic gates, multiplexers, crossbar switches, and electro-mechanical devices such as relays. Control signals cause the switching elements to configure their interconnections. A wide variety of performance characteristics can be achieved. These range from connecting a single source with a single destination through connecting all possible sources with all possible destinations.

In Figure 1.32, a multiplexer selects one register from among several for routing to register Z. To perform the operation Z←A, control signals provide the code to select register A.

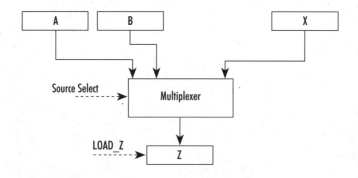

Figure 1.32   Use of a Multiplexer for Register Interconnection Routing

A crossbar switch may also be used as the basis of interconnecting a group of registers and processing elements. In Figure 1.33, registers A through Z are interconnected via a crossbar switch. By properly selecting the interconnection points among the rows and columns of the switch, the contents of any register may be moved to any other register. Simultaneous data transfers may also be made. Any source register may be connected to any destination register not already committed to a source register. Connection structures of this type are known as non-blocking. A control signal to the crossbar switch specifies the row/column interconnection points.

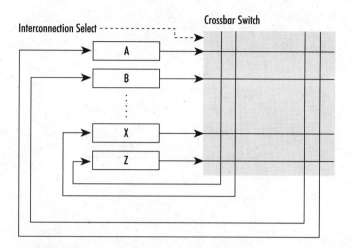

Figure 1.33   Use of a Crossbar Switch for Data Routing

Serial Data Channels

When large distances separate the source and destination, serial channels, as shown in Figure 1.29D, may be used. Data is formed into information units called packets. Packets may include additional information identifying the source and the destination. The packet is created at the source end, sent over a serial data channel, and interpreted at the destination end. Information from the packet is extracted and delivered to the selected destination. Serial channels require a substantial investment in hardware to create, interpret, and route packets. Because of the additional processing overhead and serial transmission, performance is usually lower than that obtained by the other techniques. Topics related to serial data channels will be covered in Chapter 12, "Digital Communication and Networking."

1.14   Interconnecting Registers and Arithmetic and Logic Units

Computing structures involve the interconnection of registers and data conversion elements such as arithmetic and logic units. Although the interconnection mechanism may be chosen arbitrarily, several highly regular structures using buses are often used. A more extensive treatment of this subject is included in Chapter 3.

Single-Bus Structure

Figure 1.34 shows a structure in which a single bus, ABUS, interconnects the storage and processing elements. Each register, except TEMP_1 and TEMP_2, has a single data port used for both input and output operations.

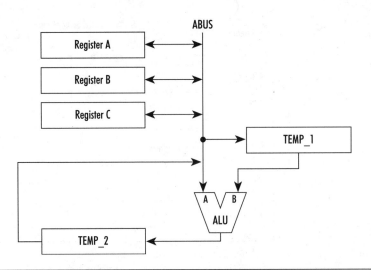

Figure 1.34   A Single-Bus Datapath Structure

Register TEMP_1 is a temporary storage location for one of the ALU operands. Register TEMP_2 provides temporary storage for data produced by the ALU. Because the bus can accommodate only one data word in any clock cycle, three clock cycles are required to perform an ALU operation. For example, to add the contents of register A and register B and store the results in register C requires the sequential generation of control signals for the registers and ALU to perform the register transfer operations in Table 1.14. The three successive clock cycles are identified as t_0 through t_2.

Clock Cycle	Register Transfers
t_0	TEMP_1←B;
t_1	TEMP_2←TEMP_1 + A;
t_2	C←TEMP_2;

Table 1.14

Two-Bus Structure

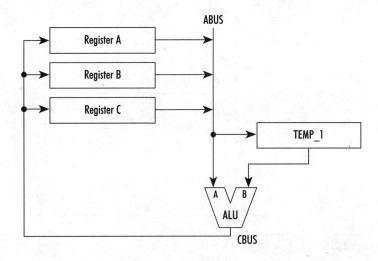

Figure 1.35   A Two-Bus Datapath Structure

Higher system performance can be attained by increasing the number of buses. In Figure 1.35, CBUS transports results from the ALU back to the destination registers. This eliminates the need for TEMP_2. The data processing operation shown in Table 1.14 can now be performed in two clock cycles.

Clock Cycle	Register Transfers
t_0	TEMP_1←B;
t_1	C←TEMP_1 + A;

Table 1.15

Three-Bus Structure

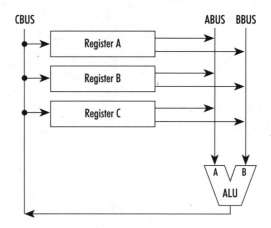

Figure 1.36   A Three-Bus Datapath Structure

Three data buses, as shown in Figure 1.36, provide two paths for operands and a path to the destination register. Each of the registers requires tri-state outputs capable of putting the contents on either ABUS or BBUS or both. If the registers are of the clocked synchronous type discussed in previous sections, the entire data processing operation can be performed in a single clock cycle.

Clock Cycle	Register Transfers
t_0	$C \leftarrow A + B$;

Table 1.16

Systems using only a single data bus for routing internal information flow may suffer from low performance due to the need to use sequential clock periods to transfer information among registers. As additional interconnection paths and buses are added, the ability to transfer information simultaneously among registers improves. Improvement continues until all information transmission required in any step of the processing algorithm can be performed concurrently. Additional buses and data interconnection structure, at this point, would provide no improved performance. The designer must choose the best balance between the higher performance possible with complex interconnection structures and the lower cost of simple interconnection structures.

These topics will be considered in greater detail in Chapter 3, "Hardware Structures for Data Processing."

1.15   Controller/Datapath Functional Decomposition

It is possible to devise small digital systems without a disciplined design methodology. The designer envisions the interrelationship of the data, hardware, and algorithm, and using skill and experience creates a functioning device. Data and control signals are often used in complex and highly interrelated ways. But such ad hoc design procedures are burdensome to document, and modification of the design may be very difficult.

Classical design approaches using logic equations and state diagrams are also not suited for complex systems. Optimization for the device, if attempted, might use minimization techniques based on logic equations. The device is documented by means of truth tables, state diagrams, and considerable written explanations. This approach has several disadvantages when used with larger and more complex systems:

- Truth tables and state diagrams are not well suited for specifying processing algorithms.
- Traditional logic design methods are primarily suited for handling data at the bit level, rather than the word level required for data processing.
- Truth tables and state diagrams are awkard for documenting the hardware action of the system.
- Slight changes in design requirements may require extensive redesign activities.
- Traditional logic design approaches are not well suited for large and complex systems.

As the number of components increases, a more disciplined technique is needed to preserve the algorithmic aspect of the design. It should allow modification and improvements to the design. One such approach involves decomposing the design into two modules as shown in Figure 1.37.

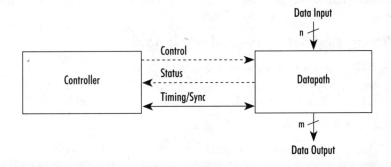

Figure 1.37   A Digital System with Controller/Datapath Decomposition

The Datapath

One module, the datapath, is comprised of the hardware associated with the data processing portion of the system. It includes registers, logic devices, memories, and communication paths for the movement, storage, and processing of data. The datapath is capable of supporting the many primitive operations that collectively implement data processing.

The Controller

The second module, the controller, is involved with the algorithmic portion of the system. It is responsible for generating control signals and sequencing the processing operations. Control signals terminate at control points associated with elements in the datapath. The control signals specify what processing actions are to be performed, where the operations are to occur, and when the actions are to begin. For example, a control signal could be used to inform a specific register to accept a unit of information at a specified clock cycle.

Typical operations involving control signals include the following:

- Loading, clearing, or complementing a register.
- Incrementing a counter.
- Routing or moving data from one location to another within the datapath.
- Specifying the logic function (from among several choices) to be performed on data by a processing element.
- Selecting the device (from among many) to participate in a data processing activity.

Status Signals

In some situations the controller requires information from the datapath before it can determine the correct control signal to provide. Status signals are derived from one or more yes-no type tests on the data. For example, one of two possible processing actions may be selected depending on whether the data in a register is positive or negative. The positive-negative information would be provided to the controller via the status signals.

For ease in referencing and documentation, status signals may also be given a descriptive symbolic name. For example, ZERO might be the name of the status signal to indicate the contents of a register has a value of zero.

Timing and Synchronization Signals

The presence of a system clock is implicit and will be omitted from a description of the controller and datapath. Additional synchronization signals, called handshake signals, are required for asynchronous transfer of data. This topic will be covered in Chapter 10.

Example 1.5:   Datapath Structures

Figure 1.38 shows the datapath structure for a small system involving two registers, two adders and a decision element. When asserted, control signals LOAD_A and LOAD_B load the registers with information from the adders. A logic block compares the value provided by register B and produces a status signal called BGT9 (B Greater Than 9). When asserted, BGT9 indicates that the contents of register B has a value greater than 9. BGT9 is returned to the controller, where it may be used to influence the sequence of the processing algorithm.

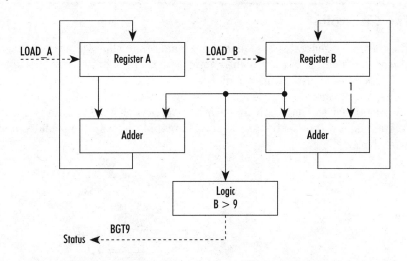

Figure 1.38 A Datapath Structure for a Small Digital Processing System

The analog to digital (A/D) converter shown in Figure 1.39 provides a second example of a datapath. Control signal CONVERT starts the conversion operation; status signal DONE indicates termination of the conversion operation. The controller can interpret the status signal to properly sequence processing actions.

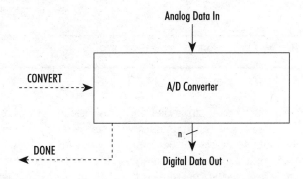

Figure 1.39 An A/D Converter with Control and Status Signals

Decomposition of a digital system into a controller and a datapath has several advantages:
- The controller and datapath portions can be designed independently.
- In some cases, changes in the processing algorithm can be accomplished by redesigning only the controller.
- Specification of the processing algorithm follows naturally from flow charts.
- System documentation proceeds naturally; using mnemonics allows easier interpretation of system actions.

1.16   The Controller

The controller generates a sequence of signals that allows the processing algorithm to be performed. This device is conveniently implemented as a finite state machine as shown in Figure 1.40. The state register contents indicate the current status of the control algorithm. Each state corresponds to one step in the processing algorithm. For each state, the output logic produces control signals for the datapath. In each clock cycle the state machine advances to a new state as determined by the next state logic. Chapter 4 presents controllers and their design more extensively.

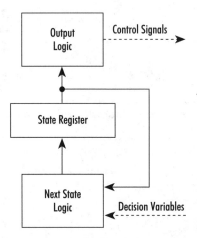

Figure 1.40   A Finite State Machine Controller

Summary

Successful computing structures use data, processing hardware, and algorithms. This chapter introduces the register as the main hardware device for storing data. Various arithmetic and logic devices allow data to be converted from one value to another. Data interconnection structures such as buses, multiplexers, and switches allow data to be routed from one location to another. Control signals asserted at control points on the devices can affect the operations of many hardware elements.

For ease of design, a complex computing structure is decomposed into a datapath and a controller. The datapath provides the basic hardware processing actions; the controller generates a sequence of control signals which cause the processing algorithm to be performed. The datapath is composed of data converters, decision elements, data routing elements, data storage elements, and interface devices. The controller is based on state machine principles. Each state corresponds to one step in the processing algorithm.

For additional information and references on the topics presented in this chapter, access the World Wide Web page provided at: http://www.ece.orst.edu/~herzog/docs.html.

Key Terms

clock cycle	design
clock skew	hardware
control points	processing algorithm
control signals	programmable logic devices (PLDs)
controller	register file
crossbar switch	Reverse Polish Notation
data	set-up time
data routing	stack
data storage	stack pointer
datapath	top of the stack
decision elements	tri-state buffer

Exercises

1.1 Give an advantage and a disadvantage of using low-level design components, such as transistors, as a starting point for a complex system design.

1.2 Give an advantage and a disadvantage of using high-level design components, such as microprocessors, as a starting point for a complex system design.

1.3 The 74LS382 arithmetic and logic unit presented in Section 1.4 is a part of the datapath shown in the Figure 1.41. Register A and register B contain 4-bit binary operands. Find a sequence of control signals and register transfer operations which will allow the following functions to be placed in Register C. It is not necessary to preserve the contents of register A and register B. You may want to use them for temporary storage of intermediate results.

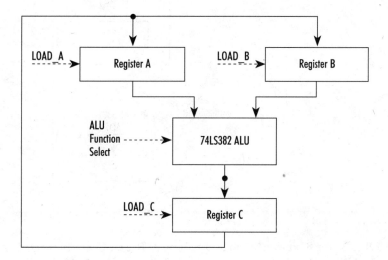

Figure 1.41

       (a) NOT A

       (b) A NAND B

       (c) B

       (d) A + 1 (Assume the carry is available as a control input to the ALU)

       (e) A − 1 (Same assumptions as in (d))

1.4   Assume that it is necessary to connect five registers, R_0, R_1, R_2, R_3, and R_4, such that the data output of any register can provide the data input for any register.

       (a) Show how this can be done using multiplexers.

       (b) Show how this can be done using a data bus.

       (c) Show how this can be done using a crossbar switch.

1.5   How many multiplexers are required to fully interconnect n registers?

1.6   Design the logic required to compare two binary numbers of 2 bits. The output should indicate if input A is greater than, less than, or equal to B.

1.7   Repeat Exercise 1.6 for the condition in which multiple compare circuits may be interconnected to provide a greater than, less than, or equal output for inputs which are a multiple of 2 bits. The GT, LT, and EQ outputs of one unit should become inputs to the next sequential unit. *Hint*: start with the most significant bits.

1.8   Three 4-bit synchronous registers are interconnected with a 74LS382 ALU as shown in Figure 1.42. Initially the register contents are X = 0001_2, Y = 0010_2, Z = 1111_2. The control signal for the ALU is 011, 100, 101, 011, 110, and 000 during successive clock pulses. What are the contents of each register after each clock pulse? Ignore overflow bits from the ALU.

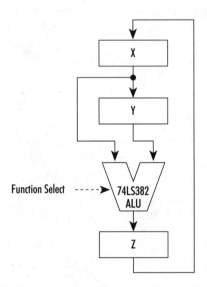

Figure 1.42

1.9   Three 8-bit synchronous registers, A, B, and C, are connected as shown in Figure 1.43. Initially, A = 000011112, B = 00110011, and C = 11110000. Clock signals for the three registers are identical. The clock has an equal time duration in its high and low value. What are the contents of A, B, and C after each of the first three clock pulses under each of the following conditions:

(a) All registers are edge triggered on the leading edge of the clock pulse.

(b) Register A is edge triggered on the leading edge of the clock pulse. Registers B and C are loaded on the trailing edge of the clock.

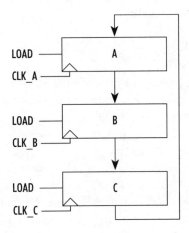

Figure 1.43

1.10 In Figure 1.43, assume that all registers are loaded with the leading edge of the clock. Initially A = 11111111, B = 11110000, and C = 00001111. The three clocks are such that CLK_B and CLK_C have significant clock skew (greater than the register set-up time) with respect to CLK_A. The sequencing is such that CLK_A appears first followed by CLK_C and then CLK_A. What are the contents of the registers after each of the first three clock pulses to register A?

1.11 The individual bits of a 4-bit register are interconnected as shown in Figure 1.44. If the initial value is 0001_2, find the repeating pattern of register contents that occurs with a continuous clock signal. Which numerical value is missing from the sequence?

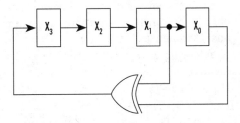

Figure 1.44

1.12 The circuitry shown in Figure 1.44 is known as a shift register generator. For an n-stage shift register generator, what is the maximum number of clock pulses that can be received without having a pattern in the register repeat? Suggest some uses for this circuitry.

1.13 A register file of 16 registers, each of 8 bits, is connected to a (double wide) 74LS382 ALU. The register file has control inputs consisting of a 4-bit A Address, a 4-bit B Address, and a LD signal which loads the contents of CBUS into the register specified by Address B. Show the sequence of control signals (address A, address B, LOAD, ALU function select, and carry input) which will perform the following actions. The contents of other registers may be changed if necessary.

(a) $R_0 \leftarrow /R_0$ (not R_0)

(b) $R_0 \leftarrow 5 * R_2$

(c) $R_0 \leftarrow R_1$ NOR R_2

1.14 What are the possible impacts of each of the following on a digital design?

(a) The system clock runs 20% slower than its design rate.

(b) The system clock runs 20% faster than its design rate.

(c) The system clock is irregular, but never has a shorter period than the clock period the device was designed for.

(d) Without changing the clock rate, less expensive, but slower, logic components are substituted into the device.

(e) The system clock is moved one meter from its original location.

1.15 The controller of a digital system occasionally produces control signals with incorrect values at the beginning of the clock cycle; after about 25% of a clock cycle, all control signals are correct. Will this cause problems with system operation?

1.16 Some of the registers in a design use the rising edge of the clock as the active transition; some use the falling edge. What effect could this have on system operation?

1.17 Assume that a supply of $1K \times 8$ (1024 words, 8 bits per word) ROM is available with control signals such as shown in Figure 1.18. How could these devices be used to create:

(a) An $8K \times 8$ memory.

(b) A $1K \times 32$ memory.

(c) An $8K \times 32$ memory.

1.18 Show a hardware configuration and sequence of control signals that will read the contents of a memory location, increment the value, and write the value back in the same memory location.

1.19 A $1K \times 8$ RAM memory contains 8-bit integers. Show a hardware configuration and sequence of register transfer statements that will move the contents of locations 1 through 8 to locations 0 through 7.

1.20 Convert the following arithmetic operations to Reverse Polish Notation for use on a four-function calculator.

(a) $(A * B) + (C * D) + (E * F)$

(b) $A * (B + C) * (D + E + F)$

(c) $A / (B + (1 / C)) / D$

1.21 Convert the following RPN statements to the equivalent arithmetic form.

(a) $A B C D E F + * - / +$

(b) $A B \cdot C D + / E F * *$

(c) $A B C D E F / / / / /$

1.22 Interconnection structures, including crossbar switches, are sometimes fabricated with contact closure devices. The binary digital switch is shown in Figure 1.45. Two transfer contacts are used. The two inputs and two outputs are interconnected in a pattern controlled by a single input variable. In one configuration, each input is connected to its corresponding output; in the other, each input is cross coupled to its opposite output. Show how several digital switches can be connected to form a multiplexer with four inputs.

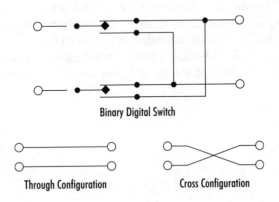

Binary Digital Switch

Through Configuration   Cross Configuration

Figure 1.45

1.23 Figure 1.46 shows an interconnection pattern for four binary digital switches. Z_{00}, Z_{01}, Z_{10}, and Z_{11} are control inputs to the two layers of switches. When the control signals are not asserted, the through connection is made; when asserted, the cross connection is made.

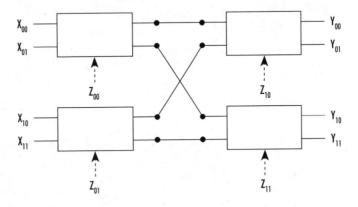

Figure 1.46

(a) If all of the control inputs are 0, what is the interconnection pattern of the switch system?

(b) If all of the control inputs are 1, what is the interconnection pattern of the switch system?

(c) Can any arbitrary assignment of inputs to outputs be made with this device? Why?

(d) If control signals are reassigned so that $Z_{00} = Z_{01} = Z_0$ and $Z_{10} = Z_{11} = Z_1$, what is the relationship between Z_0, Z_1 and the interconnection pattern?

1.24 Compare the performance of a single bus structure, two bus structure, and three bus structure.

(a) What type of operations provide the maximum improvement for the three bus structure.

(b) How much performance improvement is possible? Be as quantitative as possible.

1.25 Using an automobile as an example, explain the concept of controller and datapath decomposition. What is the controller in an automobile?

1.26 It was stated in Section 1.15 that "traditional logic design approaches are not well suited for large and complex systems." Give several arguments in support of this position.

2
The Processing Algorithm

Leonardo of Pisa

Leonardo of Pisa had queries
About rabbits, additions, and carries
Despite all the cynicism
He developed an algorithm
For producing the Fibonacci Series

Prologue

Leonardo of Pisa was a thirteenth-century Italian mathematician also known as Fibonacci. A challenging mathematical problem was presented in *Liber Abaci*, one of the journals of the day. If a pair of rabbits could produce another pair of rabbits at the end of two months and additional pairs after each additional month, and all offspring could replicate the same production schedule, how many rabbits would there be at the beginning of each month (assuming no deaths)?

The solution to this problem can be determined by counting rabbits as shown in the Figure 2.1.

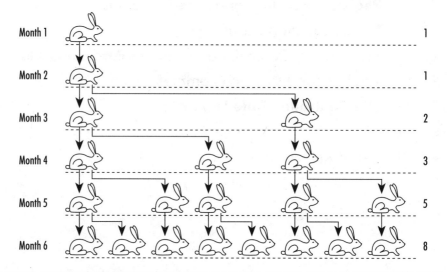

Figure 2.1   Reproduction of Rabbit Pairs

The sequence of rabbit pairs, now known as a Fibonacci sequence, has the following pattern:

{ 1, 1, 2, 3, 5, 8, 13, 21, 34, 55, 89, ... }

The first two elements in the sequence are 1. Each additional element can be obtained by adding the two previous numbers in the sequence. The pattern of the Fibonacci numbers is more than just a mathematical curiosity. Relationships exist between the Fibonacci sequence and the pattern of coefficients in the binomial theorem and even the spiral rings in the center of a sunflower.

2.1   Introduction

Chapter 1 emphasized the characteristics of digital hardware components. These devices can perform computational activities. This chapter emphasizes the role of processing algorithms. An algorithm is a sequence of simple processing actions which collectively performs a complex action. Complex systems are decomposed into a datapath and controller to aid the design process. The hardware components of the datapath perform the processing actions. The controller synchronizes and sequences the actions to perform a multi-step algorithm.

A register transfer language was introduced in Chapter 1. In this chapter, more formality for these statements will be introduced. The processing algorithm specifies the actions performed in each clock cycle.

A graphical procedure for specifying the computational algorithm, the microoperation diagram, is introduced later in the chapter. This approach offers particular advantages when the processing algorithm involves decisions and looping. The chapter concludes with several examples illustrating the development and analysis of processing algorithms.

The overall design process for a digital computing structure can be reduced to four major activities: algorithm specification, datapath structure, design of the controller, and system documentation.

1. **Algorithm Specification.** It is necessary to express the essential algorithmic operations required of the device. The processing algorithm must be based on the availability of components and data pathways of the hardware. It must allow branches in the processing algorithm. It must also preserve the basic time base of a digital system, the system clock.

   A **register transfer language** (RTL) specifies primitive processing actions. RTL statements allow a clear description of the movement, processing, and storage of digital information within the datapath.

   To express processing algorithms involving sequential processing actions, the concept of **microoperation diagram** will be developed later in this chapter. The microoperation diagram provides a graphical description of a processing algorithm while maintaining essential limitations imposed by the system clock and processing hardware.

2. **Datapath Structure.** Performance of processing operations requires a suitable set of registers, hardware processing components, and data interconnection links. The hardware structure and the algorithmic structure are related and interdependent—a change in one may require a change in the other. The hardware and interconnection structure, along with a proper set of control signals, must support the processing actions to be performed.

3. **Design of the Controller.** To fully implement a processing algorithm, a controller is needed to present signals to hardware components which cause the various processing activities to be performed in the proper sequence.

4. **System Documentation.** System design is almost always an evolutionary process. Months or years after the initial design, it may be necessary to modify the processing algorithm and change the design. The design documentation must be such that the intent of the design and the design techniques are preserved in a clear and logical manner.

This chapter focuses on the first and fourth of the designer's activities, specifying the processing algorithm and documenting the system. This specification will allow the algorithm to work with hardware components. Chapter 3 will emphasize the structure of the datapath. Controller design is presented in Chapter 4.

2.2   Digital Information

A **word** is a logical grouping of bits moved and processed as a unit. The word may represent a numerical value, a character, or it may be a set of coded information bits that conveys the status of a device. The number of bits in the word, the word width, must be consistent with the hardware and data requirements of the system. This may be related to requirements for data representation, the number of bits needed to represent a computer instruction, or the number of bits required to address system memory. Word width represents a trade off between the desire for large widths associated with data representation and more economical short widths for hardware fabrication. Common widths include 8, 16 and 32 bits. A width of 64 bits is likely to become common in the near future. In this chapter, techniques are developed for manipulating data words in much the same way that logic design is concerned with techniques for manipulating data bits.

Most numeric values considered will be in integer form. Integers can be expressed in decimal, hexadecimal, or binary formats. When necessary for clarity, subscripts such as 596_{10}, $5AB2_{16}$, or 1011_2 will be used. Numbers without any designation will be assumed to be decimal. For situations in which subscripts are impractical, such as in assembly language programs, the letters D, H, and B serve the same purpose: 596D, 5AB2H, 1011B (see Table 2.1). The symbols "/*" and "*/" will be used as delimiters of a comment; their only purpose is to define or explain an action.

25_{10} or 25D	The decimal integer 25.
34_{16} or 34H	The hexadecimal integer $34_{16} = 52_{10}$.
101_2 or 101B	The binary integer $101_2 = 5_{10}$.
48	Assumed to be a decimal integer 48_{10}.

Table 2.1

The Data Register

Words are stored in registers or in a data memory. Associated control signals clear, load, or control the register in other appropriate ways. Registers are given names, often suggesting their use, to allow ease of reference. For example:

SUM	Register named SUM.
PC	Program counter (important register in a digital computer).
MAR	Memory address register (used to hold an address for a memory device).

Table 2.2

When performing arithmetic and logic operations, data is specified by the name of the register holding the contents. For example:

SUM − 2	Value of the contents of register SUM decremented by 2.
SUM − A	The difference between the contents of register SUM and register A.

Table 2.3

The size of a register is specified by using square brackets ([ ]) to indicate the number of bits. For example, if register SUM has 8 bits, this is indicated by:

```
REGISTER: SUM[8]              /* Declares register SUM to have 8 bits.              */
```

The individual bits of a register are numbered from right to left starting with bit 0. By this convention, a numeric value in bit 0 will have a binary weighting of 2^0. Parentheses are used to indicate a specific bit or group of bits in a register. A colon (:) is used to indicate all of the bits within a specified range.

SUM (0)	The right-most (bit 0) of register SUM.
A (5)	Bit 6 (from the right) of register A.
X (9:6)	Bits 9 through 6 of register X.

Table 2.4

A group of otherwise unrelated information can be combined to form a data word through the process of **catenation**. For example:

A(7), A(5), A(3), A(1)	4-bit data unit formed from catenating odd-numbered bits of register A.
X(7:4), Y(7:4)	8-bit data unit formed from catenating the 4 high order bits of register X to the 4 high order bits of register Y.

Table 2.5

Unless otherwise indicated, reference to a register implies all of the register bits.

Data Memory

Bulk memory devices, such as RAM and ROM, provide storage locations for data words. Memory units are given a descriptive name and declared with respect to their size and type. For example:

```
ROM MEMORY: TABLE [256,16]    /* Declaring TABLE as a 256 x 16        */
                              /* ROM memory.                          */
            CHAIR [64,8]      /* CHAIR is a 64 x 8 ROM.               */
```

Individual words and bits of a **data memory** are designated by providing the word address and the bit address. When the bit address is omitted, the entire word is indicated.

TABLE(74)	Location 74 in memory TABLE.
CHAIR(56,3:0)	Bits 3, 2, 1, 0 of location 56 in memory CHAIR.

Table 2.6

2.3   Datapath—Hardware Structure for Processing Data

In information processing devices, registers and data memories are integrated with arithmetic and logic devices to create the system datapath. Several datapaths, similar to those presented in Chapter 1, are shown in Figure 2.2. Data processing operations begin with the data residing in registers. After a processing operation, data is stored in destination registers. All processing actions and associated data storage are performed in one clock cycle.

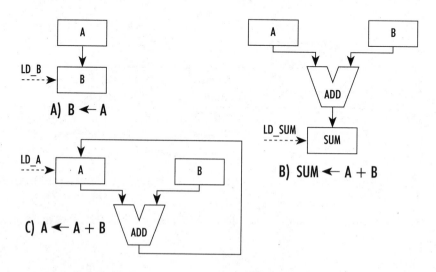

Figure 2.2   Example Datapath Structure

In Figure 2.2A a data transfer from register A to register B is performed. In Figure 2.2B, the contents of register A and register B are added and the result is placed in register SUM. Figure 2.2C shows a slightly more complex operation in which register A receives the sum of the contents of register A and register B.

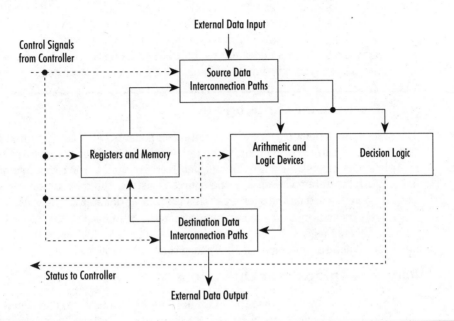

Figure 2.3   A Conceptual Model of the System Datapath

Figure 2.3 shows a more generalized representation of a data processing system which includes provisions for data routing, decisions, and possible external data sources. Data processing operations begin with data residing in the registers and memory block. This is the source of all internal data to be processed. Control signals from the controller specify register and memory operations. The registers and memory block is the only portion of the system available to permanently retain data.

The source data interconnection paths block routes information from a storage location to arithmetic, logic, and decision elements. This may involve point-to-point connections, crossbar switches, multiplexers, or data buses. Control signals interact with this block to direct the data to the proper processing elements. Data from external sources may also be introduced into the system.

Data processing involves two separate blocks. One performs arithmetic and logic operations. Results are returned for storage in the registers and memory. The second generates status signals which are sent to the controller.

The arithmetic and logic block processes data to produce new information: for example a binary adder. In some cases signals from the system controller may specify the exact nature of the processing. For example, the controller could indicate the operation for an ALU to perform on two data elements.

The decision logic block processes data to provide logical yes/no information for the system controller. A typical element is a **comparator**, which analyzes the numeric value of two data words and produces a logic signal specifying a greater-than, less-than, or equal relationship.

In the processing model of Figure 2.3, it is assumed that the arithmetic and logic blocks do not contain any memory elements or delay elements. Consequently, the result of an arithmetic or logic operation is available shortly after the proper operands are provided. The only delays are due to propagation and logic delays within the circuitry. After the data has been processed it is returned by the destination data interconnection paths for storage in system registers and memory elements. It may also be made available to an external destination.

2.4   Register Transfer Language (RTL)

The processing actions performed by digital components, such as presented in the previous section, may be described and documented by means of a register transfer language, RTL. An RTL expresses a processing algorithm in a form directly related to the registers and components of the datapath. RTL statements specify the source registers, the destination register, and the operations to be performed on the operands. This is an essential step in the overall design process.

Engineers and computer scientists use many variations of RTLs. Some are closely associated with formal languages used for simulation and design. The RTL used in this text is typical of the language structures the reader is likely to encounter in practice.

Register Replacement Operations

The arrow ($\leftarrow$) which was introduced in Chapter 1, is used to represent a replacement operation. In one clock cycle, register content specified on the tail of the arrow replaces content of the register identified on the head of the arrow. For example, the process of routing the content of register A into storage in register B would be indicated as:

```
B←A;                      /* Register B receives content of Register A.      */
C←0;                      /* Register C receives the value of 0.             */
```

In this statement, the data path may be as simple as a direct point-to-point connection of the output of register A to the input of register B as shown in Figure 2.2A. In other situations the path may be less direct and involve data buses, crossbar switches, and multiplexers. The register transfer statement indicates the operation to be performed, but it does not explicitly indicate the system resources to be used or the control signals to be asserted to initiate the action.

Simultaneous Input and Output Operations

As discussed in Chapter 1, clocked synchronous registers are used in most digital system designs. Using them assures reliable register transfer operations. If separate input and output data lines are available, the same register could be both the source and destination of information. For example:

```
A←A + 1;                  /* Content of Register A is incremented.           */
A←Ā;                      /* Content of A is complemented.                   */
```

Bit Level Operations

Operations can also be specified on the bit level. For example:

```
B(0)←A(3);                  /* A single bit operation.                      */
SUM(5:3)←A(2:0);            /* An operation moving 3 bits of data.          */
SUM(7:0)←SUM(6:0),0;        /* A shift left operation. The left-most bit of SUM  */
                           /* is lost. A 0 is shifted into the right-most  */
                           /* position.                                    */
A(2:0)←A(0),A(1),A(2);     /* The order of the 3 bits of A is reversed.    */
```

Arithmetic and Logic Operations

The arithmetic and logic operations that can be performed on the data depend on components in the datapath. Logic operations involve a bit-by-bit operation with bits of the two operands. Standard symbols will be used for their representation.

Operation	Example	Comment
AND	START←READY • GO; W←X Y Z; B←A AND C;	When the interpretation is clear, the second representation (no symbol) will be used.
OR	DONE←HALT + ERROR; STOP←t_1 OR OVERFLOW;	The "+" will be used only when the distinction from ADD is clear.
NOT	Y←$\overline{X}$ Y←X' Y←/X Y←!X	The choice of symbol is often determined by typographic convenience.
ADD	A←A + B	An addition.
SUB	A←A – B	A subtraction.
MULT	A←A * B	A multiplication.

Table 2.7

Additional logic relationships are indicated by "=," "≠," and ">" for equal, not equal, and greater than. Symbols for other operations will be presented and defined as needed.

Register Transfer Operations Performed in Parallel

Parallel processing at the register level means that more than one register transfer operation is performed simultaneously. The "," is used to delineate the parallel operations; a ";" indicates the end of the sequence of parallel operations. For example:

```
B←A, SUM←A + B;              /* These operations are performed in the same      */
                            /* clock period (separated by ",").               */

B←A,                        /* These operations are also performed in the same */
SUM←A + B;                  /* clock period (separated by ",").               */

B←A;                        /* These operations are performed in two          */
SUM←A + B;                  /* consecutive clock cycles (separated by ";").   */
```

Operations may be performed in parallel if the system structure for data routing and data processing has sufficient resources. There must be no conflicting resource demands. For example, the operations

A←B, A←C;

cannot be performed simultaneously since the same register cannot simultaneously accept the result of two inputs. The operations

A←B + C, D←A + B;

can be performed simultaneously only if two adders are available. Note that, in the latter example, the results of performing the two register transfer operations sequentially is different from that obtained by a parallel operation.

Memory Operations

Similarly, RAM memory may be used as source and destination locations.

```
A←TABLE (37);               /* Register A receives the content of             */
                            /* location 37 in RAM memory TABLE.               */

CHAIR(15)←45H;              /* Location 15 in RAM memory CHAIR                 */
                            /* receives the value 45H.                        */
```

In contrast to registers, most RAM memories use the same address and data lines for both input and output operations. This prohibits a read and write operation within the same clock cycle. Under these conditions some operations may not be allowed.

```
TABLE(25)←TABLE(0);         /* Not possible since RAM memory TABLE uses the    */
                            /* same selection mechanism for both input and    */
                            /* output.                                        */
```

Transferring information between two locations within the same memory can be done in two steps using a temporary register, TEMP.

```
TEMP←TABLE(0);              /* Use temporary register TEMP to hold data from   */
                            /* TABLE(0).                                      */

TABLE(25)←TEMP;             /* Complete the data transfer.                    */
```

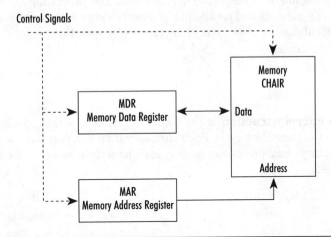

Figure 2.4   *Data Transfers between a Register and Memory*

A register provides the address associated with a memory source or destination. For example, in Figure 2.4 register MAR (memory address register) contains an address for the memory named CHAIR. Register MDR (memory data register) contains data to be stored in the memory. The system controller provides control signals for selection and read/write control. The following statements describe possible memory actions.

```
CHAIR(MAR)←MDR;          /* The memory location in RAM memory CHAIR,    */
                         /* specified by the address in register MAR,   */
                         /* receives the content of MDR. This is        */
                         /* a memory write.                             */

MDR←CHAIR(MAR);          /* MDR receives content of RAM memory CHAIR.   */
                         /* The memory location is specified by the     */
                         /* content of MAR. This is a memory read       */
```

Initializing Registers

The "=" is used with a register statement (instead of "←") to indicate assignment of an initial condition. It differs from "←" because initialization is performed during power-on or some other initializing operation. It is *not* performed in synchronism with the system clock.

```
SUM = 0;                 /* Register SUM is initialized with a value of 0.    */
```

2.5   Microoperations, Processing Primitives

A **microoperation** is a processing activity, performed during a single clock cycle, which moves or alters stored data within the datapath. Microoperations are performed under the influence of the control signals that the controller provides. The processing activity is described by a register transfer statement and results in a physical change in the system register contents. A descriptive label identifies the clock cycle of the processing operation.

For example:

```
t4:        A←A;              /* Operation is performed during clock cycle t4.    */

ONE:       B←0;              /* Operation performed in clock cycle ONE.          */
```

A **microinstruction** is the set of all microoperations which are performed *simultaneously* within the same clock cycle. Each microoperation in a microinstruction generally involves a single elementary data processing action, such as a data move or an arithmetic/logic operation. For example:

```
START:     A←TABLE(12),      /* This microinstruction contains three             */

           B←B + 1,          /* microoperations that alter the                   */

           C←0 ;             /* content of the three registers.                  */
```

Microoperations require system hardware resources such as data links and arithmetic devices. A microoperation must have exclusive use of these resources during a clock cycle. All microoperations in a microinstruction must be capable of being performed simultaneously without resource conflicts. Microoperations and microinstructions can be executed sequentially to perform more complex processing algorithms.

Example 2.1:   Hardware Structure for Fibonacci Number Generation

The hardware structure shown in Figure 2.5 has been proposed for generating a sequence of Fibonacci numbers. What microoperations can be performed by this structure?

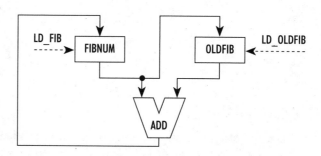

Figure 2.5   Hardware Structure for Fibonacci Number Generation

Solution:

The hardware structure in Figure 2.5 consists of two registers and two control signals to control the loading of the two registers. Register FIBNUM generates the Fibonacci sequence. There are two microoperations, labeled i and j, which are possible.

```
i:         FIBNUM←FIBNUM + OLDFIB;  /* Addition operation.                       */

j:         OLDFIB←FIBNUM;           /* Replacement operation.                    */
```

2.6   Register Transfer Language Programs

While microinstructions provide a method for specifying and documenting processing actions, sequencing and branches also must be introduced into a processing algorithm. This can be accomplished by expressing the processing actions in a program statement form using a register transfer language.

- Each RTL statement describes the microoperations to be performed in a single clock cycle.
- Each microinstruction is represented by a series of microoperations (separated by ",") that specify the microoperations to be performed in the clock cycle. A ";" is used to terminate the microinstruction.
- Microinstructions are processed sequentially unless GOTO or conditional statements indicate otherwise.

A GOTO directive specifies unconditional branching. For example, an endless looping of three processing operations associated with labels ZERO, ONE, and TWO is indicated as follows:

```
ZERO:    A←B, B←B̄;          /* Microinstruction ZERO operations.              */

ONE:     A←A + 1;            /* Microinstruction ONE operations.               */

TWO:     B←A,               /* Microinstruction TWO with                      */
         GOTO ONE;          /* unconditional jump to microinstruction ONE     */
```

An IF-THEN-ELSE statement indicates conditional branching operations. These statements have a similar structure to those encountered in a structured programming language such as Pascal or C. The following code segment shows an algorithm in which the processing action following ZERO depends on the comparison of variables A and B.

```
ZERO:    A←B, B←B̄,          /* Microinstruction ZERO operations.              */
         if (A > B) then    /* Conditional operations to determine the        */
             GOTO ONE,      /* next microinstruction.                         */
         else
             GOTO TWO,
         end if;

ONE:     A←A + 1,           /* Microinstruction ONE operations.               */
         GOTO ZERO;

TWO:     B←A,               /* Microinstruction TWO and specification         */
         GOTO ZERO;         /* of the next microinstruction.                  */
```

Register transfer language programs can be expanded to include declarations of variables, specification of registers such as width, and identification of specialized hardware components. This text will not use rigid syntactical rules for writing descriptive programs.

Example 2.2:   RTL Program to Generate Fibonacci Numbers

Given the hardware configuration shown in Figure 2.5, write an RTL program to generate a sequence of the first 13 Fibonacci numbers.

Analysis:

The first 13 Fibonacci numbers have values less than 256; 8-bit registers are sufficient to hold results. The processing algorithm can be based on the fact that each number in the Fibonacci series is generated by adding the two previous numbers in the series. Register FIBNUM holds the current number in the sequence; register OLDFIB holds the previous number in the sequence. Initial values of FIBNUM = 1 and OLDFIB = 0 allow the apparatus to start. Each of the first 13 Fibonacci numbers are generated in sequence.

Solution 1:

In this solution, straight-line programming is used. While the structure is simple, the program is long and repetitive.

```
REGISTER:     FIBNUM[8], OLDFIB[8]          /* Declare the two 8-bit registers.    */

t_0:          FIBNUM = 1, OLDFIB = 0;       /* Initialize values in registers.     */

t_1:          FIBNUM←FIBNUM + OLDFIB,       /* Generate 2nd Fibonacci number.      */
              OLDFIB←FIBNUM;                /* Update OLDFIB.                      */

t_2:          FIBNUM←FIBNUM + OLDFIB,       /* Generate 3rd Fibonacci number.      */
              OLDFIB←FIBNUM;                /* Update OLDFIB.                      */

t_3:          FIBNUM←FIBNUM + OLDFIB,       /* Generate 4th Fibonacci number.      */
              OLDFIB←FIBNUM;                /* Update OLDFIB.                      */

t_4:          FIBNUM←FIBNUM + OLDFIB,       /* Generate 5th Fibonacci number.      */
              OLDFIB←FIBNUM;                /* Update OLDFIB.                      */

                     . . .

t_11:         FIBNUM←FIBNUM + OLDFIB,       /* Generate 12th Fibonacci number.     */
              OLDFIB←FIBNUM;                /* Update OLDFIB.                      */

t_12:         FIBNUM←FIBNUM + OLDFIB,       /* Generate 13th Fibonacci number.     */
              OLDFIB←FIBNUM;                /* Update OLDFIB.                      */

t_13:         GOTO t_13                     /* Stop the processing.                */
```

In clock cycle t_0, the initial values of FIBNUM and OLDFIB are established. No processing action is performed and the values remain unchanged as the device advances to clock cycle t_1.

In clock cycle t_{13}, the GOTO statement effectively halts the program. For this solution, the content of the registers changes as shown in Table 2.8. Each clock cycle ends with the active transition of the system clock. This clock updates the content of all registers.

Clock Cycle	Content (before clock pulse)		Next Content (after clock pulse)	
	FIBNUM	OLDFIB	FIBNUM	OLDFIB
t_0	1	0	1	0
t_1	1	0	1	1
t_2	1	1	2	1
t_3	2	1	3	2
t_4	3	2	5	3
t_5	5	3	8	5
t_6	8	5	13	8
t_7	13	8	21	13
t_8	21	13	34	21
t_9	34	21	55	34
t_{10}	55	34	89	55
t_{11}	89	55	144	89
t_{12}	144	89	233	144
t_{13}	233	144	233	144

Table 2.8

Solution 2:

In this solution, looping minimizes the number of steps in the RTL program. Since the hardware lacks a counter, there is no way to stop the looping. Register overflow will occur after the 13th Fibonacci number, which has a value of 233.

```
REGISTER:    FIBNUM[8], OLDFIB[8]        /* Declare the two 8-bit registers.   */

ZERO:        FIBNUM = 1, OLDFIB = 0;     /* Initialize values in registers.    */

ONE:         FIBNUM←FIBNUM + OLDFIB,     /* Generate next Fibonacci number.    */
             OLDFIB←FIBNUM,              /* Update OLDFIB.                      */
             GOTO ONE;                   /* Repeat.                            */
```

Solution 3:

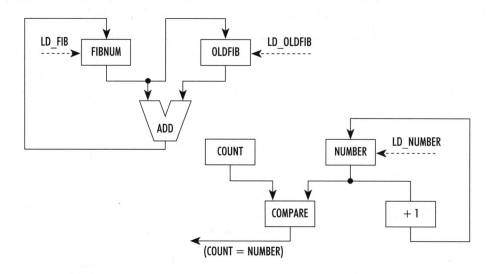

Figure 2.6   Datapath Elements for Generating Fibonacci Numbers

In this solution, the original hardware computing structure has been modified as shown in Figure 2.6. Register COUNT specifies the element to be calculated in the sequence. For example, if COUNT is 6 then the 6th element in the Fibonacci sequence is calculated. An additional register, NUMBER, has been added to count the number of elements that have been generated. When NUMBER = COUNT, the desired Fibonacci number is in register FIBNUM.

```
REGISTER:      FIBNUM[8], OLDFIB[8]      /* Declare the two 8-bit registers.   */

ZERO:          FIBNUM = 1, OLDFIB = 0,   /* Initialize values in registers.    */
               COUNT = 6, NUMBER = 1;    /* Calculate the 6th element.         */

ONE:           FIBNUM←FIBNUM + OLDFIB,   /* Generate next number.              */
               OLDFIB←FIBNUM,            /* Update OLDFIB.                     */
               NUMBER←NUMBER + 1;        /* Increment NUMBER.                  */

TWO:           if (NUMBER = COUNT) then
                   GOTO THREE,           /* Jump to THREE if done.             */
               else
                   GOTO ONE,             /* Jump to ONE if not done.           */
               end if;

THREE:             GOTO THREE;           /* End processing.                    */
```

Table 2.9 shows the sequence of microinstructions and the content of the registers for each clock cycle.

Clock Cycle	Micro Inst	Content (before clock pulse)				COUNT = NUMBER	Next Content (after clock pulse)				Next Micro Inst
		FIBNUM	OLDFIB	COUNT	NUMBER		FIBNUM	OLDFIB	COUNT	NUMBER	
t_0	ZERO	1	0	6	1	—	1	0	6	1	ONE
t_1	ONE	1	0	6	1	—	1	1	6	2	TWO
t_2	TWO	1	1	6	2	F	1	1	6	2	ONE
t_3	ONE	1	1	6	2	—	2	1	6	3	TWO
t_4	TWO	2	1	6	3	F	2	1	6	3	ONE
t_5	ONE	2	1	6	3	—	3	2	6	4	TWO
t_6	TWO	3	2	6	4	F	3	2	6	4	ONE
t_7	ONE	3	2	6	4	—	5	3	6	5	TWO
t_8	TWO	5	3	6	5	F	5	3	6	5	ONE
t_9	ONE	5	3	6	5	—	8	5	6	6	TWO
t_{10}	TWO	8	5	6	6	T	8	5	6	6	THREE
t_{11}	THREE	8	5	6	6	—	8	5	6	6	THREE
t_{12}	THREE	8	5	6	6	—	8	5	6	6	THREE

Table 2.9

2.7   Microoperation Diagram

A flow chart is a graphical method for expressing sequential events. It contains graphical symbols and descriptive narration to describe processing actions and decisions. Flow charts can express algorithms, but they cannot fully specify the time-sensitive actions of digital hardware. Register transfer language programs provide an accurate description of operations performed in each clock cycle but lack the graphic appeal of a flow chart.

A microoperation diagram is a graphical structure that combines many features of both flow charts and register transfer language descriptions. It describes algorithms performed by digital hardware, indicates microinstruction operations, and provides a visually descriptive method of the sequencing of microinstruction operations.

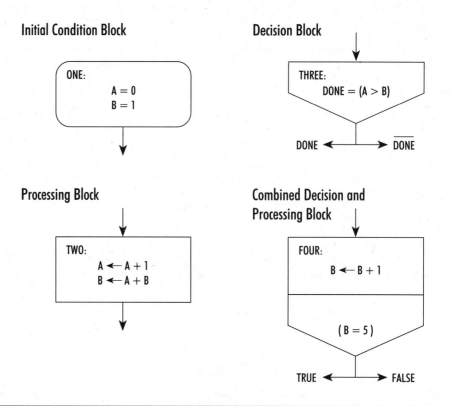

Figure 2.7   Block Elements of a Microoperation Diagram

Each block of a microoperation diagram has a unique label. Each block specifies datapath microoperations to be performed in one clock cycle. There are three types of microoperation diagram blocks: initial condition blocks, processing blocks, and decision blocks, as shown in Figure 2.7. Since processing operations and decision operations may be performed concurrently, a fourth block—combined decision and processing block—which is a combination of the two, is also possible.

Initial Condition Blocks

Initial condition blocks use a rounded rectangle shape and indicate the initial values of registers in the system. This is the state in which the system powers up (often referred to as the power-on initialize state or POI). It may also be the state to which an externally generated "initialize" or "reset" signal directs the device. No microoperations are performed in the initial state. A "=" indicates the initial condition (if known) of memory elements in the datapath. For example, the statement A = 0 in an initial condition block means that when the system is initialized, register A has content of 0.

Processing Blocks

Rectangular processing blocks specify and document the data processing operations performed by the digital hardware in one clock cycle. Blocks in the microoperation diagram are given symbolic or descriptive names such as COUNT and END. Processing blocks have one field of information to specify the microinstruction and an optional field to specify the control signals asserted.

- **Microinstruction.** The microinstruction specifies the set of register transfer operations to be performed in the clock cycle.
- **Control Set.** The control set is a set of control signals that are simultaneously asserted. For each microinstruction there is a corresponding control set that provides signals to perform the specified microoperations to the datapath control points.

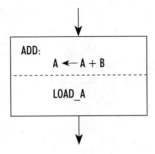

Figure 2.8   Microinstruction Processing Blocks

To perform the action A←A + B, the microinstruction is the register transfer statement A←A + B. This indicates that the next clock pulse will load register A with the sum of the current content of register A and register B. The **control set** indicates the control signals asserted to accomplish the action. In this case, a LOAD_A control signal causes register A to load data (A + B) on the next clock pulse. Figure 2.8 shows a processing block for A←A + B. For situations in which the emphasis is on the processing algorithm, the control set may be omitted from the diagram.

Decision Blocks

The pentagonal decision block is used to implement conditional logic for sequencing the execution of processing blocks. A third group of signals is used in this block.

- **Decision set.** A decision set is the set of signals generated by the decision elements. These signals are directed back to the system controller and are used to sequence the processing blocks.

Decision blocks allow processing blocks to be sequenced depending on the result of logical decisions. Decision variables generated by logic circuitry are sent to the system controller to sequence the processing algorithm. A typical decision might be based on the value in a register, comparison of two register values, or a single bit value in a register.

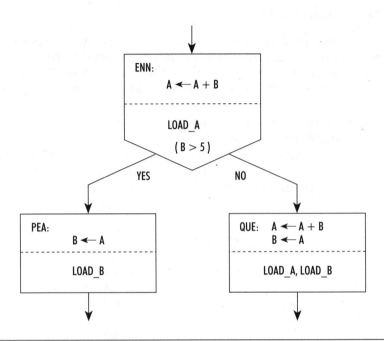

Figure 2.9   A Decision Block

Decision blocks may also include data processing activities, so full documentation of this block may require three fields of information. In Figure 2.9 block ENN has three sections; one each for the microinstruction, the control set, and the decision set. After executing microinstruction ENN, the next microoperation block to be executed depends on the system controller's logical response to decision set ENN.

When data processing activities are included in decision blocks, care must be exercised if some register content is being modified by processing elements and used by the controller to determine the next processing block. Both operations are performed simultaneously.

Example 2.3:   *Developing a Microoperation Diagram from an RTL Program*

Find the microoperation diagram for the following RTL program segments.

Program 1:

```
ZERO:      A←B, B←B̄;      /* Microinstruction ZERO operations.              */

ONE:       A←A + 1;       /* Microinstruction ONE operations.               */

TWO:       B←A,           /* Microinstruction TWO with unconditional        */

           GOTO ONE;      /* jump to microinstruction ONE.                  */
```

This program consists of three microinstructions with labels ZERO, ONE, and TWO. Each has a corresponding block in the microoperation diagram. There is an unconditional branch from block TWO to block ZERO. There is no initial condition block. The solution is shown in Figure 2.10A.

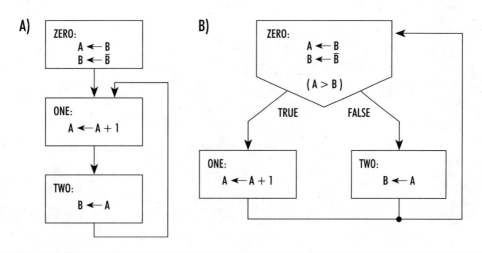

Figure 2.10   Example Microoperation Diagrams

Program 2:

```
ZERO:     A←B, B←B̄,        /* Microinstruction ZERO operations.            */

          if (A > B) then   /* Conditional operations to determine the     */
              GOTO ONE,      /* next microinstruction.                      */
          else
              GOTO TWO,
          end if;

ONE:      A←A + 1,          /* Microinstruction ONE operations.            */
          GOTO ZERO;

TWO:      B←A,              /* Microinstruction TWO and specification       */
          GOTO ZERO;        /* of the next microinstruction.               */
```

Program 2 also consists of three blocks labeled ZERO, ONE, and TWO. There is a conditional branch following block ZERO. The solution is shown in Figure 2.10B.

Example 2.4:   Analysis of a Microoperation Diagram

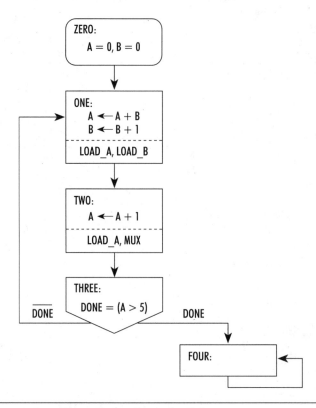

Figure 2.11   Microoperation Diagram for Example 2.4

An example microoperation diagram is shown in Figure 2.11. Initial condition block ZERO and processing blocks ONE, TWO and THREE are performed in successive clock cycles. When DONE is asserted, based on A > 5, processing block FOUR is performed; $\overline{DONE}$ returns the algorithm to block ONE. No operations are performed in block FOUR. The tight unconditional loop to itself in block FOUR effectively terminates the algorithm.

Figure 2.12 shows a possible datapath structure to support this algorithm. Arithmetic units compute A + 1, B + 1, and A + B. A single logic device produces DONE for the controller when the condition A > 5 is satisfied. Three control signals are used. LOAD_A and LOAD_B allow the respective registers to be loaded at the next clock pulse. MUX provides a signal to the selector multiplexer at the data input to register A. This allows data to be selected from either A + 1 (when MUX is asserted) or A + B (when MUX is not asserted).

An analysis of the operations specified by the microoperation diagram is included in Table 2.10. When initialized in clock cycle t_0, the device is in block ZERO with initial content of register A and register B equal to 0. There are no register transfer operations in block ZERO. The next clock pulse advances the system to microoperation block ONE with A and B retaining their value of 0. In block ONE, control signals are generated to load A with the value of A + B and to load B with the value B + 1. The tabular analysis indicates that in block ONE the next values of A and B are 0 and 1 respectively. The content of the registers changes at the end of the clock cycle with the arrival of the next clock pulse. The next microoperation block after ONE is TWO; no decisions are involved.

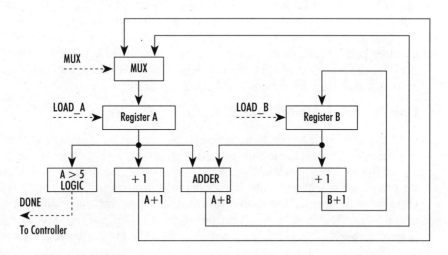

Figure 2.12   Datapath for Example 2.4

Clock Cycle	Micro Inst	Current Contents (before clock)		Decision Value	Next Micro Inst	Contents (after clock)	
		A	B	A > 5		A	B
t_0	ZERO	0	0	—	ONE	0	0
t_1	ONE	0	0	—	TWO	0	1
t_2	TWO	0	1	—	THREE	1	1
t_3	THREE	1	1	F	ONE	1	1
t_4	ONE	1	1	—	TWO	2	2
t_5	TWO	2	2	—	THREE	3	2
t_6	THREE	3	2	F	ONE	3	2
t_7	ONE	3	2	—	TWO	5	3
t_8	TWO	5	3	—	THREE	6	3
t_9	THREE	6	3	T	FOUR	6	3
t_{10}	FOUR	6	3	—	FOUR	6	3

Table 2.10

In block TWO control signals are generated to load A with the value A + 1. In this example, the source of the values to be loaded into A is specified by control signal MUX. Block THREE follows unconditionally from block TWO.

In block THREE during clock cycle t_3, a decision is performed by the controller based on the value of DONE, the output of the A > 5 compare circuit. If DONE is not asserted, the controller causes the system to return to block ONE; if asserted it advances to block FOUR.

Note that the decision is based on the current content of register A. This is the value before the clock pulse. After several loops involving blocks ONE, TWO, and THREE, DONE is asserted in t_9 and the device advances to block FOUR. When the system reaches block FOUR it loops continuously, retaining the current content of the registers.

Combined Decision and Processing Blocks

Blocks containing both decision and processing operations can easily lead to erroneous interpretation. This type of block contains both a processing operation and a decision that affects selection of the successor block in the microoperation diagram. It is easy to mistakenly assume that the processing operation is completed before the decision is made. This is not the case. Although the processing operation is placed first in the listing, both operations are performed simultaneously. The decision is based on current register values, not the values produced by the current processing block. The next three examples will attempt to illustrate this point.

Example 2.5:   ***Analysis of a Microoperation Diagram with a Complex Decision Block***

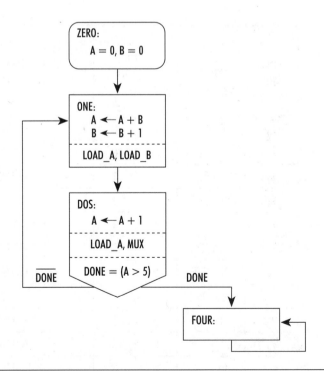

Figure 2.13   Microoperation Diagram for Example 2.5

Figure 2.13 shows a microoperation diagram similar to that considered in Example 2.4. The same datapath structure is assumed. In this figure, processing block TWO has been combined with the decision block THREE to produce a new block labeled DOS. Although the microoperation diagrams are similar, the analysis of the system, as shown Table 2.11, shows some major differences.

Clock Cycle	Present Block	Current Contents		Decision Value	Next Block	Next Contents (after Clock)	
		A	B	A > 5		A	B
t_0	ZERO	0	0	—	ONE	0	0
t_1	ONE	0	0	—	DOS	0	1
t_2	DOS	0	1	F	ONE	1	1
t_3	ONE	1	1	—	DOS	2	2
t_4	DOS	2	2	F	ONE	3	2
t_5	ONE	3	2	—	DOS	5	3
t_6	DOS	5	3	F	ONE	6	3
t_7	ONE	6	3	—	DOS	9	4
t_8	DOS	9	4	T	FOUR	10	4
t_9	FOUR	10	4	—	FOUR	10	4
t_{10}	FOUR	10	4	—	FOUR	10	4

Table 2.11

In clock cycle t_6, register A has a value of 5. Control signals generated in block DOS perform an increment operation to increase the value to 6. The decision in this block is based on the current content of the register, which is 5. The test result is false and the system returns to block ONE. Note that even though processing block DOS causes the next content of register A to change, the decision logic is based on the current content of the register. The register does not receive a new value until the clock transition at the end of the clock cycle. The test is based on the current content and is not affected by the new value to be loaded into register A. Chapter 1 and Appendix A present a more thorough analysis of the timing of register transfer operations.

Although the analysis of Example 2.5 is straightforward, the results may not be intuitive to a person accustomed to flow charts. Our intuition tends to suggest that the processing be performed first, and then the decision made based on the new value. Actually, the variable is simultaneously processed and used as a decision variable. The following table summarizes the problem.

Condition in Decision Block	Analysis	Comments
Block contains only decision operations.	The next block is determined by the current value of the decision variable.	The analysis is straight-forward.
Block contains decision operations and processing operations which *do not* affect the decision variable.	The next block is determined by the current value of the decision variable.	The analysis is straight-forward. The processing actions have no influence on the decision.
Block contains decision operations and processing operations which *do* affect the decision variable.	The next block is determined by the current value of the decision variable.	The processing action and decision actions are performed independently.

Table 2.12

Including processing operations in decision blocks may complicate the interpretation of the processing algorithm. In some situations, however, the combination produces algorithms that are performed in fewer clock cycles.

Example 2.6:   *Splitting a Microoperation Diagram with a Complex Decision Block*

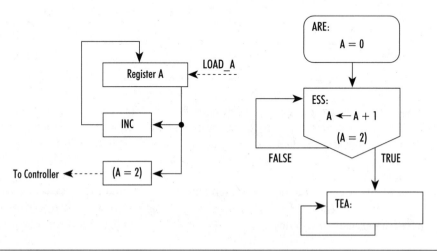

Figure 2.14   Example Datapath and Microoperation Diagram

Consider the example system shown in Figure 2.14. In block ARE of the microoperation diagram, the value of A is initialized to 0. In block ESS the value of A is incremented and (simultaneously) a test is made to determine if A (before it is incremented) has a value of 2. The increment and the test are both performed on the value of A stored in register A. Block TEA performs no operations and ends the short algorithm by endlessly looping. Only one control variable, LOAD_A, is used.

The following table analyzes the operational sequence in the controller and system datapath as they respond to clock pulses.

Clock Cycle	Micro Inst	Present Content Register A (before clock)	Decision $(A = 2)$	Next Micro Inst	Next Content Register A (after clock)
t_0	ARE	0	—	ESS	0
t_1	ESS	0	F	ESS	1
t_2	ESS	1	F	ESS	2
t_3	ESS	2	T	TEA	3
t_4	TEA	3	—	TEA	3

Table 2.13

Clock pulse 0, arriving at the end of clock cycle t_0, advances the device to block ESS where register A retains its initial value of 0. Clock pulse 1 increments the value of A to 1 and loops the device back to block ESS again. In t_2, clock pulse 2 increments the value of A to 2. The decision circuitry (of block ESS) in clock cycle t_2 uses the original value in register A, which is 1, in making the decision for the next block. In this case since the value is 1, the loop back to ESS is made a second time. During clock cycle t_3, the value of A is 2 and the decision variable is true. Clock pulse 3, at the end of clock cycle t_3, increments A to a value of 3. The next block is TEA, which is entered with A = 3.

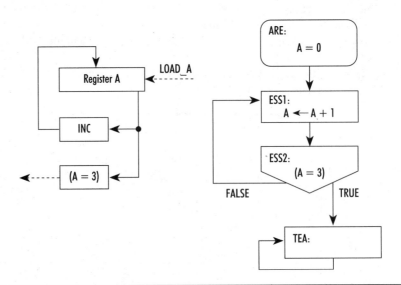

Figure 2.15   Separate Processing and Decision Blocks

One way of avoiding this potentially confusing interpretation of the algorithm is to split data processing operations and decision activities into separate microoperation blocks. For this example, Figure 2.15 shows a possible microoperation diagram producing the same processing results. The operations performed in block ESS have been split into two sequential blocks: ESS1 and ESS2. The algorithm proceeds as shown in the following table.

Clock Cycle	Micro Inst	Present Content Register A (before clock)	Decision (A = 3)	Next Micro Inst	Next Content Register A (after clock)
t_0	ARE	0	—	ESS1	0
t_1	ESS1	0	—	ESS2	1
t_2	ESS2	1	F	ESS1	1
t_3	ESS1	1	—	ESS2	2
t_4	ESS2	2	F	ESS1	2
t_5	ESS1	2	—	ESS2	3
t_6	ESS2	3	T	TEA	3
t_7	TEA	3	—	TEA	3

Table 2.14

By splitting the operations, the data processing is performed first, followed by the decision in the next clock period. This approach makes the algorithmic logic more intuitive at the expense of increasing the effective processing time of the device.

The idea of associating algorithms to state machines was originally developed by Christopher Clare[1]. He called the concept an algorithmic state machine, ASM. The microoperation diagram is an adaptation of many of the ASM ideas.

2.8 Supervisory Control Signals in the Processing Algorithm

Digital computing structures may receive various signals from manually operated sources or from external devices. These signals are assumed to be synchronous with the system clock. Techniques for synchronizing them will be considered in a later section.

Manual RESET and START

The RESET signal, whether generated manually or during power-on, forces the system into a known state prior to beginning the data processing algorithm. This is an important hardware consideration since some memory elements may energize in a random condition and could cause unplanned system operations.

The START signal, as shown in Figure 2.16, allows control over the actual start of the processing algorithm. Following power-on or RESET, the initialized system will remain in block READY until START arrives. Asserting START allows the system to proceed with the processing algorithm. To assure proper processing, the START signal must be synchronized with the system clock.

To terminate the processing action, an END processing block can be used. As shown in Figure 2.16A, the END block loops endlessly to itself and performs no processing actions. Logically interrupting the system clock to terminate processing actions is generally not a good idea. Any logic involvement with the system clock may cause delays. Clock glitches, which may cause unpredictable behavior, are also possible.

When the processing algorithm concludes, other actions are also possible. In Figure 2.16B the processing algorithm returns to the READY block, where it remains until another START signal is asserted. Note that there is a potential problem if the START signal persists for a long period of time, perhaps generated by a manual push button. The system may finish the processing algorithm, return to the READY block and again detect that START is asserted. It would then again repeat the processing algorithm, perhaps with undesired results. As a possible solution, START can be restricted to last only one clock cycle. In Figure 2.16C the algorithm remains in the END block until a RESET signal is received. Another alternative is to have RESET equivalent to the release of START.

---

1. Clare, C.R. *Designing Logic Systems Using State Machines*. New York: McGraw-Hill Book Co. 1973.

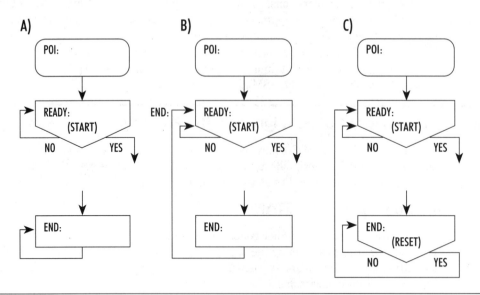

Figure 2.16   Use of START and RESET Signals in a Microoperation Diagram

PAUSE, RESUME, and ABORT

Other supervisory external signals may also participate in the processing algorithm. These signals are generated manually or by other electronic systems. Their implementation details will be presented in Chapter 4, "The Controller."

- **PAUSE** stops the processing algorithm and may be used for various purposes. It can delay execution of one system to allow coordinated action with another system. For example, PAUSE may delay a processing algorithm until another system provides an essential piece of data. It may also be used as a method to share components among several systems.

- **RESUME** may be an independent external signal or it may be implemented as the logical complement of PAUSE. In either case, RESUME allows the system to continue its processing operations during the next clock cycle. PAUSE and RESUME should not affect the content of the system registers or memory. They should affect only the relative timing of the processing algorithm.

- **ABORT** causes the system to abandon the processing algorithm and return to a known condition. If used for safety purposes, ABORT may initiate a sequence of operations that leaves the system in a power-down and safe condition. Following an ABORT, the system requires initialization before processing operations can resume.

2.9   Analysis and Design Examples

The following examples illustrate many of the analysis and design concepts discussed in this and the previous chapter.

Example 2.7:   Analysis of an RTL Program

Given the following RTL program description, find the sequence of microinstructions and the content of registers A and B while in each clock cycle.

```
REGISTER: A[8], B[8];          /* Declaration of registers.                    */

POI:          A = 1, B = 3;   /* Initialize values for registers.             */

UNE:          A←A + B,        /* Perform data processing operations.          */

              B←A;

DUE:          A←A + 2,        /* Processing and conditional branch.           */

              if (A > 7) then

                  GOTO TRE,

              else

                  GOTO DUE,

              end if;

TRE:          B←B + A,        /* Processing and unconditional branch.         */

              GOTO TRE;
```

Analysis:

This algorithm uses four processing microinstructions. POI is a power-on-initialize microinstruction for registers A and B. UNE performs arithmetic operations on the content of A and B. DUE involves both data processing and a decision and must be carefully analyzed. Depending on the value of A, either TRE is executed or DUE is processed again. We must be careful to assure that decisions are based on values residing in registers, not future values specified by the micro-operation.

TRE involves a repeated arithmetic operation. The accompanying table analyzes the behavior of the system. Values of the registers change simultaneously at the end of the clock cycle. Decisions are based on the current content of the registers.

Clock Cycle	Micro Inst	Current Conents		Decision	Next Micro Inst	Next Contents	
		A	B	A > 7		A	B
t_0	POI	1	3	—	UNE	1	3
t_1	UNE	1	3	—	DUE	4	1
t_2	DUE	4	1	F	DUE	6	1
t_3	DUE	6	1	F	DUE	8	1
t_4	DUE	8	1	T	TRE	10	1
t_5	TRE	10	1	—	TRE	10	11
t_6	TRE	10	11	—	TRE	10	21

Table 2.15

Following an initialization in POI, the next clock pulse advances the device to UNE; no changes in A and B are involved in the transition to UNE. While in UNE, values of A and B remain as they were in POI. Arithmetic circuitry computes A + B and control signals are generated to load registers A and B with values of 4 and 1 respectively. Note that these operations occur simultaneously. Since there are no decision variables involved in block UNE, the next microinstruction is DUE.

When DUE is entered as a result of the clock at the end of clock cycle t_1, the new values of A and B are also stored in the registers. The "next contents" from UNE become the "current contents" to be used in DUE. While in DUE, the decision, based on the current values of A and B, specifies that the next microinstruction again be DUE. The next values of A and B are also computed to load with the next clock pulse.

Microinstruction DUE is again executed after receiving the clock at the end of clock cycle t_2. Again the decision (A > 7) is false. The current value of A, which is 6, is used in this decision. The next microinstruction is again DUE.

DUE is entered a third time at the end of clock cycle t_3. During clock cycle t_4, the result of the decision is true and the next microinstruction is TRE. TRE is finally entered at the end of clock cycle t_4. The next values of A and B from TRE are registered in A and B. The device executes microinstruction TRE for all additional clock cycles. Register B continues to be modified with each succeeding clock cycle.

Example 2.8:   *Finding the Largest Number in a ROM Memory*

TABLE, a ROM, is organized as 256 words by 8 bits and contains positive integers. Design a system to search TABLE to find the largest number stored in its memory. Place the first occurrence of this number in register LARGE, and put the address of this first occurrence in register ADLARGE. Generate a DONE signal when the algorithm is finished. Prepare an appropriate datapath and microoperation diagram for the device. Do not include any supervisory signals (START, PAUSE, ABORT, RESET). Do not design the controller.

Analysis:

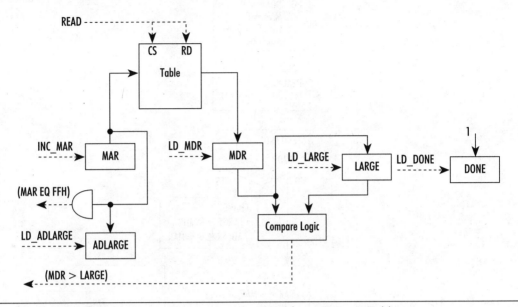

Figure 2.17   Example Datapath Structure for Modifying a Data Table

A possible datapath is shown in Figure 2.17. Since only one ROM is used, CS and RD (CHIP SELECT and READ) are tied to the READ control signal. If power consumption is not important, they can both be permanently asserted. The current largest value is to be stored in LARGE.

The ROM is read sequentially starting with location 0. After each read operation the data is compared to the content of LARGE. If the value read from the ROM is larger than the value stored in LARGE, the content of LARGE and ADLARGE are updated. MAR, the address register for the ROM, is a counter which can be incremented by control signal INC_MAR. A comparator produces an appropriate decision variable when the content of the MDR, the value from the ROM, is greater than LARGE. An AND gate determines when the value of MAR is FFH. This indicates that all data in the ROM have been examined.

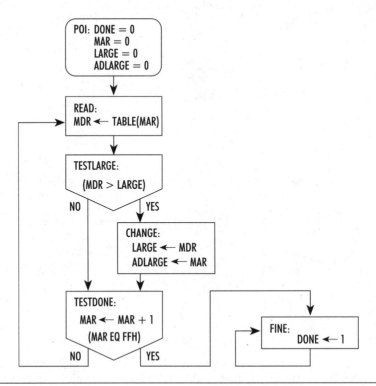

Figure 2.18   Microoperation Diagram for Table Search Example

A microoperation diagram is shown in Figure 2.18. All critical registers are initialized during power-on in POI. Microinstruction TESTDONE reads the next value from TABLE and simultaneously increments the address register for the ROM. The value of MAR (the value *before* being incremented) is checked to see if the total ROM has been examined. Memory locations from 00H through FFH will be sequentially examined before the algorithm terminates by advancing to block FINE.

TESTLARGE compares MDR and LARGE. If changes are necessary LARGE and ADLARGE receive their updated values in CHANGE. The process continues until the contents of the ROM are exhausted.

Example 2.9:   An Example Processor to Compute A Modulo B

As an example of applying a microoperation diagram, consider the design of a system with three 32-bit registers OPERAND, MODULUS, and RESULT as shown in Figure 2.19. After power-on initialization of register values, a manually entered START signal begins the processing. The device computes the modulus of OPERAND with respect to MODULUS and places the result in RESULT. This involves repeated subtractions of MODULUS from OPERAND. The operation stops when one additional subtraction would produce a negative result.

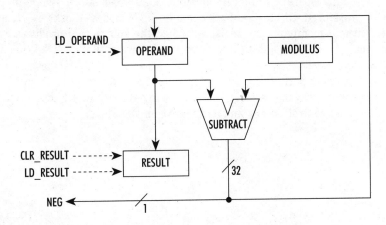

Figure 2.19   Datapath for Computing A Mod B

Analysis:

Registers OPERAND and MODULUS have previously been initialized. Bit widths of the registers and interconnection paths are 32 bits. The SUBTRACT circuitry works with two's complement arithmetic. A negative result from subtraction produces a "1" in bit 31 (the most significant bit) of the subtracter output. This is a characteristic of negative numbers represented in two's complement. Bit 31 serves as a decision variable called NEG.

Solution:

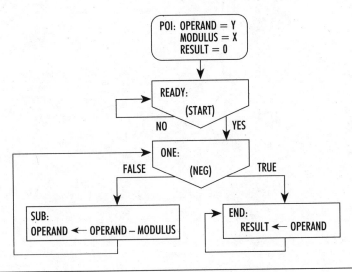

Figure 2.20   Microoperation Diagram for Computing A Mod B

The microoperation diagram of Figure 2.20 presents a possible processing algorithm. Block POI is initialized with the values for computing Y mod X. Block ONE tests the result of a subtraction without modifying any registers. When the subtraction produces a positive result, the subtraction is completed and the algorithm cycles back to block ONE. When the subtraction produces a negative result, the current register content provides the desired result and block END is entered without performing the subtraction. A register transfer language representation provides an alternative way of describing the processing algorithm. Additional formality has been included to document control signals and hardware components.

```
REGISTERS:  OPERAND[32],              /* Holds the initial operand.      */
            MODULUS[32],              /* Holds the modulus.              */
            RESULT[32];               /* Holds result of calculation.    */

CONTROL:    LD_OPERAND,               /* Load OPERAND register.          */
            LD_RESULT,                /* Load RESULT register.           */
            CLR_RESULT;               /* Clear RESULT register.          */

ARITHMETIC/LOGIC: SUBTRACT;          /* A 32 Bit Subtracter.            */

DECISION: NEG;                        /* Indicates result of             */
                                      /* subtraction is negative.        */

POI:        RESULT = 0;               /* Initialize RESULT.              */

READY:      if (START) then
                GOTO ONE,             /* Wait for START.                 */
            else
                GOTO READY,
            end if;

ONE:        if (NEG) then             /* Test if finished.               */
                GOTO END,
            else
                GOTO SUB,
            end if;

SUB:        OPERAND←OPERAND − MODULUS, /* Reduce OPERAND.                */
            GOTO ONE;

END:        RESULT←OPERAND,           /* All Done.                       */
            GOTO END;
```

Both the microoperation diagram of Figure 2.20 and the RTL statements above are representations of the same algorithm. Both contain five microinstructions.

Example 2.10: An Example Processor for Multiplication

Multiplication is an important arithmetic operation that can be performed by a large number of processing algorithms. Consider the multiplication of two positive integers, MPLIER and MCAND, to produce a product, PROD. If an arithmetic device to perform the multiply is available, the algorithm is quite uncomplicated. If a multiplier is not available or unsuitable for other reasons, for example cost, the multiply operation can be decomposed and performed by less complex components.

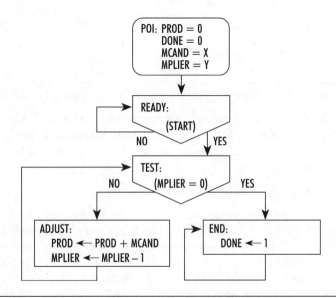

Figure 2.21   Microoperation Diagram for Repeated Addition Multiply

The microoperation diagram of Figure 2.21 and the datapath of Figure 2.22 illustrate a decomposition of the multiplication operation that allows it to be performed by sequential addition. Several higher performance algorithms for multiplication will be considered in Chapter 5, "Data Representation and Arithmetic." The microoperation blocks perform add and decrement operations. The decision is based on a decremented counter that counts the number of times the multiplicand has been added to itself. The results of this decision are made available to the controller through decision variable ZERO.

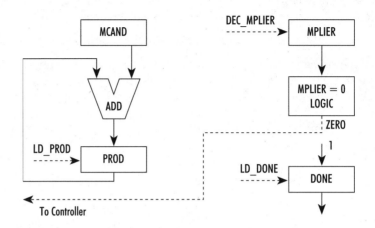

Figure 2.22   Datapath Structure to Perform a Repeated Addition Multiply

Block TEST determines if an appropriate number of ADDs have been completed. If not, ADJUST performs an ADD of MCAND to the accumulated product. The value of MPLIER is also reduced. When MPLIER is found equal to zero, block END is entered, DONE is asserted, and the algorithm terminates.

Example 2.11:   BCD to Binary Conversion

A RAM with 256 words, each of 8 bits, contains integer data obtained from digital instrumentation. The values are represented with 8 bits using binary-coded decimal. In this representation a group of 4 binary bits represents each of the decimal values in the range 0–9. The total data range is 00 to 99. Design a device to convert the BCD data to binary form. The converted data is to be stored in-place in the RAM. Select a suitable set of hardware components and prepare a microoperation diagram. Supervisory signal START, assumed synchronous, is to begin the conversion. When completed, the device is to halt. Use a ROM look-up table to perform the BCD to binary conversion.

Solution:

Figure 2.23 shows a datapath to perform the required action. The system is organized with a single data bus that links the major components. DATA is a RAM that contains the initial data. TABLE is a ROM organized as a look-up table. BCD values read from the RAM are used as addresses for the ROM. The contents of the memory location specified by the address are the required binary values. Some of the memory locations are unused since BCD does not use all possible binary bit patterns.

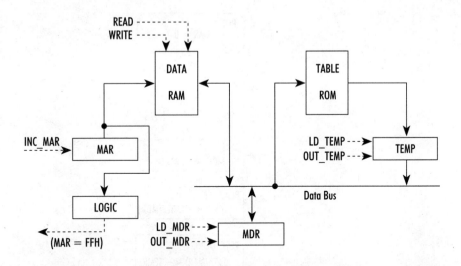

Figure 2.23   Datapath Structure to Support Data Conversion Example

MAR, a binary counter, points to the current data location in DATA. A logic block (AND gate) detects the "FFH" condition and indicates that the complete range of DATA has been processed. MDR receives the data from memory DATA and provides an address to the look-up ROM memory, TABLE. TEMP provides temporary storage of the converted data from the look-up table. It is needed to avoid conflicts on the data bus. The result in TEMP is then written to DATA.

The microoperation diagram of Figure 2.24 shows a possible algorithm for performing the required actions. In block TEST_DONE the MAR is incremented and also tested. The test is always performed on the current content of the MAR, the value before the register is incremented.

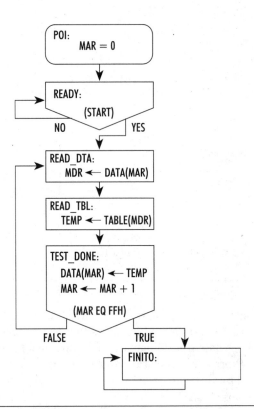

Figure 2.24   Microoperation Diagram for Data Conversion Example

2.10   The Controller State Diagram

The actual control and sequencing of the processing algorithm is performed by the controller. As indicated in Chapter 1, the controller is a finite state machine. A set of control signals is produced for each state of the controller.

Although an RTL program or a microoperation diagram can fully describe the processing action, more traditional state machine approaches are also useful. State diagrams describe finite state machines and indicate the sequence of states and system outputs as a function of system inputs. The **controller state diagram**, which is used for describing a controller, differs in several ways from the classical state diagram. This is because the controller state diagram describes and documents the controller actions rather than describing the logical status of the controller. First, the states are labeled with mnemonic names, such as CALCULATE and UPDATE, rather than binary state designations. Second, the output associated with a state is a set of control signals directed to control points within the datapath. The individual bits of the output are also given a mnemonic, such as LD_R_1 and WRITE_RAM, to suggest the action to be performed when the control signals are asserted.

There is a one-to-one correspondence between each microoperation block in the microoperation diagram and each state in the controller state diagram. To emphasize this close relationship, the same names are used to label processing blocks in the microoperation diagram and states in the controller state diagram.

The controller state diagram identifies output only when the corresponding control signals are asserted. When an output is not indicated, it is assumed to be in its passive condition; no register transfer operations related to the signal are performed. The concept of assertion avoids confusing references to positive and negative logic.

Descriptive mnemonics indicate the conditions causing state transitions. The datapath uses logic decisions based on stored data values to provide this information to the controller. The START, RESET, and other supervisory signals are external inputs to the controller. They originate from either manual entries or other cooperating systems.

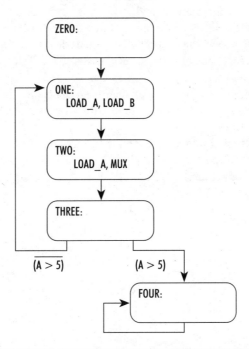

Figure 2.25   Controller State Diagram for Datapath and Microoperation Diagram of Example 2.4

Figure 2.25 exhibits the controller state diagram associated with the datapath and microoperation diagram of Example 2.4 (Figure 2.11). Each processing block of the microoperation diagram has a corresponding state in the state machine and is represented by a rounded box. Each state is given a symbolic name the same as that used in the microoperation diagram. The state block lists all of the control signals asserted when the controller is in that state. An arrow terminates on the next state. Decision variables, such as (A > 5), are used with arrows when the next state depends on a decision variable. The design of state machine controllers is the major topic of Chapter 4.

Summary

This chapter presents techniques to express the datapath processing algorithm. Data is stored in registers and data memory. It is processed by arithmetic and logic devices and routed via interconnection paths. Control signals from the system controller influence all aspects of the processing activity. The specification of the data processing activity is stated by register transfer operations. These mathematical statements indicate the source of information, the arithmetic/logic operation performed, and the destination of the results.

Register transfer language programs describe the operation of a digital device. This is a hardware descriptive program that specifies the register operations during each clock cycle.

As an alternative, the microoperation diagram specifies the register transfer operations to be performed in each clock cycle. The graphical structure is similar to a flow chart. An appropriate microoperation diagram involving processing blocks, initialization blocks, and decision blocks allows complex algorithms to be specified using available datapath components. The microoperation diagram allows the content of all registers to be determined for each clock cycle.

Supervisory control signals, such as RESET, START, PAUSE, and ABORT are also important to system operation.

The combination of the microoperation diagram, the controller state diagram, and the hardware configuration of the datapath provides a means of both specifying and documenting the computing structure.

For additional information and references on the topics presented in this chapter, access the World Wide Web page provided at: http://www.ece.orst.edu/~herzog/docs.html.

Key Terms

ABORT	microoperation
catenation	microoperation diagram
comparator	PAUSE
control set	processing blocks
combined decision and processing block	register transfer language (RTL)
controller state diagram	
data memory	RESUME
decision blocks	supervisory control signals
initial condition blocks	word
microinstruction	

Readings and References

Hennessy, John L. and David Patterson. *Computer Architecture: A Quantitative Approach.* San Mateo, CA: Morgan Kaufmann Publishers, 1990.

This is a fine second-level textbook emphasizing quantitative analysis of design implementations. Chapter 5 gives a good overview of microoperations.

Gibson, Glenn A. *Computer Systems Concepts and Design*. Englewood Cliffs, NJ: Prentice-Hall, 1991.

   Chapter 7 includes some useful material on using microoperations to describe processing actions.

Clare, Christopher R. *Designing Logic Systems Using State Machines*. New York: McGraw-Hill Book Co., 1973.

   This is the original source of the algorithmic state machine concept.

Hill, Frederick J. and Gerald R. Peterson. *Digital Logic and Microprocessors*. New York: John Wiley & Sons, 1984.

   Chapter 10 includes good reference material on algorithmic state machines and register transfer languages.

Exercises

2.1   A, B, and C are 16-bit registers. Write the register transfer statements to perform the following operations:

  (a) Put the lower byte of A into the lower byte of B.

  (b) Interchange the two bytes of register A.

  (c) Shift the content of register B two positions to the left. Bits emerging from the most significant bit position are introduced in the least significant bit position.

  (d) Complement the upper byte of register A.

  (e) Put the lower byte of A into the upper byte of A, put the upper byte of A into the lower byte of B, put the lower byte of B into the upper byte of B, put the upper byte of B into the lower byte of A.

2.2   Show a hardware organization capable of performing each of the data movements of Exercise 2.1.

2.3   A, B, and C are 16-bit registers. Write register transfer statements to perform the following operations:

  (a) Replace A with the sum of B and C.

  (b) Replace the lower byte of A with the exclusive or (XOR) of the upper bytes of B and C.

  (c) Replace A, B, and C with the sum of all three registers.

  (d) Reverse the order of the bits in register A.

2.4   Show a hardware organization capable of performing each of the data movements of Exercise 2.3.

2.5   DATA is a 256×16 RAM memory. A, B, and C are 16-bit registers. Write register transfer statements to perform the following operations:

  (a) Move the content of memory location 25_{16} to register A.

  (b) Read the content of two consecutive memory locations specified by register A into registers B and C.

  (c) Register A is used to provide an address and data to DATA. Write 0 into location 0, 1 into location 1, 2 into location 2, and n into location n.

2.6   Show a hardware organization capable of performing each of the operations specified in Exercise 2.5.

2.7   The initial value of register A = 1 and B = 2. What is the difference between the register transfer statements in *(a)*, *(b)*, and *(c)*?

(a) B←B + 1, A←A + B;

(b) B←B + 1; A←A + B;

(c) A←A + B − 1; B←B + 1

2.8   Develop a microoperation diagram, including control signals, which will store numbers in memory TABLE, a 256 × 8 RAM. 0 is to be stored in location 0, 1 in location 1, 2 in location 2, etc. Show a hardware structure to support the microoperations.

2.9   Develop a microoperation diagram and supporting hardware to clear DATA, a 64K × 8 memory, and then set a CLEAR bit to indicate the operation is complete. Include the control signals in the microoperation diagram.

2.10   Develop a microoperation diagram and supporting hardware to examine the content of DATA, a 256 × 8 RAM memory. If bit 7 of a data word is 1, the contents are to be replaced with 00_{16}. Include control signals in the microoperation diagram.

2.11   Develop a microoperation diagram with control signals and supporting hardware that will count the number of blank locations (content 00H) in DATA, a 256 × 8 ROM, and store the value in register B.

2.12   Develop a microoperation diagram with control signals and supporting hardware that will examine DATA, a 256 × 8 memory. Remove all values of 00_{16} from the memory and move up the following data to take its place. When finished, the algorithm is to stop. *Hint:* Use two address registers; one for reading, the other for writing.

2.13   In the Figure 2.26, the registers are initialized with values A = 0, B = 0. Clock cycle 0 advances the controller from block POI to block ONE. Show the content of registers A and B at the beginning and at the end of clock cycles t_0 through t_6.

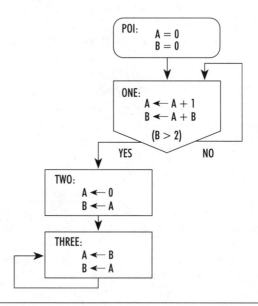

Figure 2.26

2.14 Registers A, B, and C contain positive integers and are connected on a single bus as shown in Figure 2.27. TEMP1 and TEMP2 are connected to a compare circuit. Derive a microoperation diagram that will order the content of A, B, and C so that A contains the smallest value and C contains the largest value.

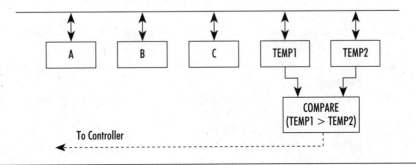

Figure 2.27

2.15 Assume that devices to perform integer multiplication and addition are available. Given an integer in register N, show a datapath and microoperation diagram that will compute N! [N factorial = N (N − 1) (N − 2) ... (2) (1)]. Assume that all registers and arithmetic devices are sufficiently long to accommodate the length of the result.

2.16 Figure 2.28 describes a digital device. The device is initialized in block POI. It advances to block ONE at the conclusion of clock cycle t_0. Determine the content of registers A, B, and C at the beginning and end of each of the clock cycles t_1 through t_5. Prepare a table to show the values.

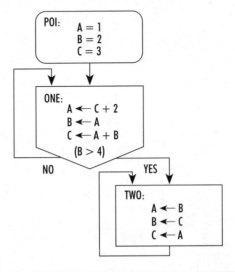

Figure 2.28

2.17 Redesign the microoperation diagram of Figure 2.18 so that the operations in blocks READ and TESTLARGE are combined into a single block.

2.18 Write a register transfer language program to describe the microoperation diagram from Exercise 2.8.

2.19 Write a register transfer language program to describe the microoperation diagram from Exercise 2.9.

2.20 Write a register transfer language program to describe the microoperation diagram from Exercise 2.10.

2.21 Write a register transfer language program to describe the microoperation diagram from Exercise 2.11.

2.22 Write a register transfer language program to describe the microoperation diagram from Exercise 2.12.

2.23 In Example 2.7, microinstruction DUE contains both a processing action and a decision based on the same variable. Prepare an RTL program for the example in which the processing and decision operations are separated into two microinstructions. The algorithmic action is to be the same.

2.24 Some digital systems are designed using a two-phase clock. Phase A is used to clock the datapath; phase B is used to clock the controller. Both clocks have the same frequency with the active transition of phase A occurring before the active transition of phase B. What are the advantages and disadvantages of this timing?

2.25 In Example 2.2, in Solution 3 of the Fibonacci number generator, two registers, COUNT and NUMBER, have been added to allow a test of the looping. Show how this could have been accomplished with a single register. Show the modification of the RTL program needed.

3
Hardware Structures for Data Processing

Alan Turing

An Englishman named Alan Turing
Found computing to be quite alluring
He found, it would seem
That a simple machine
Would make his name quite enduring

Prologue

Alan Turing was born in England in 1912. After a childhood marked by burning curiosity and dogged determination to complete any task he started, Turing won several awards for his mathematical ability. At the age of 15 he read and critiqued Einstein's book on the theory of relativity.

While at Cambridge University, Turing, at age 24, formulated a conceptual model of a computational device, now called a **Turing Machine**. The model consists of a tape of unbounded length, a processor with a finite memory, and a mechanism for reading and writing symbols on the tape. Turing showed that his machine was capable of performing all computational operations that can be performed in a finite length of time. The Turing Machine will be examined in greater detail later in this chapter.

Turing completed graduate studies and earned a doctorate from Princeton University. John von Neumann offered him a position as an assistant working on computer-related research. With the clouds of war gathering, he chose instead to return to England where German invasion was thought to be inevitable

The German air force used a highly sophisticated mechanical encryption device for its central radio communications. Turing and an encryption expert interviewed a Polish engineer with first-hand knowledge of the machine, known as Enigma.

Turing was successful in breaking the code and supervised the design and construction of a computer capable of deciphering the Enigma messages. One of the first decoded messages indicated that the English town of Coventry was to be bombed. Winston Churchill, the British Prime Minister, chose not to alert his air force to defend the town, an action that deprived the enemy of the knowledge that its code had been broken but resulted in a large loss of life.

The strain of these events and personal problems had their toll on Turing. Using his home chemistry laboratory, he produced potassium cyanide through the electrolysis of a solution obtained from seaweed. He coated an apple with the mixture, went to bed, ate the apple, and died.

3.1   Introduction

While Turing's work does not necessarily aid in designing high-performance processors, it does indicate that simple processors can perform complex tasks if given enough time. The designer's role is to make reasonable trade-offs between device complexity (the number of components) and performance.

In systems that process information, trade-offs between the system datapath and the controller are possible. It is usually possible to perform processing operations faster (higher performance) with less control involvement if highly specialized datapath hardware is used. Simpler and less costly datapath hardware can be used if the controller algorithm is allowed to become more complex.

This chapter takes a closer look at processing information. In particular it will examine the organization of the processing elements in the datapath and the effect of this organization on system performance.

Many of the following sections will use a processing action of converting a binary number to a binary-coded decimal number. Though of limited practical importance, this example illustrates many of the trade-offs and compromises which are possible in datapath design.

3.2   Structure of the Datapath

Figure 3.1 presents a model for an information processing architecture. This figure is a modification of the model presented in Figure 2.3. The source data operands originate in registers or memory devices (M). Data is routed or switched through a data interconnection structure (S) to the processor (P), which contains the arithmetic and logic elements. Processed data returns to storage (M) via additional data routing and switching circuitry (S). Each datapath element has appropriate control points that allow the controller to sequence a processing algorithm.

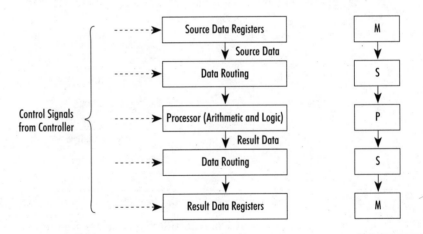

Figure 3.1   Datapath Structure

The quantity of data simultaneously processed and switched depends on the datapath structure. For example, an adder can combine two operands selected from two source data registers to produce a sum that can then be stored in a single result data register.

It is assumed that elements in the datapath can produce the desired results in a known length of time. In most situations, results are produced fast enough for data to be selected, routed, processed, and returned to a storage location in one clock cycle. Operations of this type can be conveniently described by a microoperation diagram.

High performance requires that the total processing activity take a minimum length of time. Delays associated with logic gates and setup times for registers must be small. P must have a relatively small number of AND-OR logic levels. If results are not available for storage within one clock cycle, a slower clock or techniques involving multiple clock cycles must be used.

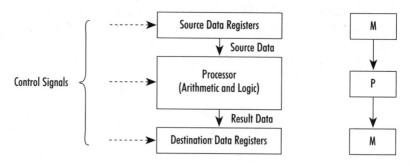

Figure 3.2   A Simplified Datapath Structure with Fixed Routing

For situations in which there is fixed data routing circuitry, for example point-to-point routing, the less complex organization shown in Figure 3.2 is appropriate. No control is exerted over the data routing. When data is moved between storage elements with no data processing operations, the simplified organization in Figure 3.3 is appropriate.

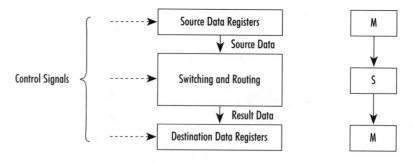

Figure 3.3   Simplified Datapath Structure for Data Movement with No Processing

In some situations the source and destination registers are the same, as shown in Figure 3.4. This is called **in-place processing** and allows data to be processed in a repetitive or iterative manner and returned to the source registers.

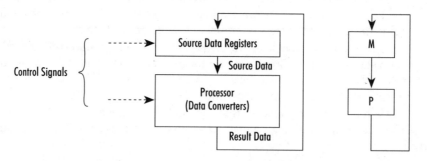

Figure 3.4   In-Place Processing

3.3   Datapath Processing

A processing activity, P, can be performed in many ways. One method is to implement the processing activity in specially designed digital logic. When provided with input data, the processor uses arithmetic and logic hardware to produce the output data. Even for operations as complex as multiplication of 8-bit integers, the results can be expressed in a truth table. There are 16 input variables to consider; 8 from the multiplier and 8 from the multiplicand. The truth table has 2^{16} rows. It would be formidable and unrealistic to fabricate the resulting two-level AND/OR circuitry. A read-only memory holding the 64K (2^{16}) possible 16-bit products would be reasonable for some applications. Expanding the technique to accommodate the multiplication of 32-bit integers would require a look-up table with 2^{64} entries. Its fabrication would not be reasonable with current technology.

An alternative processing approach decomposes large and complex processing activities into several less complex processing steps. If possible, this method can lead to processing techniques offering trade-offs between the quantity of processing hardware and processing performance. The overall processing is done by performing the simpler activities sequentially.

If the original processing activity is P, each member of the decomposition set will be referred to by p_i, where i references the sequential ordering of the p's. Each p_i is performed by a different processing element in the datapath structure. If P produces the same results as the sequential processing of a series of p's, this will be indicated as:

$$
P = \begin{bmatrix} p_1 \\ p_2 \\ \dots \\ p_n \end{bmatrix}
$$

In-Place Processing

In-place processing subjects a data set to a sequence of processing actions in successive clock cycles. After each clock cycle processed data is returned to the original registers. Figure 3.5 shows a structure in which each of the processing actions p_1 through p_n are made available to the data registers through a multiplexer switch. The data is processed first by p_1, second by p_2, etc. Some of the processors may be used multiple times if processing activities are repetitious; fewer than n processors may be sufficient. The controller selects the multiplexer settings to process the data sequentially through the processor elements.

A processing algorithm involves the interaction of data with processors. This can be illustrated with a scheduling table. Each location in the table corresponds to a clock cycle for one of the processors. Assume that P is decomposed into three sequential processing operations, p_1, p_2, and p_3. If there are two data sets, D_1 and D_2, to be processed, six time periods are required.

In this example only one processor is active in each time slot; two are idle. High performance systems try to use all available resources during each clock cycle. **Datapath processor efficiency** is defined as the ratio of the number of used processing clock cycles to total number available. In the schedule shown, 6 of the 18 available time slots are used for an efficiency of 6 / 18 or 0.33.

Processor	Clock Cycle					
	t_0	t_1	t_2	t_3	t_4	t_5
p_1	D_1	—	—	D_2	—	—
p_2	—	D_1	—	—	D_2	—
p_3	—	—	D_1	—	—	D_2

Table 3.1

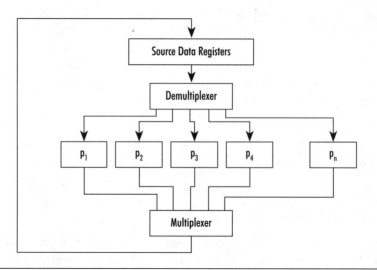

Figure 3.5   In-Place Processing

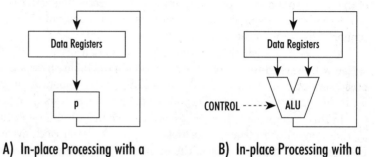

A) In-place Processing with a
Single Processor

B) In-place Processing with a
Programmable ALU

Figure 3.6   Iterative Processing in a Datapath with a Single Processing Element

For the special case where all of the processing actions are identical, a single processor can be used, as shown in Figure 3.6A. More complex repetitive actions can be performed using an ALU that can use a control signal to select the arithmetic and logic operation. This is shown in Figure 3.6B. The schedule table for processing two data sets with three processing actions is the same for both situations, as shown in Table 3.2. Although six time units are still required, the processor is used with every clock cycle. The datapath processor efficiency is 1.00.

Processor	Clock Cycle					
	t_0	t_1	t_2	t_3	t_4	t_5
p	D_1	D_1	D_1	D_2	D_2	D_2

Table 3.2

When using in-place processing, assume that the processing time for a single data set is n clock cycles. After each cycle the processed data is returned to the original registers. Since no registers are available, a new data set cannot be introduced until the first has completed its processing. Each additional data set also requires the full n clock cycles.

Example 3.1:   In-Place Processing

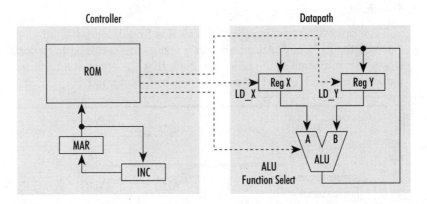

Figure 3.7   An Iterative Processing Unit Using an ALU Chip

Figure 3.7 shows an example system in which the processor is a 74LS382 arithmetic and logic unit (ALU). This device, as shown in Table 3.3, can produce eight different arithmetic and logic operations specified by the code provided to the three function select lines. Two registers, X and Y, are connected to the ALU and initially contain operands. Demonstrate that this system can be used to compute the functions:

X←(3 * Y) + X,
Y←0

Function Select			Function	
S_2	S_1	S_0		
0	0	0	$F = 0000$	ZERO
0	0	1	$F = B + \overline{A} + C_{in}$	SUBA
0	1	0	$F = A + \overline{B} + C_{in}$	SUBB
0	1	1	$F = A + B + C_{in}$	ADD
1	0	0	$F = A\ XOR\ B$	XOR
1	0	1	$F = A\ OR\ B$	OR
1	1	0	$F = A\ AND\ B$	AND
1	1	1	$F = 1111$	ONE

Table 3.3

Solution:

The processing function $P = (3 * Y) + X$ is first decomposed into a sequence of functions provided by the ALU unit. All intermediate results are returned to the original registers. Table 3.4 shows the utilization of the processor. Note that the effect of a processing action is not observed until the next clock cycle. Processor datapath efficiency is 1.00.

	Clock Cycle				
	t_0	t_1	t_2	t_3	t_4
Processor Control (function select)	ADD	ADD	ADD	ZERO	—
Register Control	LD_X	LD_X	LD_X	LD_Y	—
Register Transfer	X←X + Y	X←X + Y	X←X + Y	Y←0	NO OP
Contents of X (based on original contents)	X	X + Y	X + 2 * Y	X + 3 * Y	X + 3 * Y
Contents of Y (based on original contents)	Y	Y	Y	Y	0

Table 3.4

$$P = \begin{bmatrix} p_1 \\ p_2 \\ p_3 \\ p_4 \end{bmatrix} = \begin{bmatrix} A - B \\ A + B \\ A + B \\ 0000 \end{bmatrix}.$$

The datapath in Figure 3.7 performs a sequence of processing operations on the data in registers X and Y. By providing the proper sequence of control signals, the ALU generates an arithmetic or logic function using the contents of X and Y as operands. The results are loaded into either X or Y for further processing.

The control signals for the registers and the function specification for the ALU are stored as a memory word in the ROM. The memory address register is incremented after each data processing operation. The combination of the ROM, its address register, and the increment circuit functions as a control unit. No provision has been made for terminating the processing algorithm.

Pipeline Processing

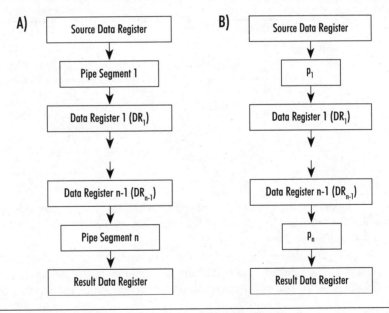

Figure 3.8   Pipeline Processing

The structure for **pipeline processing** is shown in Figure 3.8A. Data operands originate in one or more source data registers. The single datapath is decomposed into several smaller datapaths called **pipe stages** or **pipe segments**. Data is processed in each pipe stage for one machine cycle. A **machine cycle** consists of a fixed number of clock cycles, usually in the range of one to four. The datapath can perform processing actions on the data within a machine cycle. At the conclusion of the machine cycle, the data is passed via a data register to the next pipe segment. In Figure 3.8B, a machine cycle is limited to a single clock cycle. Only one processing action can be performed. The pipe segment in this configuration has no storage. After each stage of processing, the results are stored in intermediate data registers. In the first time interval, p_1 processes data set D_1, and the results are stored in intermediate data register DR_1. In the next time unit p_2 processes D_1 with the results stored in DR_2. If each processing action requires a single clock period, the entire processing action for D_1 requires n clock cycles, with the final processed results stored in the Result Data Register.

After the initial data, D_1, has progressed from the source data register to DR_1, it is possible to load new data, D_2, into the source data register. In this manner p_2 processes D_1 while p_1 processes D_2. Once the pipeline is filled, each of the processing elements can simultaneously perform a different portion of the processing algorithm on different data sets. The data is processed in an assembly line manner.

With an n-stage pipeline, the first full processing action requires n machine cycles. At the end of n machine cycles the pipeline is filled with partial results from processing other elements in the data stream. An additional processed result emerges from the pipeline after each machine cycle. If a machine cycle consists of a single clock cycle, each clock cycle produces a new set of processed information.

The pipeline schedule for the case of three processors and two data sets is shown in Table 3.5.

Processor	Clock Cycle					
	t_0	t_1	t_2	t_3	t_4	t_5
p_1	D_1	D_2	—	—	—	—
p_2	—	D_1	D_2	—	—	—
p_3	—	—	D_1	D_2	—	—

Table 3.5

The processing action is completed in four time units, t_0 through t_3, with a processor efficiency of $6 / 12 = 0.5$. Clock cycles t_4 and t_5 are not included in the efficiency calculation since the processing action terminates at t_3.

Table 3.6 shows the device utilization when processing four data sets. Six time units are required. The schedule uses 12 of the 18 available processor time cycles for an efficiency of 0.67 ($12 / 18 = 0.67$). The efficiency continues to increase with the number of data sets to be processed.

Processor	Clock Cycle					
	t_0	t_1	t_2	t_3	t_4	t_5
p_1	D_1	D_2	D_3	D_4	—	—
p_2	—	D_1	D_2	D_3	D_4	—
p_3	—	—	D_1	D_2	D_3	D_4

Table 3.6

Since all stages in the pipeline are used simultaneously, pipelining is a form of parallel processing. Pipeline processing can process multiple data sets very efficiently.

Cascade (Ripple) Processing

Figure 3.9 shows **cascade processing**, also known as **ripple processing**—an alternative for implementing sequential processing activities. In contrast to pipelined processing, the results from the first processor are immediately applied to the second processor. The results from successive processors are immediately provided to their successors without intermediate storage. No looping of data from later processing elements to earlier elements is allowed.

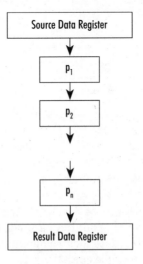

Figure 3.9   A Cascade Processor

Since each processor cannot produce valid results until it receives valid information from its predecessor, there is a ripple of valid results which begins at p_1 and proceeds to p_n. If the exact delay of each processing element is known, the clock period can be adjusted to assure that the result data register receives valid information before the next clock pulse arrives. As an alternative, multiple clock cycles can be allocated to the processing activity.

Since the exact time delay of each processing element is not generally known, the portion of the cascade processors producing valid results cannot be determined in advance. Often with arithmetic operations, the processing time is data dependent. Addition of numbers with many internal carry operations, for example, are inherently slower than additions with no carries. In a cascaded system, a worst-case processing time for each component must be known. For an adder, assuming that a carry was generated in the lowest bit position and rippled to the most significant bit would allow us to determine the total amount of time that must be allocated to finish the cascaded computation.

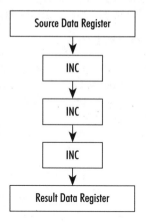

Figure 3.10   An "Add Three" Device Using Cascaded Increment Units

For example, suppose a processing element is needed which accepts an integer value and increases its value by three (ADD_3). Assume that incrementers are available to perform the increment (add one) operation within one clock cycle (worst case). Assuming that no overflow occurs, the cascaded system shown in Figure 3.10 performs the ADD_3 operation. A single clock cycle may not be sufficient, however, for valid results to be available for storage at the result data register.

Cascade processors, because of the relatively large number of layers of logic, require a longer time period to produce a valid result than a single processing unit requires. The system clock can be slowed down to accommodate long computation periods so that its period exceeds the time required by the worst-case ripple operations. In the above ADD_3 example, a clock period with a duration three times that required for a single processor would allow enough time for all ripple effects to end. By extending the clock cycle and slowing the clock rate, the performance of all operations decreases, even those not requiring an extended processing period.

Logic designers in the dark ages used an experimental procedure to determine the proper clock rate by constructing a prototype in which the clock frequency was gradually increased until processing errors were detected. The clock rate was then decreased by 20% to account for component variability. Using modern simulation tools, it is possible to determine accurately the worst-case delays due to logic, path propagation, and register set-up times.

Instead of changing the clock rate, the microoperation diagram can be modified to support the slow ripple operation. Figure 3.11 shows a portion of a microoperation diagram to accommodate the ADD_3 hardware. In block ONE the source data register receives the operand. Blocks TWO, THREE, and FOUR allow time for the processed data to propagate through the processors. In block FOUR the now completed result is captured in the result data register. Three microoperation blocks (TWO, THREE, and FOUR) have effectively been combined to produce one slow microoperation block.

Microoperations require stable control signals for the time period before the clock pulse arrives. When processing operations exceed a single clock cycle, continuity to control signals affecting elements in the datapath for longer than one clock cycle must be assured. Discontinuities in the control signals may occur during state transitions of the control unit. With multi-period control signals, the controller must be designed to avoid any control signal discontinuities. If a programmable processor's control signal has a glitch in the midst of a cascaded processing operation, data could be corrupted.

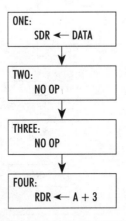

Figure 3.11   Microoperation Diagram to Support Slow Cascade Processing

Using multiple clock cycles for slow operations maintains a high rate for the system clock while still accommodating a few slow operations. It requires that the worst-case ripple time be known in advance. In general, an n-stage ripple processor will produce a result in fewer than n clock periods. This is true because intermediate registers require no set-up time.

Example 3.2:   Pipeline and Cascade Processing

Two's complement is a coding scheme that allows a representation of both positive and negative binary numbers. When using two's complement, a negation can be performed by complementing all of the bits and adding 1.

For example, in an 8-bit representation, the integer +5 is represented as 00000101. The representation of –5 is performed by first complementing all the bits. This gives 11111010. The addition of "1" then produces the two's complement representation of –5, 11111011.

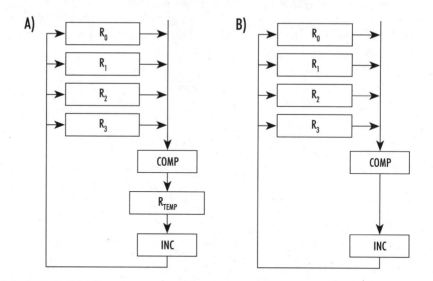

Figure 3.12   Datapath for Pipeline and Cascade Processing of "Negate"

Figure 3.12 shows two possible datapaths to perform a two's complement negate operation. A 33 MHz clock (clock cycle of 30 nanoseconds) is to be used. Table 3.7 shows the worst-case delays for system components.

Component	Time Required	Comment
Complement Logic	5 nanoseconds	Logic, propagation delay.
INC Logic	10 nanoseconds	Logic, carry propagation delay.
Register	20 nanoseconds	Register, setup time.

Table 3.7

Find a microoperation diagram for pipeline processing the contents of registers R_0 through R_3 using the datapath structure of Figure 3.12A. Results are to be returned to the source registers. Find the total processing time. Repeat using ripple processing of the registered data, as shown in Figure 3.12B.

Solution 1, Pipeline Processing:

In the pipeline processor, complement followed by a load into R_{TEMP} requires $5 + 20 = 25$ nanoseconds. INC followed by a load back into one of the source registers requires $10 + 20 = 30$ nanoseconds. Each of the operations can be performed within one clock cycle. The microoperation diagram for pipeline processing the data is shown in Figure 3.13A. A total of 5 microinstruction blocks are required. The processing of data in the 4 registers is completed in 5 clock cycles. The total processing time is 150 nanoseconds.

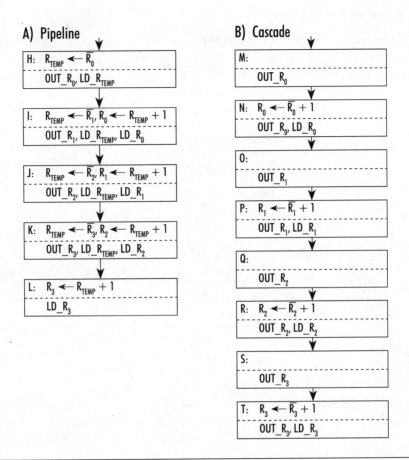

A) Pipeline

H: $R_{TEMP} \leftarrow \bar{R}_0$
OUT_R_0, LD_R_{TEMP}

I: $R_{TEMP} \leftarrow \bar{R}_1$, $R_0 \leftarrow R_{TEMP} + 1$
OUT_R_1, LD_R_{TEMP}, LD_R_0

J: $R_{TEMP} \leftarrow \bar{R}_2$, $R_1 \leftarrow R_{TEMP} + 1$
OUT_R_2, LD_R_{TEMP}, LD_R_1

K: $R_{TEMP} \leftarrow \bar{R}_3$, $R_2 \leftarrow R_{TEMP} + 1$
OUT_R_3, LD_R_{TEMP}, LD_R_2

L: $R_3 \leftarrow R_{TEMP} + 1$
LD_R_3

B) Cascade

M:
OUT_R_0

N: $R_0 \leftarrow \bar{R}_0 + 1$
OUT_R_0, LD_R_0

O:
OUT_R_1

P: $R_1 \leftarrow \bar{R}_1 + 1$
OUT_R_1, LD_R_1

Q:
OUT_R_2

R: $R_2 \leftarrow \bar{R}_2 + 1$
OUT_R_2, LD_R_2

S:
OUT_R_3

T: $R_3 \leftarrow \bar{R}_3 + 1$
OUT_R_3, LD_R_3

Figure 3.13 *Microoperation Diagram for Example Datapath*

Solution 2, Ripple Processing:

In the ripple processor, both the complement and addition are performed in cascade. The required time is 5 + 10 + 20 = 35 nanoseconds. This cannot be performed in a single 30-nanosecond clock period; two must be used.

A microoperation diagram for the ripple solution is shown in Figure 3.13B. Since the combination of complement and add cannot be performed in one clock cycle, two blocks of the microoperation diagram, for example M and N, must be used. Each operation requires the continuity of the OUT_R_i control signal for two clock periods. Eight processing blocks are required; processing time is 8 * 30 nanoseconds = 240 nanoseconds.

If it were possible to slow the clock down to a period of 35 nanoseconds, the cascaded operations could be performed in one clock period. Four processing blocks would be required (a total processing time of 4 * 35 nanoseconds = 140 nanoseconds). This processing time is better than that required using pipeline processing. The disadvantage is that all additional processing actions would also be slowed down.

Parallel Processing

In Figure 3.14 P is decomposed so operations p_i through p_n can be performed simultaneously to produce result equivalent to P. This type of decomposition is not always possible. If each of the p_i can be performed in one clock cycle, the total operation can be completed in one clock cycle. For example, if vector A = $[a_1\ a_2\ a_3]$ and vector B = $[b_1\ b_2\ b_3]$, the sum of the two vectors can be computed with three processors in one clock cycle: A + B = $[a_1 + b_1, a_2 + b_2, a_3 + b_3]$.

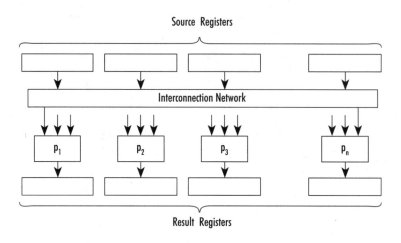

Figure 3.14   A Parallel Datapath Structure

Because parallel computation can perform extensive computation in a short period of time, it has been of great interest to computer designers. It should be noted that not all computational problems can be effectively solved using parallelism. A more extensive discussion of this topic is included in Chapter 11.

3.4   Comparison of Processing Structures

Consider the problem of converting an 8-bit binary integer to a binary-coded decimal (BCD) representation. In BCD a group of 4 bits is used to store the binary equivalent of a decimal digit. For example:

$$11000101_2 = C5_{16} = 197_{10} = 0001\ 1001\ 0111\ \text{(BCD)}$$

The conversion algorithm is not trivial since each bit of the BCD representation may be a function of multiple bits of the binary representation.

In the shift and add three algorithm, the original integer n-bit binary number is assumed to have been divided by 2^n. This is equivalent to aligning the binary point to the left of the most significant bit in the binary number. The algorithm reconstructs the value of the original number by multiplying the number by 2 and repeating the process n times; this is equivalent to shifting the n bits across the binary point.

In the reconstruction, corrections are made so that the resulting number after each of a series of shifts is a BCD number rather than a binary number. This is done by dividing the area to the left of the binary point into BCD fields for units, tens, and hundreds. Before a number is shifted, each of these fields is examined to determine if the contents *after the shift* would be 10 or greater. A number greater than 10 would be an incorrect BCD representation.

If the number were 10 or greater after the shift, it would be 5 or greater *before the shift*. It is this condition that is detected. A correction is made by adding 3. This is equivalent to adding 6 after the shift, which skips over the unused BCD codes of 1010 through 1111. The correction is made in each of the fields of the BCD number. The algorithm is demonstrated in Example 3.3.

Example 3.3: *Convert the 8-bit binary number 11111111 (255_{10}) to its BCD form.*

Description	100s	10s	1s	Binary
Initial Alignment				11111111
shift			1	1111111
shift			11	111111
shift			111	11111
add 3 (1s position)			1010	11111
shift		1	0101	1111
add 3 (1s position)		1	1000	1111
shift		11	0001	111
shift		110	0011	11
add 3 (10s position)		1001	0011	11
shift	1	0010	0111	1
add3 (1s position)	1	0010	1010	1
shift	10	0101	0101	
Decimal Equivalent	2	5	5	

Table 3.8

The binary to BCD conversion algorithm uses the add three data conversion extensively. In the following examples, this function will be used as a building block for several alternative processor designs. For a single decade, Table 3.9 shows the truth table of the add three operation.

Inputs				Outputs			
A_3	A_2	A_1	A_0	B_3	B_2	B_1	B_0
0	0	0	0	0	0	0	0
0	0	0	1	0	0	0	1
0	0	1	0	0	0	1	0
0	0	1	1	0	0	1	1
0	1	0	0	0	1	0	0
0	1	0	1	1	0	0	0
0	1	1	0	1	0	0	1
0	1	1	1	1	0	1	0
1	0	0	0	1	0	1	1
1	0	0	1	1	1	0	0

Table 3.9

It is not necessary to consider input conditions greater than 1001 since such inputs cannot occur. A field with value 0101 or greater would be corrected (ADD_3) before the shift. The result after the shift would be no greater than 1001.

Several possible implementations of this algorithm will be examined. They will indicate some of the possible trade-offs between processing performance and the complexity of the datapath processing elements.

Single Processing Element

The complete conversion operation for an n-bit binary number can be specified by a truth table with n input variables and 2^n rows. For binary to BCD conversion the number of output variables will be greater than or equal to the number of input variables. An 8-bit binary number requires three decades and 10 bits to represent 255, the largest 8-bit integer.

By conventional logic design, the truth table can be converted into a two-level AND OR logic circuit. In counting levels of logic, NOT gates are ignored. A circuit so designed would indeed be very fast, since only two levels of processing gate delay would be required to produce the result. This circuit would also be very large, requiring several hundred gate-level components.

When using a single processor element with no more than two levels of logic, high speed is obtained at the cost of a large amount of specialized hardware. Each component plays a unique role in the conversion. The complexity of single processor elements often grows rapidly as a function of the number of input variables. While such a system may not be feasible because of the high cost of components, its performance, requiring only one clock cycle, represents a benchmark for judging performance of other forms of processor organization.

Instead of logic components, a ROM with 2^n memory locations can be used. The original binary number provides the address; the contents of the memory cell are the BCD representation. While a table look-up ROM for an 8-bit number requires only 2^8 or 256 words of memory, the memory requirement doubles for each additional bit in the operand. ROM table look-up for converting a 32-bit number would require 2^{32} words of storage. While ROM implementation may be feasible, it will not be considered in detail in this discussion.

Pipelined Processor

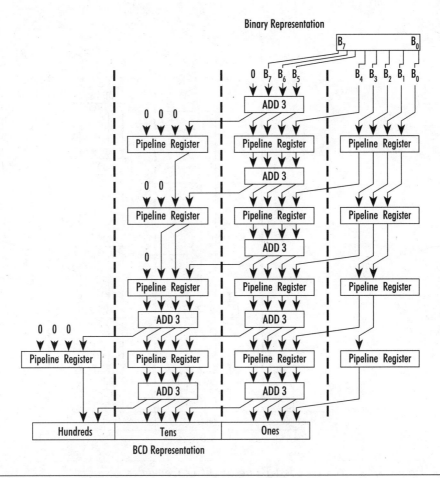

Figure 3.15   A Pipelined Processor Implementation of the Binary to BCD Conversion Algorithm

Figure 3.15 illustrates a pipelined implementation for the binary to BCD data conversion algorithm. Each machine cycle for a pipe segment is a single clock cycle. The ADD_3 circuitry is replicated seven times. There is both a horizontal and a vertical repetition of the circuitry. Shifting is accomplished by routing the outputs from each stage of the processing one position to the left before entering the next stage. To capture intermediate results, pipeline registers are provided at the output of each processing level. Five clock cycles are required to produce the BCD results. If pipeline features are used, additional conversion results can be obtained in each additional clock cycle.

Cascade Processor

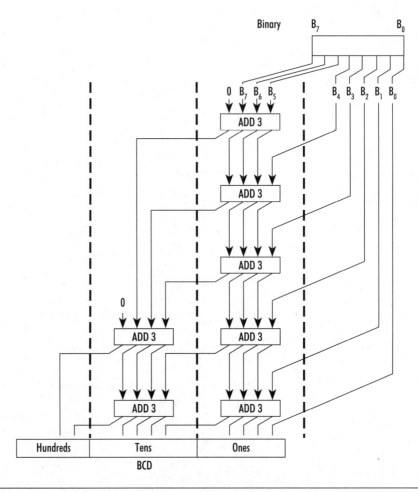

Figure 3.16   A Cascaded Binary to BCD Converter

Figure 3.16 shows a cascade processor implementation of the algorithm. The circuitry is similar to the pipeline processor of Figure 3.15 with the removal of the internal pipeline storage registers. Using the ripple processor is slower than using a single processor because of the delay caused by multiple layers of logic. Processing requires long clock cycles or multiple clock cycles. A trade-off has been made between performance and complexity; to reduce hardware complexity, performance is degraded.

Performance of the cascaded processors can also be compared to the pipelined implementation. The pipelined approach requires five clock cycles to process one binary number. The cascaded approach, although it has similar processing hardware, does not have internal registers. Since register set-up time for internal registers is not required, the processing time for one binary to BCD conversion is the sum of the delay times for each of the sequential processing gates. It may correspond to several clock cycles. For performing a conversion on a single piece of data, the time duration is less than that required by the pipeline. With the cascaded design, it is not possible to introduce new data with each clock cycle. Therefore pipelined processing offers better performance for multiple blocks of data.

In-Place Processing

A single register serves as both a source register and destination in the in-place processing structure shown in Figure 3.17. Only one level of ADD_3 devices is required. This reduces the quantity of processing elements. Because the maximum number in 8-bit binary is 255, an ADD_3 is not required for the 100s position. The shift and ADD_3 operations are combined by spatially routing all outputs from the ADD_3 unit one position to the left before performing the parallel load to the destination register. The least significant bit of the BCD result register is loaded with a shift from the binary number register.

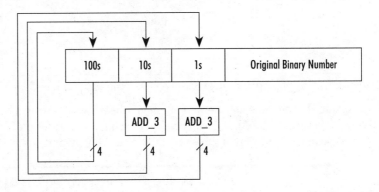

Figure 3.17   Binary to BCD Conversion Using In-Place Processing

In this implementation, the registers and ADD_3 devices are used several times in an iterative manner to produce the final results. Fewer components are needed, but it is necessary to use them more than once in the processing algorithm. The minimum-length clock cycle for this system depends on the time required to produce all of the output bits from the ADD_3 circuits. The maximum clock rate is the same as that for the pipelined processor. The processing time for a single data unit is eight clock cycles. In contrast to the pipelined processor, this method requires an additional eight clock cycles for each data element processed after the first. For processing a block of data, this system is slower than both the pipeline and cascaded processors.

By time sharing a single level of processing hardware, the saving of hardware has been considerable. Only two of the ADD_3 units are required. Hardware savings are even more dramatic for processing longer bit length BCD numbers. This savings has been gained at the expense of longer processing time required by time-sharing the group of ADD_3 devices.

Nibble Serial Processor

With in-place processing of the previous section, separate ADD_3 devices were used for each decade. This can be reduced by sharing a single ADD_3 device among the three groups of 4 bits (one-half of a byte or a **nibble**) representing each decade. Each decade shares the correction circuitry sequentially. The result, as shown in Figure 3.18, is a system in which a single ADD_3 corrector is used. Performance is again degraded due to the extra time required to sequentially share the ADD_3 corrector.

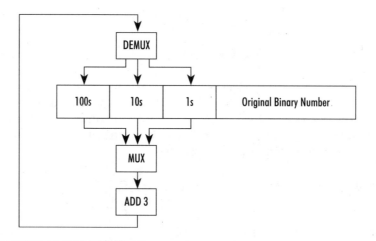

Figure 3.18   Binary to BCD Conversion Using a Nibble Serial Processor

Note that there is no interaction among the decades during a correction. No overflow carry is generated during the ADD_3 correction of each decade. A total of seven ADD_3 correction cycles are required (in addition to the eight shift cycles) for the conversion process.

	Number of ADD_3 devices	Internal Register Required	Time Required for First Result	Time for Additional Results
Single Processor	NA	0	1 clock cycle.	1 clock cycle.
Pipelined Processor	7	4	5 clock cycles.	1 clock cycle.
Cascade Processor	7	0	Depends on logic delays.	Depends on logic delays.
In-place Processing	2	0	8 clock cycles.	8 clock cycles.
Nibble Serial Processor Solution	1	0	15 clock cycles.	15 clock cycles.

Table 3.10

The multiple solutions presented in this section demonstrate that it is possible to have many viable trade-offs between performance and hardware complexity. The performance and complexity of each of the solutions is shown in Table 3.10. The greater the amount of unique hardware used, the less time required to perform the operation.

3.5   Bit Serial Processors

Processing actions discussed up to present have been performed on multiple-bit data words. Some processing operations, because of their repetitive nature, can be performed by a sequence of bit-level operations. This is known as **bit serial processing**. These operations are often encountered in performing arithmetic. In the serial binary addition of two numbers $A = A_n \ldots A_2 A_1 A_0$ and $B = B_n \ldots B_2 B_1 B_0$, for example, A_0 and B_0 are first combined in a 1-bit binary adder to produce S_0, the first bit of the sum, and C_1, a carry. This is followed by successive operations involving the remaining bits in additional 1-bit binary adders.

Figure 3.19 shows how the addition can be accomplished with a conventional ripple adder. Data from register A and register B are simultaneously provided to the adder inputs. A carry signal, which may originate at any of the bit positions, ripples to the left, where it may affect the sum and carry bits of other adder segments. None of the adder segments can produce the correct sum bit until the carry input is known. At the conclusion of all the carry ripples, the results can be placed in register SUM. Because of the potential carry propagation from the lowest to the highest bit positions, ripple adders may require a long processing time.

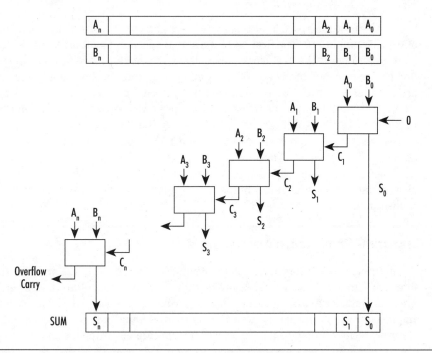

Figure 3.19   A Ripple Adder

Since each of the processor units in the ripple adder is identical, an alternate, bit serial implementation, shown in Figure 3.20 can be used. Instead of completing the add operation in one clock cycle with n adder sections, a bit serial adder uses n clock cycles with a single adder. A shift register is used to present the operands 1 bit at a time. The sum bits are collected, one per clock cycle, by register SUM; SUM←A + B.

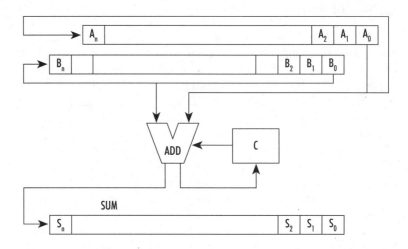

Figure 3.20   A Bit Serial Adder

In the first time period bits A_0 and B_0 are presented to the 1-bit adder unit. The initial value of the carry input is 0. The first bit of the sum (least significant bit) is routed to the SUM shift register for storage at the next clock pulse. The carry bit generated is saved in the C register. Bits A_0 and B_0 are returned to the end of the A and B registers. In the second time period, bits A_1 and B_1 are combined with the carry produced in time period 1. The sum and carry bits are stored as before. The process proceeds through n time periods to produce the sum in the SUM register. The overflow carry, if any, is stored in C. Registers A and B are unchanged.

Several early computers used bit serial techniques to implement arithmetic because of the economy of components. These techniques are very slow and can be justified only when circuitry must be minimized or when long time delays are acceptable. Many pocket calculators use extensive serial processing since human operators are insensitive to delays up to about 200 milliseconds.

3.6   Iterative Processing Techniques

As an alternative to directly implementing a hardware datapath, indirect methods can be used effectively in some situations. Many of the useful techniques are based on iteration. An initial guess is made for a desired result. Based on computations and decisions, the value of the guess is improved until a sufficiently accurate solution is found. For a modest number of iterations, the iterative technique can be a good alternative for implementing a complex computation.

Iterative methods are based on techniques developed to find the solution to an algebraic equation. Consider, for example, the following equation:

$$\sin(x) = 0.5$$

A solution to this equation is a value of x, in radians, for which $sin(x) = 0.5$. An equivalent problem is to solve:

$$f(x) = sin(x) - 0.5 = 0$$

$f(x)$ is plotted in Figure 3.21 for several values of x. A solution for this equation is a value of x for which $f(x) = 0$ or $sin(x) = 0.5$. It can be seen from the figure that the solution is between the values $x = 0.4$ and $x = 0.6$.

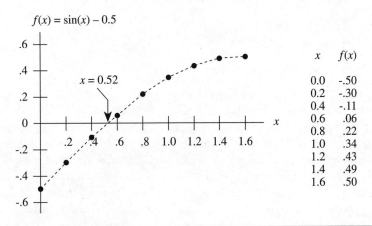

x	$f(x)$
0.0	-.50
0.2	-.30
0.4	-.11
0.6	.06
0.8	.22
1.0	.34
1.2	.43
1.4	.49
1.6	.50

Figure 3.21   An Equation to be Solved by Iterative Techniques

An exact solution for this equation is:

$$x = sin^{-1}(0.5)$$

Techniques to calculate the inverse sine directly, using specialized hardware, a table look-up, or Taylor series expansion may not be economically justified. Calculation of the sine, while of similar complexity, is more likely to be available. The following techniques involve iterative solutions to computational operations.

Iterative Solution

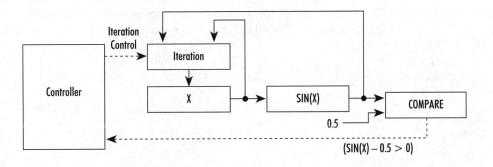

Figure 3.22   Implementation of an Iterative Computation Structure

Figure 3.22 shows an implementation of a processing structure with a method for examining various values of x to see if they satisfy the requirements for a solution. x is measured in radians. The 8-bit representation of x is scaled such that the binary point is to the left of the most significant bit. This gives x a range of:

$$.00000000 \leq x \leq .11111111$$

When a value of $f(x_i) = 0$ is produced, x_i is the root of the equation. With discrete values of x_i it may not be possible to find a value for which $f(x_i)$ is exactly zero. If two values of x are such that:

for $x = x_p$, $f(x_p) < 0$

  and

for $x = x_q$, $f(x_q) > 0$,

then a root of the equation lies between x_p and x_q. Iterative techniques attempt to find values of x_p and x_q that are close together and bound the range of the solution within an acceptable value.

The performance of iteration algorithms is strongly influenced by the technique for generating iteration values for x_i. The next value of the iteration is based on the value produced by the compare circuit, the current value of x_i, and perhaps the value of $f(x_i)$.

Note that the exact solution requires the computation of $\sin^{-1}(x)$. The iterative solution requires the computation of $f(x) = \sin(x) - 0.5$. The computation of one function $(\sin^{-1}(x))$ has been replaced with the computation of its inverse $(\sin(x))$. This substitution may be an advantage if the inverse function is already available or easier to calculate.

Assuming a solution exists and $f(x)$ is continuous in the search region, several possible algorithms can be used to adjust the value of x to search for a solution.

Trial and Error

A solution can be found by sequentially exploring the entire range of values of x. This can be done by implementing x as a counter. A value of $x_0 = 0$ can initially be selected as a solution. The value of $f(x)$ is computed and the sign of $f(x)$ noted. x is then incremented and $f(x)$ can again be computed and compared to 0. The process continues until there is a change in the sign of $f(x)$. This indicates that the function has passed through zero and a solution is within the range of the two preceding values.

This process may be slow, especially if the number of bits associated with x is large. The exact number of iterations is data-dependent. To find a root very close to $x = 00000000$ requires only a few iterations; a root close to $x = 11111111$ would require 255 iterations. Implementing the iteration technique requires only a counter. Table 3.11 shows a solution using this technique for $\sin(x) - 0.5 = 0$. For ease of interpretation, a decimal representation of $f(x)$ has been used. x has been scaled to a range of 0.00000000 to 0.11111111.

Iteration	x_i	$f(x) = \sin(x) - 0.5$	$f(x) > 0$	Comment
0	00000000	−.5000	N	
1	00000001	−.4961	N	
...	...	...	...	
132	10000100	−.0069	N	
133	10000101	−.0001	N	
134	10000110	.0032	Y	Change of sign. A solution is found.

Table 3.11

Binary Search

The **binary search technique** is similar to the trial-and-error method, except it is much more efficient. The bits of x are adjusted, one at a time, beginning with the most significant bit. For example, in an 8-bit system, assume that x has been scaled and it is known that a root of the equation exists for $0.00000000_2 < x < 0.11111111_2$. Assume also that $f(0.00000000) < 0$ and $f(0.11111111) > 0$.

The algorithm starts with the most significant bit set to 1; the remaining bits are 0. $f(x)$ is evaluated to determine if the result is greater than or less than zero. If it is greater than zero, x is too large a number and the most significant bit is returned to a value of 0. If $f(x)$ is less than zero, the bit retains its value of 1. In both cases the process repeats with the trial value of the next bit set to 1. The following table shows a binary search for the solution of $f(x) = \sin(x) - 0.5 = 0$.

Iteration	x_i	$f(x) = \sin(x) - 0.5$	$f(x) > 0$	x_{i+1}	Comment
0	10000000	−.0206	N	11000000	MSB is 1.
1	11000000	.1816	Y	10100000	2nd bit is 0.
2	10100000	.0851	Y	10010000	3rd bit is 0.
3	10010000	.0333	Y	10001000	4th bit is 0.
4	10001000	.0066	Y	10000100	5th bit is 0.
5	10000100	−.0069	N	10000110	6th bit is 1.
6	10000110	−.0001	N	10000111	7th bit is 1.
7	10000111	.0032	Y	10000110	LSB is 0.
8	10000110	−.0001			Done.

Table 3.12

The binary search technique locates a solution in eight iterations. The number of iterations is always equal to the number of significant bits in the data registers. One additional bit of the solution is determined in each iteration cycle. Because of the general usefulness of this process, specialized circuitry, called a **successive approximation register**, is available to perform the required bit manipulation.

Example 3.4:   BCD to Binary Converter

As an example of the binary search technique, consider the design of a BCD to binary converter. Several methods of implementing the inverse function, a binary to BCD converter, were presented earlier in this chapter. These can be used as the basis for computing the BCD to binary conversion. Figure 3.23 shows how the previous technique can be used to perform this action.

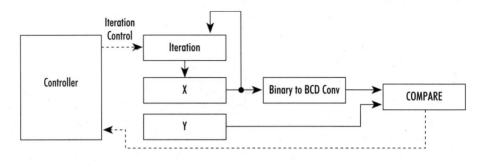

Figure 3.23   A BCD to Binary Converter Using Iteration Techniques

The BCD operand is placed in register Y. X_0, an initial value to be used in the algorithm is placed in register X. The value in Register X is converted using the binary to BCD algorithm. The resulting BCD value is compared with the value in register Y. (It might be noted that a binary compare circuit will also work with BCD values). The value in register X is adjusted according to the iteration algorithm until the required binary result is achieved.

For example, assume that the system in Figure 3.23 is to find the binary representation of the BCD number 0001 0101 1001 (159_{10}). The successive approximation using a binary search is to be used in the iteration. Table 3.13 shows how the values of the registers would be changed.

Iteration	x	f(x)	Y	f(x) < Y
0	1000 0000	0001 0010 1000	0001 0101 1001	Yes
1	1100 0000	0001 1001 0010	0001 0101 1001	No
2	1010 0000	0001 0110 0000	0001 0101 1001	No
3	1001 0000	0001 0100 0100	0001 0101 1001	Yes
4	1001 1000	0001 0101 0010	0001 0101 1001	Yes
5	1001 1100	0001 0101 0110	0001 0101 1001	Yes
6	1001 1110	0001 0101 1000	0001 0101 1001	Yes
7	1001 1111	0001 0101 1000	0001 0101 1001	Yes
8	1001 1111			

Table 3.13

In this example, the correct value of the binary representation of 0001 0101 1001 is found in eight iterations. One iteration is required to adjust each bit sequentially in the result.

Newton-Raphson Technique

The Newton-Raphson technique is a classical numerical analysis technique for finding the root of an algebraic equation. The root of an equation $f(x) = 0$ can be found using the following iterative process.

$$x_{i+1} = x_i - \frac{f(x)}{f'(x)}$$

For the above example $f(x) = \sin(x) - 0.5$. $f'(x)$, the derivative of $f(x)$, is $\cos(x)$. Table 3.14 shows the convergence of this algorithm based on an initial value of $x = 0.00000000$.

Iteration	x_i	$f(x) = \sin(x) - 0.5$	$f'(x) = \cos(x)$	$x_{i+1} = x_i - f(x)/f'(x)$	Comment
0	00000000	−.5000	1.0000	10000000 (.5000)	First guess.
1	10000000	−.0206	0.8776	10000110 (.5235)	
2	10000110	−.0001	0.8661	10000110 (.5236)	No change.

Table 3.14

The rapid convergence of this algorithm is readily apparent. The disadvantage of using it is the need to perform more complex computations to determine the replacement value. Circuitry is required to calculate $\cos(x)$ and perform a binary division.

3.7   Bit Slice Processors

In studying datapaths, registers and ALUs are normally considered as modular devices that are interconnected to make them useful. An alternative view is the **bit slice processor**, which is a modular device consisting of a narrow slice of a general assortment of registers, data buses, and arithmetic and logic units. Common widths are 2 bits and 4 bits. An assortment of bit slice components may be interconnected to make a general-purpose datapath of arbitrary width. Each unit processes a portion of the full data word width. Each of the processor units has data connections to its neighbor to allow shift operations and the propagation of carry and borrow information.

Prior to the development of modern techniques for fabricating application-specific integrated circuits (ASICS) from part libraries, bit slice units provided an alternative to constructing a datapath from selected components such as registers and logic. Studying such devices still provides insight to modular computing structures.

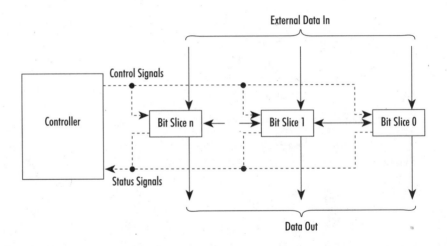

Figure 3.24   A Bit Slice Processing Architecture

Bit slice processors can be used, as shown in Figure 3.24, to construct the datapath of a digital processor. To complete the design, a controller provides control signals for programmable datapaths, arithmetic and logic functions, and register operations. Because many of the components and interconnections have been chosen in advance, the user's influence is limited to selecting the data width, providing external data paths, and implementing the controller.

Historically, bit slice processors have been useful for applications requiring performance greater than that available from a conventional microprocessor. The internal electronics of bit slice components are inherently faster than the type used in most general-purpose microprocessors. By designing the controller, the processing algorithm can be customized for high performance in a specific application. Bit slice components also allow moderately complex systems to be constructed with a relatively small number of parts.

Figure 3.25 shows a conceptual organization of the AMD 2901, which was originally introduced by Advanced Micro Devices Corporation (AMD). Although not of modern vintage, this device is representative of the bit slice approach to system design. In addition to the bit slice units, the 2900 family contains supporting devices including control units and memory devices. The idea of modular datapath devices still exists in libraries available for computer aided design.

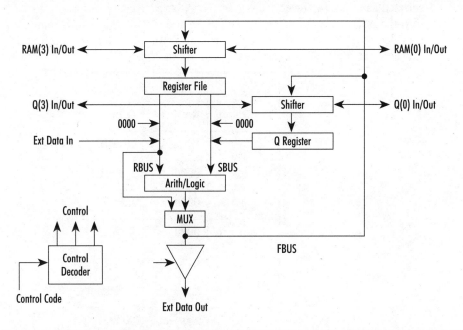

Figure 3.25   Conceptual Model of the AMD 2901 Bit Slice Microprocessor

The 2901 has a word width of 4 bits. The major memory elements are 16 registers, organized as a register file, and a 4-bit Q register, which acts as an accumulator. These registers are combined in a three bus interconnection architecture. Two buses, RBUS and SBUS, provide inputs to the arithmetic and logic unit. A third bus, FBUS, routes results from the ALU back to the registers. Provisions also exist to accept external input data and to provide external data output from the ALU. External control inputs provide addresses and a selection code to specify the registers to be placed on the buses.

To facilitate shifting operations among bit slice units, the input path to the register bank and the Q register passes through a shifter with external data pins. This allows a register to accept information which has been shifted one bit to the right or left across a chip boundary.

The arithmetic and logic unit can perform three arithmetic operations and five logic operations as specified by a 3-bit control signal. A carry input and a carry output facilitate arithmetic operations among multiple units. Status signals generated by the ALU include bits to indicate overflow, the sign of the result (positive or negative), and zero (the result of the operation was 0).

The sixteen registers in the register file are interconnected on two data buses, ABUS and BBUS. Two addresses, each of 4 bits, are used to select source registers from the register file to be placed on ABUS and BBUS. The BBUS address also is used to specify a destination register within the register file. Additional coding, via a 9-bit instruction, selects the connection between the internal ABUS and BBUS, and the RBUS and SBUS connected to the ALU.

Rather than provide direct control signals to the individual registers and ALU, a 9-bit control code, I(0:8), is provided. (The reversing of the usual bit order has been done to agree with the literature published by AMD.) The code is decoded on the bit slice chip to allow a choice of operations selected as useful by the chip designer. The 9-bit code may be broken into three fields. I(0:2) specifies the information to be placed on RBUS and SBUS. The RAM[A] and RAM[B] addresses are provided as separate information to the chip. Eight possible pairs of information can be provided. The accessibility of 0000 is useful for several processing actions.

I_0	I_1	I_2	RBUS	SBUS
0	0	0	RAM[A]	Q Register
0	0	1	RAM[A]	RAM[B]
0	1	0	0000	Q Register
0	1	1	0000	RAM[B]
1	0	0	0000	RAM[A]
1	0	1	Ext Data	RAM[A]
1	1	0	Ext Data	Q Register
1	1	1	Ext Data	0000

Table 3.15

I(3:5) specifies the ALU function to be performed.

I_3	I_4	I_5	Function Performed by ALU	
0	0	0	ADD	R plus S plus C_{in}
0	0	1	SUBTRACT	S minus R minus $\overline{C}_{in}$
0	1	0	SUBTRACT	R minus S minus $\overline{C}_{in}$
0	1	1	OR	R OR S
1	0	0	AND	R AND S
1	0	1	COMP/AND	$\overline{R}$ AND S
1	1	0	EXCLUSIVE OR	R XOR S
1	1	1	EQUIVALENCE	R EQUIV S

Table 3.16

I(6:8) specifies the destination of results from the ALU and Q register. In the following table, F is the output of the ALU. F/2 and 2F are the output of the ALU shifted to the right and to the left respectively. Similar shift operations can be performed with the entry into the Q register. Shift operations carry information between different slices of the multi-slice structure.

I_6	I_7	I_8	ALU Destinations and Shift Operations		
			DATA	OUTRAM[B]	Q
0	0	0	F	—	F
0	0	1	F	—	—
0	1	0	RAM[A]	F	—
0	1	1	F	F	—
1	0	0	F	F/2	F/2
1	0	1	F	F/2	—
1	1	0	F	2F	2F
1	1	1	F	2F	—

Table 3.17

By properly choosing the three control fields, a wide variety of register transfer operations involving ALU functions can be performed. The following are several examples of register transfer operations that can be implemented with the 2901 bit slice unit.

R(2)←R(2) + R(1)

ABUS Address:	0001	; Choose Register 1 as data source.
BBUS Address:	0010	; Choose Register 2 for source/destination register.
$I_0 I_1 I_2$:	001	; Place RAM[A] on RBUS and RAM[B] of SBUS.
$I_3 I_4 I_5$:	000	; ALU performs an ADD.
$I_6 I_7 I_8$:	011	; F is routed back to RAM.
C_{in} :	0	; No carry in.

R(15)←R(6)

ABUS Address:	0110	; Choose Register 6 as data source.
BBUS Address:	1111	; Choose Register 15 for destination.
$I_0 I_1 I_2$:	100	; Place 0000 on RBUS and RAM[A] on SBUS.
$I_3 I_4 I_5$:	011	; ALU performs an OR.
$I_6 I_7 I_8$:	011	; F is routed back to RAM.

R(7)←2 * R(7) Shift Left

```
ABUS Address:          0111  ; Choose register 7 as data source.
BBUS Address:          0111  ; Choose register 7 for destination.
I₀I₁I₂ :               100   ; Place 0000 on RBUS and RAM[A] on SBUS.
I₃I₄I₅ :               011   ; OR operands to produce register 7.
I₆I₇I₈ :               111   ; 2 * F is routed back to register 7 for storage.
```

Bit slice processors offer a method for implementing custom designs with a minimum of hardware fabrication. The structure of the datapath units makes it possible to fabricate digital computers with arbitrary word width and custom instruction sets.

3.8   Processor Performance

Processor performance measures the ability of the processing unit or computer to perform processing activities. It is related to the system clock rate, the number of clock cycles required to perform a processing activity, and the amount of processing activity that can be performed in a clock cycle.

Two measures of processor performance are of interest. The **response time** is the time between the start and finish of a computational task. This is of primary importance to an individual user. The **throughput** is a measure of the total amount of computational work accomplished in a unit time. This is of primary importance to the system manager of a computer system.

Processor performance is directly affected by the rate at which the clock may run and still achieve reliable operation. Since the clock is assumed to run at a constant rate, processor speed is determined by the slowest of the processing operations performed in one clock period. Enough time must exist within one clock period to route data to a processing element, perform an arithmetic or logic operation on the data, route the data to a storage register, and load the storage register. Delays are generated by:

- **Logic delay.** Each logic gate introduces a delay of 1 to 10 nanoseconds. (1 nanosecond = 1×10^{-9} seconds).
- **Propagation delays.** Electric signals on a wire travel at the speed of light, about one foot in a nanosecond.
- **Capacitive effects.** The capacitance of interconnecting wires limits the rate at which voltages on interconnection lines can be changed. This affects both control signals and data.
- **Set up times.** Registers require data to be present at their input for a specified time before it can reliably by captured.

The system designer controls the number of clock periods required for a processing element and the rate at which data can be accepted. The choice of architectural configuration will strongly influence performance. Using parallel processing, when possible, can dramatically increase the amount of data processing that can be performed in a single clock cycle. Data processing constraints must be balanced by the system designer. At the extremes, systems may be designed with fast clock rates that do little per clock cycle, or with slow clock rates to accommodate slow and complex processing elements.

3.9   The Turing Machine

The history of computational structures can be traced to Alan Turing. Turing proposed his abstract model of a computational device in 1936. It consisted of a state machine controller and a memory unit formed from a length of recording tape. The state machine is restricted to a finite number of states; the tape is of infinite length. Figure 3.26 illustrates a Turing Machine.

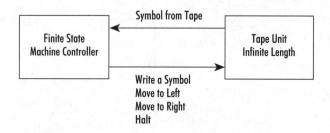

Figure 3.26   A Representation of a Turing Machine

Four operations are possible. A symbol from a finite alphabet can be written on the tape. The tape can be moved one position to the left, or one position to the right, or it can be halted. During each time unit a symbol is read from the tape and applied as an input to the state machine controller.

Turing proved that any computable problem could be solved using a Turing Machine. Although not a practical computing device, the Turing Machine represents a type of atomic component. It is the least complex device still capable of performing meaningful computations. The following example was inspired by John P. Hayes in his book *Computer Architecture and Organization*. It indicates some of the techniques used to perform calculations with this primitive device.

Example 3.5:   Addition Using a Turing Machine

Consider a natural number system using two symbols, 0 and 1. The value of a number is represented by a repetition of 1s delimited by 0s. For example, the number 4 would be represented by 011110; the number 7 would be represented by 011111110. Find a Turing Machine algorithm which will add two natural numbers separated by a single 0 delimiter.

Analysis:

The numbers are initially placed on the tape with perhaps some leading 0s. To perform the addition it is necessary to find the string of 1s and search for the delimiter separating the two numbers. The 1 at the beginning of the sequence must be changed to a 0. The delimiter must be changed from a 0 to a 1. A possible algorithm follows:

```
0001111011111110000      ; Original data, 4, 7.

0000111011111110000      ; Change the first 1 to a 0.

0000111111111110000      ; Find the delimiter and change it to 1.

0000111111111110000      ; Go to beginning of tape and halt.
```

Solution:

The following table shows a possible solution. States are assigned to each of the conditions triggering a change in the processing algorithm. Inputs from the tape are restricted to 0 and 1. The controller asserts control signals to guide the processing actions. The actions are abbreviated as R (move to right), L (move to left), 0 (write a 0 on the tape), 1 (write a 1 on the tape) and H (halt). It is assumed that the tape read head is initially positioned to the left of the two numbers. At the conclusion of the algorithm, the solution is on the tape.

Present State	Input	Next State	Action	Comment
S_0	0	S_0	R	Move to right, search for first 1.
S_0	1	S_1	0	First 1 found, change to a 0.
S_1	0	S_2	R	This is 0 just written, move right, look for 0.
S_1	1	—	—	Not possible, a 0 was just written.
S_2	0	S_3	1	Delimiter found, change to a 1.
S_2	1	S_2	R	Keep searching for delimiter.
S_3	0	—	—	Not possible, a 1 was just written.
S_3	1	S_4	L	Start moving back to the beginning.
S_4	0	S_0	H	Beginning of sequence found, Halt.
S_4	1	S_4	L	Keep searching for beginning.

Table 3.18

The above table describes the state machine controller for this device. With five states it can be implemented using three state variables. Five control signals corresponding to the five processing actions (R, L, H, 0, 1) must be provided. The value read from the tape presents a single status input to the controller.

Summary

This chapter is concerned with the structure and performance of the datapath. The datapath is a composite of arithmetic devices, memory elements, and a system interconnection structure. All respond to control signals from the system controller.

There are many ways to accomplish datapath processing. With sufficient hardware, even complex operations can be performed in a single clock cycle. Most processing actions require a sequence of simple operations on the data set.

Pipeline processing achieves high processing rates and efficiency by processing multiple data sets with an assembly line of processing elements separated by storage registers. Cascade processing allows the ripple of information, without storage, between processors. Both methods have advantages and disadvantages.

Time-sharing hardware resources at the word, byte, nibble, or bit level allow system implementation with fewer parts. This economy is achieved at the expense of additional clock cycles for the processing and additional complexity in the processing algorithm.

Iterative processing techniques are used for specialized situations in which lower performance can be tolerated, or for operations that are seldom performed. The successive approximation binary search technique and the Newton/Raphson iteration techniques are used, especially in situations in which inverse functions, such as analog to digital conversion and digital to analog conversion, are to be generated.

Bit slice processors offer a standardized datapath element that can be specialized using a controller to perform data processing activities. The Turing machine is the most generalized datapath element. Although it is of only theoretical interest, the Turing Machine is capable of obtaining a solution to any computable problem.

For additional information and references on the topics presented in this chapter, access the World Wide Web page provided at: http://www.ece.orst.edu/~herzog/docs.html.

Key Terms

binary search technique	pipe stages
bit serial processor	pipeline processing
bit slice processing	processor performance
cascade processing	response time
datapath processor efficiency	ripple processing
in-place processing	successive approximation register
iterative processing	throughput
machine cycle	trial and error
nibble serial processor	Turing machine
pipe segments	

Readings and References

Hayes, John P. *Computer Architecture and Organization*. New York: McGraw-Hill Book Co., 1988. This book has a good treatment on bit slice devices and the Turing Machine.

Exercises

3.1 What is the difference between ripple processing, pipeline processing and parallel processing? What criteria would you use to choose among these approaches?

3.2 In Example 3.2, what clock rate should be selected to achieve highest performance in the pipeline datapath? What clock rate should be selected for ripple processing?

3.3 Under the conditions of Example 3.2, how many data sets must be processed in order for pipeline processing performance to exceed ripple performance?

3.4 Consider the ripple processor solution to Example 3.2. Since the processing delay is 35 nanoseconds, why not provide the LD_R_0 control signal in processing block M?

3.5 In Example 3.2, repeat the analysis and solution if a 50 MHz clock is used.

3.6 In Example 3.2, repeat the analysis and solution if the components require 5 nanoseconds for a complement, 15 nanoseconds for an INC, and 10 nanoseconds for a register load.

3.7 Using the datapath structure of Figure 3.7, indicate a sequence of register transfer operations which will produce the following. It is not necessary to retain the contents of the register.

   (a) $X \leftarrow 2 * Y + X, Y \leftarrow 0$

   (b) $X \leftarrow \overline{X}, Y \leftarrow 0$

   (c) $X \leftarrow \overline{X}$ OR $\overline{Y} \leftarrow 0$

   (d) $X \leftarrow XY$ OR $\overline{X}\,\overline{Y}$

3.8 An n-stage pipeline is to process m data sets, $m > n$. What is the processor datapath efficiency?

3.9 Assume that a sequential processing activity p_1, p_2, p_3, p_4, p_5 can be performed by either cascaded processing elements or pipelined processing elements. Assume that the combined processing gate delay and propagation delay for each processing element is 0.7 microseconds and that the set-up time for each register is 0.3 microseconds. A 1.0 microsecond clock period is to be used.

   (a) Show the system hardware configuration and microoperation diagram for each.

   (b) What is the advantage and disadvantage of each method?

   (c) What is the maximum processing rate of each method?

3.10 Memory M_1 contains 1024 bytes of data. It is desired to take the contents of memory M_1, negate the value (take two's complement), and store it in memory M_2. Negation can be performed by complementing the value followed by an increment. It is known that the hardware components to perform various operations require the following amount of time. Assume control signals and data from registers are available instantaneously.

Read from M_1	40 nanoseconds
Complement	10 nanoseconds
Increment	25 nanoseconds
Write to M_2	25 nanoseconds
Write to a register	10 nanoseconds

   (a) If no registers are used between M_1 and M_2 (cascade processing) what is the fastest clock rate possible? How much time is required to transfer the complete contents of M_1 to M_2?

   (b) If additional registers are used between M_1 and M_2 (show your choice of quantity and location), what is the fastest clock rate possible? What is the shortest time required to complete the complete data transfer?

3.11 With the ADD_3 device used in the binary to BCD conversion, verify that an input greater than 1001 is not possible.

3.12 Use don't-care conditions to design minimum complexity logic to implement the ADD_3 logic of Table 3.9.

3.13 To accommodate a slow processing action, three approaches can be taken.

   (a) Faster (and more expensive) components can be acquired.

   (b) The system clock can be slowed to accommodate the slow device.

(c) The slow operation can use multiple clock cycles.

Under what circumstances would you use each of the above techniques?

3.14 A processing action uses two processors: p_1 and p_2. Each data set requires processing by the sequence p_1, p_2, 2. If pipelining is used, find the processor datapath efficiency if large numbers of data sets are to be processed.

3.15 For a three-stage pipeline, determine the datapath processor efficiency for 1, 2, 5, 10, 100, and 1000 data sets. Sketch a curve of efficiency vs. number of data sets. What conclusions can be drawn?

3.16 For five data sets, determine the datapath processor efficiency for pipelines of length 1, 2, 5, 10, 100, and 1000. Sketch a curve of efficiency vs. length of pipeline. What conclusions can be drawn?

3.17 Consider looping using a processor more than once in a processing action. Can this be done with a cascade connection? A pipeline connection?

3.18 Estimate the size and types of logic gates needed to perform a 4-bit addition using two-level logic. Compare this with the requirement when using ripple adders such as shown in Figure 3.19.

3.19 Show the structure of a pipelined adder using 1-bit adders to add a sequence of 8-bit integer pairs.

3.20 If the adder in Exercise 3.19 requires 10 nanoseconds for an add and 20 nanoseconds for register loading, compare the time required for a ripple adder and a pipelined adder for 100 pairs of integers.

3.21 A 1-bit adder requires 10 nanoseconds to produce a sum and carry and a register load requires 20 nanoseconds. Compare the performance of a ripple adder and a carry save bit serial adder when adding two 8-bit operands. Show the datapath structure for each.

3.22 For the function

$$f(X) = e^X - 9_{10} = 0$$

use a binary search technique (and a scientific calculator) to find a binary value of X to satisfy the equation. Note that the solution should be Ln 9_{10}. Confine your search to the region $00.000_2 \leq X \leq 11.111_2$

3.23 For the function in Exercise 3.22, use a trial-and-error search to find a root. Use an initial value of 10_2.

3.24 For the function in Exercise 3.22, use the Newton-Raphson iteration to find a solution. Perform the operations in decimal with a calculator.

3.25 Assume that for binary integer values from 0 through 11111111 (255_{10}), a ROM contains the value of the squares of the input value. Show a possible hardware configuration and microoperation diagram to find the square root of a 16-bit integer presented in register NUM. Use trial and error. The final solution is to be placed in register RESULT.

3.26 Show the register transfer operations to be performed in each clock cycle in a bit slice processor to implement the following register transfer operations. Contents to the register do not need to be preserved.

(a) $R_7 \leftarrow R_0 + 2R_1 + R_2 / 2$

(b) $R_5 \leftarrow R_5$ NAND R_4

(c) $R_3 \leftarrow !R_3$

3.27 Determine the values that must be provided for the A address (4 bits), B address (4 bits), and I_0–I_9 in each clock cycle to allow an AMD 2901 bit slice processor to perform the actions indicated in Figure 3.27. Registers R_4 through R_{15} may be used as scratchpads.

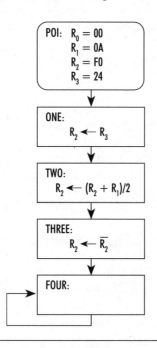

Figure 3.27

3.28 Using bit slice processors, show a 16-bit datapath structure that will allow a 16-bit integer from external register A to be multiplied by 3.5 and then stored in external register B. Assume there is no overflow.

3.29 Two numbers, *A* and *B*, with $A \geq B$, are represented by their natural form with a single 0 between them. Develop an algorithm for a Turing Machine to compute $A - B$.

3.30 Develop an algorithm using a Turing Machine to convert a 2-bit binary number to its natural representation.

4
The Controller

Maurice Wilkes

An Engineer by the name of Maurice
At Manchester sought some release
From control unit confusion
With a microprogram solution
Made from a diode matrix

Prologue

Maurice Wilkes at Manchester University in Great Britain in 1951 was one of the first to methodically organize the design of a controller. His approach, now called microprogramming, created an environment in which controllers for complex data processing actions could be systematically designed, documented, and modified.

Wilkes proposed a new method to fabricate a control unit. The contents of a memory cell, rather than digital logic, provided control signals, which were generated as a function of the value in an address register. Each memory location could contain any desired combination of the available control signals.

With Wilkes' procedure, an address register of n bits was decoded to produce 2^n output lines. Using diodes, the output lines were formed into a type of crossbar switch whose outputs were the control signal fields and the next address field. To allow conditional branching, decision variables interacted logically with the next address information to determine the next **microinstruction**. The use of jumps and conditional branching is similar to the actions performed by a computer program. Since the actions are at a lower level, (within the control unit) the term *microprogramming* describes this process.

4.1   Introduction

Throughout this book, complex digital systems have been designed based on decomposition into a datapath to perform arithmetic, logic, and data movement; and a controller to direct the algorithmic flow of the processing activity. In computing structures, the controller is central in affecting overall performance. This chapter considers techniques and strategies for implementing the controller.

Controllers are constructed using clocked sequential logic components. Figure 4.1 shows a finite state machine implementation of a controller. The state register is a synchronous, clocked memory device that maintains the current state of the controller between clock pulses. Each state must have a unique, binary state representation. Each microoperation block of the microoperation diagram is associated with one state of the controller.

For each state of the controller, the **output logic** provides a fixed set of control signals for the datapath. These may be signals for setting register modes, establishing data interconnection paths, or specifying the action performed by arithmetic and logic devices. This action occurs at the end of the clock cycle with the arrival of the clock pulse. All transitions of information between registers take place at the end of the clock cycle.

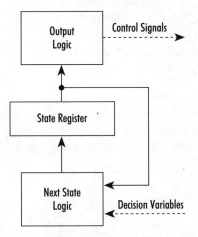

Figure 4.1   A Finite State Machine Controller

The **next-state logic** determines the state register's value for the next clock cycle. For unconditional sequential processing actions, the next state is a function of only the present state. For conditional situations, there are several possible successor processing microoperations. One or more decision variables, in conjunction with the present state, determine the next state.

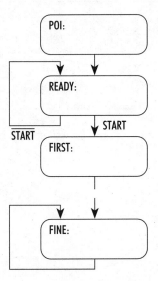

Figure 4.2   Generalized Form of a Controller State Diagram

The **controller state diagram** documents the sequence of states as a function of the decision signals from the system datapath. This diagram also lists the asserted control signals for each of the states. Figure 4.2 shows a generalized form of a controller state diagram. The controller is initialized in the power-on initialize (POI) state. It advances to the READY state where it waits for a START signal before beginning the processing algorithm. After the processing algorithm, it remains in the FINE state.

4.2   Timing of Controller Operations

As shown in Figure 4.3, a **clock cycle** consists of a sequence of cause-and-effect actions affecting both the controller and datapath. Initially, the controller is in a state specified by its state register; the datapath has a set of data values in its registers and memory. Table 4.1 describes the sequence of interactions between the controller and datapath.

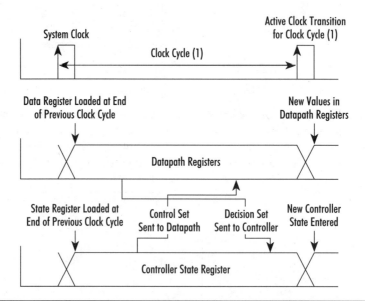

Figure 4.3   *Datapath and Controller Timing during Microoperation Execution*

Time Sequence	Controller Actions	Datapath Actions
0	The state register assumes its value from the previous clock cycle.	The datapath registers assume their values from the action of the previous clock cycle.
1	A control set, derived from the current content of the state register, is generated and sent to the datapath. It selects data interconnection paths, gates data to data buses, specifies arithmetic/logic operations, and informs registers of their appropriate action. All data transfers occur at the arrival of the next active clock transition.	A decision set, derived from the current contents of registers in the datapath, is generated and sent to the controller. The decision set, along with the current contents of the state register, determines the value the state register will assume at the next active transition of the system clock.
2	At the end of the clock cycle, the active transition of the system clock arrives and the new value of the state register is loaded.	At the end of the clock cycle, the active transition of the system clock arrives and the new values of the registers in the datapath are loaded.

Table 4.1

The maximum clock rate depends on the amount of time required for the control set to be generated, information to propagate along the data interconnections, processing delays in the data converters and decision elements, information to propagate back to storage elements, and set-up time for storage elements. A clock cycle must be allocated enough time to accommodate all processing operations and allow processed data to return to registers for storage.

Example 4.1: *Analysis of a Controller and Datapath Structure*

Figure 4.4 shows the circuitry for a controller and processor datapath. At power-on, register A (4 bits) is initialized to 0010_2, register B (4 bits) is initialized to 0101_2, register C (1 bit carry register) is initialized to 0_2. The two flip-flops in the controller are both initialized to 0. Analyze the device to determine the register contents through the first five clock cycles.

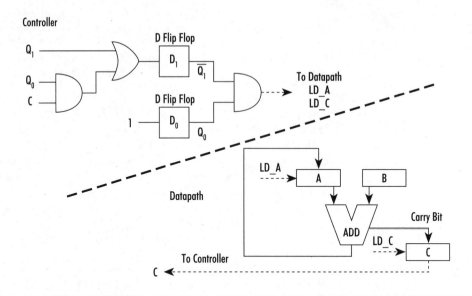

Figure 4.4   *Controller and Datapath Structure*

Analysis:

The circuitry shown is comprised of a controller section and a datapath. The controller consists of two flip-flops and miscellaneous logic. It produces control signals, LD_A and LD_C, which are sent to the datapath. The datapath has three registers; two for operands are 4 bits each. C, a 1-bit register, stores the carry from the add operation. A 4-bit adder produces a sum and a carry. The sum is stored in register A, which acts as an accumulator. The carry bit is used as a decision variable and is sent to the controller.

a) Find the controller state diagram for this device.

There are two state variables, Q_1 and Q_0, for the controller. The device is initialized with both equal to 0. Although there are four possible states, only three of them can be entered due to the logic circuitry. Figure 4.5 shows the resulting controller state diagram. The labels used are the decimal equivalent of binary representations of Q_1 and Q_0. The two control signals are produced only in state ONE (Q_1Q_0=01). The device remains in this state until the carry bit in register C becomes 1 and the corresponding decision variable is asserted.

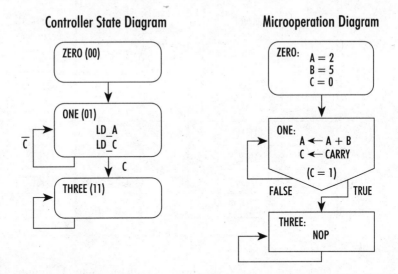

Controller State Diagram   **Microoperation Diagram**

Figure 4.5   State Diagram and Microoperation Diagram

b) Find the microoperation diagram.

    The microoperation diagram for the device is also shown in Figure 4.5. There are as many microoperation blocks as there are states in the state diagram. The first block, ZERO, involves the initialization process. In block ONE, register transfer operations affecting registers A and C are performed. There is a looping operation until register C has a value of 1 and the decision variable is asserted. The device then advances to block THREE, where it remains in an infinite loop.

c) Find the values of registers A, B, and C through the first five clock cycles.

    The present and next state of the controller and the present contents and next contents of each of the registers is specified in Table 4.2. Decimal numeric values are used. Note that in clock cycle t_3, there is an overflow of 4-bit register A. A carry is produced, which is loaded into C at the end of clock cycle t_3. The decision to advance to state THREE is based on the contents of register C, which does not affect the controller until t_4.

Clock Cycle	Current Controller State	Current Register Contents			Decision	Next Controller State	Next Register Contents		
		A	B	C	C = 1		A	B	C
t_0	ZERO	2	5	0	—	ONE	2	5	0
t_1	ONE	2	5	0	N	ONE	7	5	0
t_2	ONE	7	5	0	N	ONE	12	5	0
t_3	ONE	12	5	0	N	ONE	1	5	1
t_4	ONE	1	5	1	Y	THREE	6	5	0
t_5	THREE	6	5	0	—	THREE	6	5	0

Table 4.2

Note that in clock cycle t_4 the values of registers A and C change. This is because the device is still in block ONE of the microoperation diagram. The transition to block THREE is accompanied by the register transfer operations specified by block ONE. The problems associated with performing both processing actions and decisions in the same processing block are considered in more detail later in this chapter.

4.3   Controller Design

There are several different approaches to the design of controllers; each is especially effective for certain design environments.

- **Discrete logic.** This approach results in a classical form of a state machine, uses AND/OR/NOT gates to implement the output logic and next-state logic and uses standard logic design techniques. Using gate-level devices is convenient for uncomplicated controllers, but may be burdensome and require many components for moderately complex designs. Discrete logic allows very little flexibility for redesign. Appendix A presents techniques for designing state machines with discrete logic.
- **Programmable logic devices.** PLDs allow the designer to implement the controller, including the state register, next-state logic and output logic, within a single component.
- **Special controller chips.** Several manufacturers currently make special-purpose IC controller chips. These chips incorporate programming structures that are especially suitable for microprogramming.

Independent of the means of implementing the controller, several common requirements must be addressed.

Initializing the System Controller

The controller's first responsibility is to initialize itself in a known initial state when system power is applied. It is also responsible for initializing registers and memory elements in the datapath. This is conveniently accomplished using special circuitry to generate a glitch-free RESET signal during system power-up.

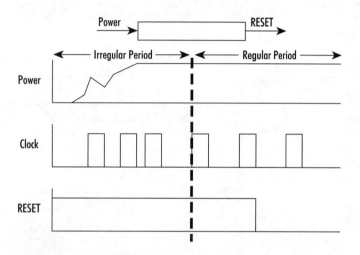

Figure 4.6   The Power-On RESET Signal

The characteristics of power-on RESET circuits are shown in Figure 4.6. At power-on, moderate transients are likely to be introduced into the system circuitry. The RESET signal is maintained in its asserted condition long enough, commonly 10 to 20 clock pulses, to allow the clock to be fully functional and the power to be stable. RESET causes the controller to initialize in the power-on initialize (POI) state. Clocked register transfer operations usually are not performed in the POI state. The RESET signal overrides any clocked actions of the flip-flops registers. It allows initialization of the state register and datapath registers upon system start up.

The controller remains in the power-on initialize (POI) state for as long as RESET is asserted. The exact duration is determined by timing constants associated with electronic circuit components. When the RESET is removed, the next clock pulse causes the controller to advance to its next processing state, READY, as shown in Figure 4.2.

Asynchronous Inputs—Manual Start

Following initialization, an optional READY state allows an operator to manually start the processing. The controller remains in state READY until given permission to advance by assertion of the START signal. Since the number of clock cycles the controller remains in state READY is dependent on an external signal, any microoperations performed in READY may be repeated multiple times. START may be generated manually or electronically from other components. As with RESET, the START signal must be glitch free.

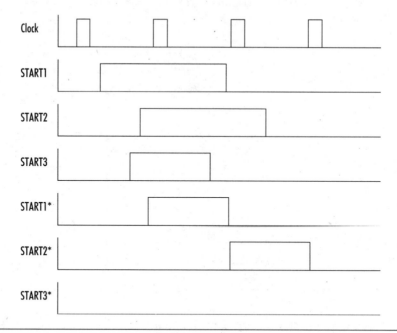

Figure 4.7   Synchronizing Input Signals

Figure 4.7 illustrates a potential problem if START is not synchronous with the system clock. START1, START2, and START3 are three variations of the START signal and are asynchronous with the system clock. They may have been produced manually or originated in another digital system. START1 arrives early in the clock cycle. There is ample time for the controller next-state logic to prepare to advance to FIRST, the correct next state. The device will perform correctly.

If the START signal arrives too late in the timing cycle, such as START2, there may not be enough time for the logic to produce signals allowing the state register to correctly advance to FIRST. For example, some of the faster flip-flops in the state register may assume their correct next state while the slower flip-flops remain in their current state or enter an unexpected state, which would produce an incorrect sequence of control signals.

START3, like START2, arrives too late in one clock cycle for reliable detection. It also terminates before the next clock arrives. With this timing, it is possible that the signal will not be detected and the controller will never advance beyond READY.

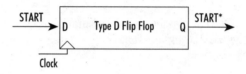

Figure 4.8   A Circuit for Synchronizing an Asynchronous Input

To eliminate these problems, convert START to a synchronous signal with transitions controlled by the system clock. This is easily done by applying START to the input of a clocked flip-flop as shown in Figure 4.8. The flip-flop output is START*. START* begins and ends in synchronism with the system clock. As shown in Figure 4.7, START1* and START2*, the synchronized versions of START1 and START2, differ in assertion by one clock period, which is normally of little consequence.

A start signal of short duration, such as START3, arrives too late for detection in the first clock cycle and terminates before detection by the following clock cycle. It may miss detection by the synchronizer altogether. To assure detection, an asynchronous signal must span at least one full clock cycle. This requires START to have a duration of at least two clock periods.

4.4   The Sequencer

A **sequencer** is a device that performs a limited subset of controller operations. Sequencers are capable of performing power-on initialization, sequential operations, and unconditional jumps. They do not respond to status information from the datapath. Their application is limited to processing algorithms requiring only unconditional sequencing of microinstructions. Even with this restricted ability, sequencers find a wide range of applications.

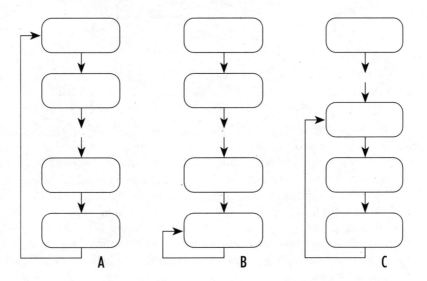

Figure 4.9   Three Variations of Sequencers

Figure 4.9 shows three controller state diagram configurations suitable for a sequencer. No control signals are shown. In circuit A, a sequence of microinstructions is performed in a non-ending loop. In circuit B, the sequence is performed once. A halt is accomplished by continuously repeating a microinstruction. In the third variation, circuit C, a sequence of microinstructions is performed once followed by a continuously looped sequence.

Synchronous, Parallel Load Counter Sequencers

Because of their relative simplicity, sequencers can be designed using medium-scale integrated circuit (MSI) chips, such as synchronous counters and shift registers. Counters with parallel load capability, such as the 74LS163 shown in Figure 4.10, are especially useful. Control signal CLEAR clears the count to 00-0. LOAD causes the value from the RELOAD register to be introduced to the counter at the next clock pulse. This chip is a 4-bit binary counter that allows designs of up to 16 states. Multiple chips can accommodate a larger number of states. The counter register is the state register for this device. The internal state transition sequence is synchronous and avoids the potential problems of ripple counters.

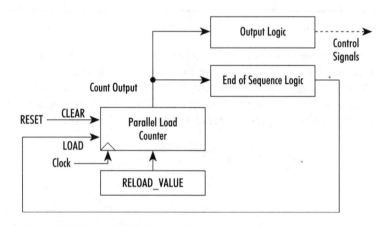

Figure 4.10   Controller with Unconditional Jump Capability

The sequencer of Figure 4.9A can be implemented with a modulo n counter, where n is the number of states in the controller. Values of n that are powers of 2 are especially easy to create with binary counters. For other values of n, the counter can be reset by appropriate logic.

The controller structure of Figure 4.10 can implement all of the controller structures of Figure 4.9. The unconditional jump address is stored in the RELOAD_VALUE register. If the reload value is 00-0, the unit loops to its initial state, as shown in the state diagram of Figure 4.9A. If the reload value is the state value of the last state, the unit effectively halts, the condition in Figure 4.9B. An intermediate reload value causes the unit to loop to an intermediate location, as shown in Figure 4.9C. With a parallel load counter, the POI state can be entered using RESET at the CLEAR control point.

Example 4.2:   Controller Design Using a Parallel Load Counter

Use a parallel load counter to design a controller as specified by the controller state diagram of Figure 4.11. The device is to have five states and provide output control signals C_0 and C_1.

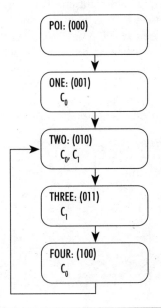

Figure 4.11 Controller State Diagram

Analysis:

The controller has five states and is in the form of a sequencer as shown in Figure 4.9C. Only 3 bits of the counter are used.

Solution:

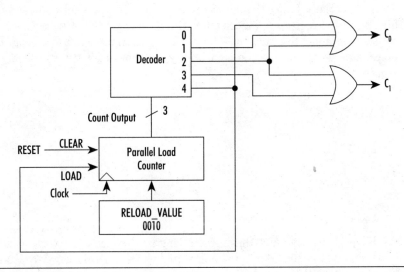

Figure 4.12 A Parallel Load Sequencer

Figure 4.12 shows one possible solution. The counter will start in state 0000 using the RESET signal applied to the CLEAR input terminal. The states are assigned in ascending sequential order from 000 through 100. Only 3 bits of the counter are required. Each of the five states is detected by a decoder. The outputs of the decoder are ORed together to generate control signals C_0 and C_1.

An unconditional jump is required from state FOUR (100) to state TWO (010). This is accomplished using the decoded signal to indicate when the device is in state FOUR. This signal causes the next clock signal to load 0010, the contents of the RELOAD register.

Shift Register Sequencers

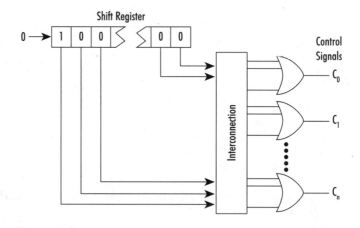

Figure 4.13   A Shift Register Sequencer

An especially simple sequencer can be formed using a shift register as the state register. Each flip-flop in the register is associated with one state of the state diagram. In Figure 4.13, a shift register initializes to a value of 100-00. The serial input is a 0. With each clock pulse, the single 1 propagates down the register and finally is dropped off the end. This corresponds to a state sequence of:

100-00
010-00
...
000-10
000-01
000-00
...
000-00

By properly choosing the register length, this sequence corresponds to Figure 4.9B. The power-on state is 100-00 and the end state, which is continuously repeated with no microoperations, is 000-00. Because each of the possible states has only a single 1 in its representation (with the exception of the 000-00 state), no decoding of the output states is required to generate control signals. The appropriate group of flip-flop outputs are ORed together to generate the control signals.

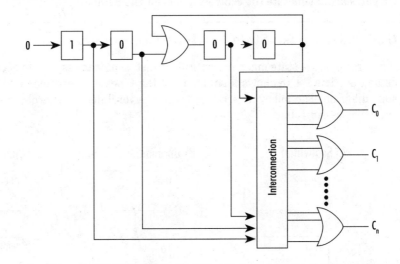

Figure 4.14   A Shift Register Sequencer with an Unconditional Jump to an Intermediate State

An unconditional jump, such as required in the sequencers in Figure 4.9A and 4.9C can be implemented using feedback on the last stage of the register. Figure 4.14 shows a configuration to support continuous looping to an intermediate state.

Example 4.3:   Design Using a Shift Register Sequencer

Repeat the design of the sequencer from Figure 4.11 using a shift register sequencer.

Analysis:

A design based on Figure 4.14 will satisfy the conditions of the problem.

Solution:

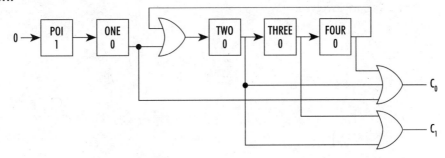

Figure 4.15   A Shift Register Sequencer Solution

One possible solution is shown in Figure 4.15. Five flip-flops are used; one for each block of the controller state diagram. RESET initializes the shift register to 10000. The unconditional jump is accomplished by ORing the output of flip-flop FOUR with the input to flip-flop TWO. The OR gates again generate the control signals for the datapath.

4.5   General-Purpose State Machine Controllers

A general-purpose state machine controller must be able to respond to decision variables from the datapath. Figure 4.16 shows a summary of the types of transitions encountered in a controller state diagram. The following sections present several design alternatives for implementing general-purpose controllers.

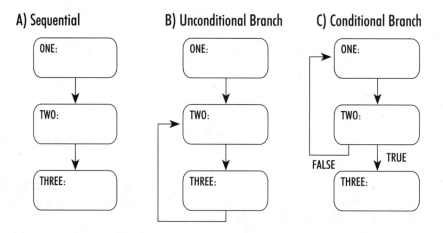

Figure 4.16   State Transitions in a Controller State Diagram

One-Hot Controller

The major disadvantage of shift register controllers is their limitation to state diagrams without status inputs from the datapath. The one-hot controller is an intuitive and straightforward method of implementing shift register controllers when conditional jumps and other controller structures are involved.

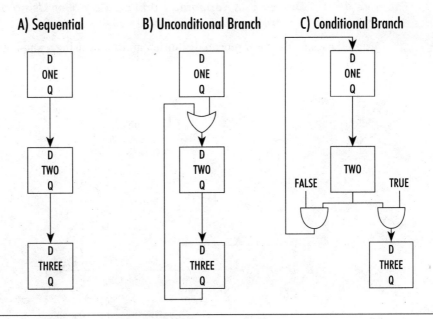

Figure 4.17   One-Hot Controller Structures

The structure of a one-hot controller is similar to that of a shift register controller. One flip-flop is assigned to each state of the device. The state of the system is determined by the single-hot flip-flop with a state value of 1. The controller design establishes logic to pass the hot-bit from the flip-flop corresponding to the present state to the proper next-state flip-flop. Figure 4.17 shows the one-hot structures associated with each controller structure in Figure 4.16.

Example 4.4:   Design of a Repeated Addition Multiplier Using a One-Hot Controller

Design the controller for a repeated addition multiplier using a one-hot controller.

Analysis:

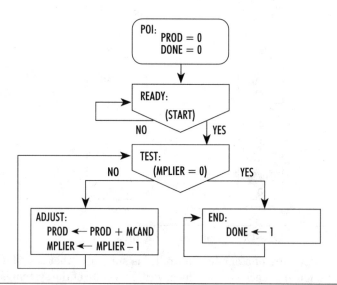

Figure 4.18   Microoperation Diagram for a Repeated Addition Multiplier

Figure 4.18 shows the microoperation diagram for the repeated addition multiplier presented as Example 2.10 in Chapter 2. The corresponding controller state diagram and datapath for this device are shown in Figures 4.19 and 4.20.

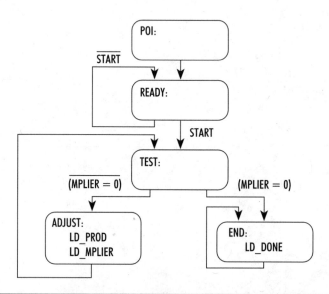

Figure 4.19   Controller State Diagram for a Repeated Addition Multiplier

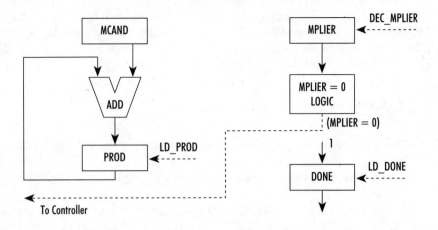

Figure 4.20   Datapath Structure for a Repeated Addition Multiplier

Solution

Figure 4.21 illustrates the design of a one-hot controller for the repeated addition multiplier. Five D flip-flops, one for each of the states, are used. The structures of Figure 4.17 pass the 1 between the flip-flops determined by status variables START and (MPLIER=0).

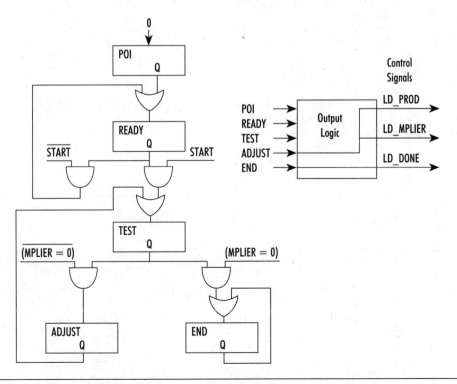

Figure 4.21   One-Hot Controller for a Digital Multiplier

Initialization with a RESET signal clears PROD and DONE. It also places the hot-bit 1 in flip-flop POI and a 0 in all other flip-flops. The first clock pulse advances the hot-bit to flip-flop READY; a 0 is loaded into POI. With each clock pulse the hot-bit is repeatedly loaded back into READY through the OR gate for as long as START is not asserted. Asserting START passes the hot-bit to flip-flop TEST. If the MPLIER register has not counted down to 0, the hot-bit is passed to flip-flop ADJUST. This state generates control signals to cause

PROD←PROD + MCAND,
MPLIER←MPLIER − 1;

The multiplicand is added to the product register and the multiplier register is decremented. The hot-bit is again passed to TEST. The sequence of TEST and ADJUST repeats until the multiplier register has been decremented to 0. The hot-bit is then passed to END, where it recirculates continuously.

A logic block produces control signals based on the state of the controller. In this example the logic is of trivial complexity. LD_PROD and LD_MPLIER are asserted in state ADJUST; LD_DONE is asserted in state END. In more complex situations, AND/OR/NOT logic might be involved in this block.

One-hot design is intuitive. There is a one-to-one correspondence between states and flip-flops in the state register. Modifications to the microoperation diagram may dramatically change the required logic components. The number of flip-flops required, one per state, exceeds the number required if all state assignments were allowed.

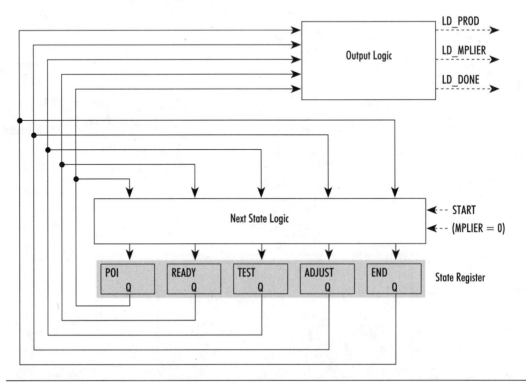

Figure 4.22   One-Hot Controller in "Conventional" State Machine Format

The one-hot controller for the multiplier has been redrawn in Figure 4.22 to suggest a more regular structure. The state flip-flops become the state register. The blocks of logic may be grouped into a next-state logic and an output logic as shown. In this form the device is easily implemented using discrete logic, ROMs, or programmable logic devices.

Finite State Machine Controller

In the general form of a finite state machine, logic is required to generate the next-state value for the state register and the output control signals. In the following example, AND/OR/NOT logic is used in the implementation.

Example 4.5:   A Controller Design Using AND/OR/NOT Logic

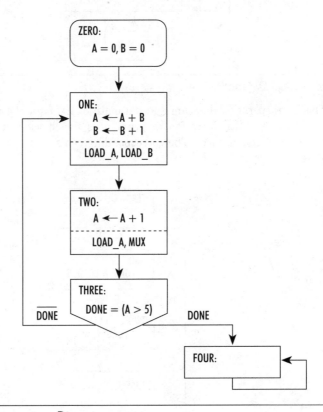

Figure 4.23   Microoperation Diagram

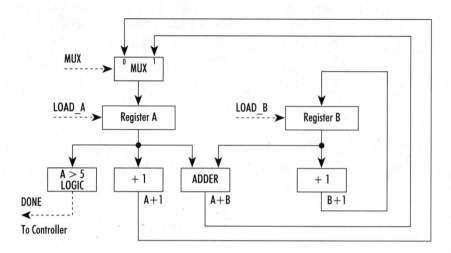

Figure 4.24   Datapath

The microoperation diagram and datapath of Example 2.4 are repeated as Figure 4.23 and Figure 4.24. The microoperation diagram uses five blocks: an initial block, three processing blocks, and a decision block. The corresponding controller state diagram, Figure 4.25, requires five states.

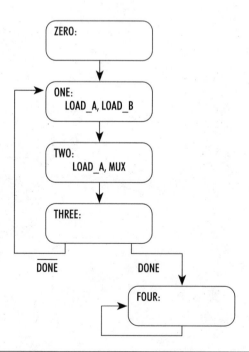

Figure 4.25   Controller State Diagram

Each of the five states is assigned a binary state variable designation. The following table lists all of the state transitions as a function of the decision variable (A > 5). For each state, the next-state and controller outputs are presented. A "—" in the table represents a don't-care condition; the next state does not depend on a decision signal.

Present State of Controller		A > 5	Next State of Controller		Control Signals		
					LD_A	LD_B	MUX
ZERO	000	—	ONE	001	0	0	0
ONE	001	—	TWO	010	1	1	0
TWO	010	—	THREE	011	1	0	1
THREE	011	No	ONE	001	0	0	0
THREE	011	Yes	FOUR	100	0	0	0
FOUR	100	—	FOUR	100	0	0	0

Table 4.3

Completion of the controller design requires implementing the state machine using appropriate devices for the next-state logic and output logic. This can be accomplished using logic design techniques. It can be verified (Exercise 4.10) that the controller can be implemented using three type D synchronous flip-flops with outputs Q_2 Q_1 Q_0 to represent the five required states. The three unused states can be used as don't-care conditions in designing the logic. The inputs to the flip-flops D_2, D_1, and D_0, can be obtained from the following logic equations. The symbol "S" represents the A > 5 decision signal.

$S = A > 5$
$D2 = Q_2 + Q_1 \, Q_0 \, S$
$D1 = Q_1 \, \overline{Q}_0 + \overline{Q}_1 \, Q_0$
$D0 = \overline{Q}_2 \, \overline{Q}_0 + S * Q_1$

The three control signals, LOAD_A, LOAD_B, and MUX can be obtained from the following output logic:

$\text{LOAD\_A} = Q_1 \, \overline{Q}_0 + \overline{Q}_1 \, Q_0$
$\text{LOAD\_B} = \overline{Q}_1 \, Q_0$
$\text{MUX} = Q_1 \, \overline{Q}_0$

ROMs and programmable logic devices offer a convenient method to implement logic in a more structured manner. A controller configuration using two ROMs to implement the next-state and output logic is shown in Figure 4.26. Power-on initialization (POI) is accomplished using the system RESET signal to initialize the state register. Conditional and unconditional changes of state are generated by logic circuitry incorporated within the next-state ROM. Because of the ease of reprogramming, the ROM also allows some modifications to the controller with a minimum of circuitry changes.

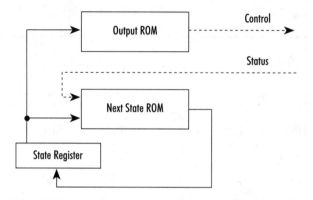

Figure 4.26   A State Machine Controller Using ROM Logic

Once the number of controller states *(N)*, state variables *(n)*, status variables *(m)*, and control outputs *(p)* are known, appropriately sized ROMs can be selected. There must be *n* address lines for the output ROM, one for each state variable. The ROM has 2^n words of width *p*.

The next-state ROM requires address bits from the *n* state variables and *m* status variables. The ROM has 2^{m+n} words of width *n*.

Example 4.6:   A Controller Design Using AND/OR/NOT and ROM Logic

Design a controller with characteristics specified by the controller state diagram of Figure 4.11.

Analysis:

Since in this example there are no status signals, both the next-state and control output are a function of only the present state. The required logic can be specified by a truth table as follows.

Name of State	Present State $Y_2 Y_1 Y_0$	Next State $Y_2^* Y_1^* Y_0^*$	Control Signals	
			C_0	C_1
POI	000	001	0	0
ONE	001	010	1	0
TWO	010	011	1	1
THREE	011	100	0	1
FOUR	100	010	1	0

Table 4.4

Solution 1, Discrete Logic:

Three type D flip-flops are used to store $Y_2 Y_1 Y_0$, the three state variables. In a type D flip-flop the next state, Y_i^* is equal to the value at the D_i input terminal. The following relationships can be derived by analyzing the truth table. Unused states provide don't-care conditions allowing some logic minimization.

$$D_2 = Y_2^* = Y_1 Y_0$$
$$D_1 = Y_1^* = \overline{Y}_1 Y_0 + \overline{Y}_1 Y_0 + Y_2$$
$$D_0 = Y_0^* = \overline{Y}_2 \overline{Y}_0$$

$$C_0 = \overline{Y}_1 Y_0 + Y_1 \overline{Y}_0 + Y_2$$
$$C_1 = Y_1$$

Solution 2, ROM Logic:

The contents of the ROM are determined directly from the truth table. Since no decision variables are used, the next-state logic and the output logic use only the present state as input variables. For the next-state ROM and the output ROM the address is specified by the present state. The next-state ROM requires storage for at least five words (for the five state representations), each of at least three bits (for the three next values of the state variables).

The output ROM requires storage for at least five words of 2 bits. At least 25 bits of storage are required.

Since no status signals are involved, the next-state ROM and output ROM can be combined into a single memory device with at least five words and 5 bits per word. This topic is treated in greater detail in the next section.

Single ROM Controller

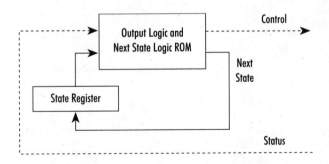

Figure 4.27　A State Machine Controller Using a Single ROM

In Figure 4.27 the next-state logic and the output logic for a general-purpose controller have been combined in a single ROM. The outputs are a function of both the state and the input status signals. This offers a potential problem. If a status signal changes while executing a microinstruction, this will cause one of the address lines of the ROM to change. This in turn will lead to a momentary glitch in the output. A change in the address will cause the output to be obtained from a different memory cell. Even if both cells have the same contents, a glitch will occur during the transition. Since the output to the datapath is a control signal, even a brief glitch in the middle of the clock cycle can lead to unexpected results.

Registered Outputs

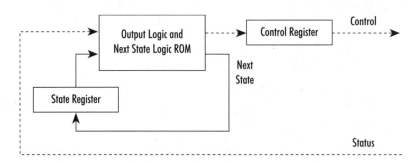

Figure 4.28   Use of an Output Register for Control Signal Continuity

The problem of glitches in the output can be controlled using additional registers. In Figure 4.28 an output register for the control signals has been included. This assures that the value of the control signals will remain constant throughout the clock period, regardless of transitional changes in the status variables. This solution slightly changes the manner in which the control signals must be stored. The information stored in the ROM must be the *next* value of the control signal; the value to be asserted when the controller assumes its next state at the end of the current clock cycle.

Registered Inputs

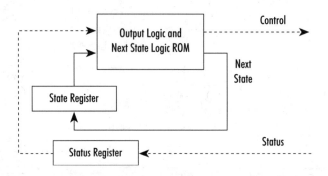

Figure 4.29   Use of an Input Register to Assure Control Signal Continuity

As an alternative to registering the output control signals from the ROM, the input status signals from the datapath may be registered as shown in Figure 4.29. This solution, while quite satisfactory in many applications, introduces an additional delay in the use of status signals. A status signal generated in one clock cycle must have the result registered for use during the next clock cycle.

Example 4.7:   Design Using Registered Controller Outputs

Design a controller to implement the controller state diagram of Figure 4.11. Use a control register for the control output.

Analysis:

The solution using an output register requires that the control output for the *next* state be determined. Since the outputs will be registered, they will not be available to the datapath until the next clock cycle; this is a delay of one clock period. To compensate for the delay, the contents of the control field must be the control output needed for the *next* state rather than for the present state.

Except for the initial power-on initialize (POI) state, the sequence of output control signals will be identical for both this example and Example 4.5. Since POI is entered as the result of a RESET signal, the exact number of clock cycles in this state is not known. Determining the next control output for state POI is potentially problematic. Although the controller will eventually advance from POI to state ONE, the device will remain in POI until RESET is no longer asserted. This will cause a repetition of actions caused by control signals associated with the next state, ONE.

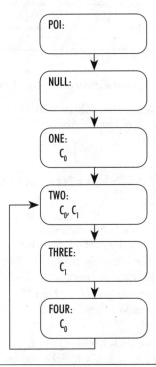

Figure 4.30   Introduction of a NULL State for Use with Registered Outputs

As a possible solution for this problem, an additional state, NULL, with no control outputs, has been added between POI and ONE, as shown in Figure 4.30. This ensures that the control actions asserted by C_0 in state ONE are performed only when the device is actually in state ONE.

Solution:

The truth-table description of the controller is shown in Table 4.5. The logic may be implemented with a single ROM without concern for glitches in the control signals.

Name of State	Present State $Y_2 Y_1 Y_0$	Next State $Y_2^* Y_1^* Y_0^*$	Next Control Signals $C_0 C_1$
POI	000	001	00
NULL	001	010	10
ONE	010	011	11
TWO	011	100	01
THREE	100	101	10
FOUR	101	011	11

Table 4.5

Example 4.8: A ROM-based Controller Design for a Repeated Addition Multiplier

The microoperation diagram for the repeated addition digital multiplier from Chapter 2, Example 2.10 has been reproduced as Figure 4.18. The corresponding controller state diagram is shown in Figure 4.19; the datapath is shown in Figure 4.20. Design a ROM-based controller for this datapath.

Method 1, Use of Separate ROMs for Next-state Logic and Output Logic

Analysis:

The controller will be designed using separate ROMs for the output logic and next-state logic. This avoids potential glitches in the control signals since the control signals are a function of the present state only; status values are not involved. Since the state diagram has five states, three state variables are needed for the state register.

Three control signals are needed. DEC_MPLIER decrements register MPLIER. LD_PROD loads the interim value of the product into register PROD. LD_DONE turns on the DONE flip-flop when the operation is finished.

The output logic ROM requires three inputs from the state register. Three outputs from the ROM provide the three control signals for the datapath. The output logic ROM requires five words, one for each of the five states. A total of 15 bits is necessary.

The next-state ROM uses five inputs. Three come from the state Register, one comes from the decision logic (MPLIER=0), one comes from the supervisory signal START. Three outputs from the ROM provide the next state for the state register. The next-state ROM requires 20 words; five states with $2^2 = 4$ decision possibilities for each state. A total of 60 bits is required. Unless custom fabrication were used, it is unlikely that a ROM of this size could be found.

Solution:

The initial state, POI, is assigned the value 000 with the remaining states of the controller arbitrarily assigned values READY = 001, TEST = 010, ADJUST = 011, and END = 100. The following tables present the address and contents of the ROMs. In Table 4.7, "—" is used as a don't-care; both values of the status variables produce the same result.

Name of State	ROM Address	ROM Contents		
	Present State	DEC_MPLIER	LD_PROD	LD_DONE
POI	000	0	0	0
READY	001	0	0	0
TEST	010	0	0	0
ADJUST	011	1	1	0
END	100	0	0	1

Table 4.6   Output Logic ROM

Name of State	ROM Address			ROM Contents	
	Present State	START	(MPLIER = 0)	Next State	
POI	000	—	—	READY	001
READY	001	0	—	READY	001
READY	001	1	—	TEST	010
TEST	010	—	0	ADJUST	011
TEST	010	—	1	END	100
ADJUST	011	—	—	TEST	010
END	100	—	—	END	100

Table 4.7   Next-state Logic ROM

When using a ROM implementation for logic, don't-care conditions are not considered. In the next-state logic ROM, four addresses, corresponding to the four logic combinations of START and (MPLIER=0), are associated with each of the states.

Method 2, Implementation Using a Single ROM and Registered Control Outputs

Analysis:

In the previous solution, separate ROMs were used for the next state and the control outputs. This was necessary because the output control signals needed to remain constant during the clock period. This was accomplished by making the control signals a function of only the state of the system.

As an alternative, a control register can be used to hold the control signal information. Under these conditions, a single ROM can generate both the next-state and next-output information. C_2, C_1, and C_0 represent DEC_MPLIER, LD_PROD, and LD_DONE respectively.

When a single ROM implements the above logic, five address lines are required. Six output lines are required: three for the next state and three for the next output. At the end of the clock cycle, the three-bit next-state signal is transferred to the state register; the three-bit next control signal is transferred to the output control register.

Solution:

The next-state and next-control outputs must be stored in the ROM. Since the READY state does not involve processing actions, an additional NULL state is not necessary.

Name of State	ROM Address			ROM Contents		
	Present State	START	(MPLIER $= 0$)	Next-state		Next Output C_2 C_1 C_0
POI	000	—	—	READY	001	0 0 0
READY	001	0	—	READY	001	0 0 0
READY	001	1	—	TEST	010	0 0 0
TEST	010	—	0	ADJUST	011	1 1 0
TEST	010	—	1	END	100	0 0 1
ADJUST	011	—	—	TEST	010	0 0 0
END	100	—	—	END	100	0 0 1

Table 4.8

4.6   Multiplexing Status Variables

Logic design is simplified when fewer input variables are considered. In implementing state machine controllers, the next state is a logical function of the n present state variables and the m status/decision variables.

When a ROM implements the logic, the ROM storage capacity is a function of both the number of address bits and the width of the output. The address bits are the n state variables and m status variables. The memory capacity of the next-state ROM is 2^{n+m} words; each word has n bits. For each additional status variable the word capacity of ROM storage doubles. If logic gates implement the next-state logic, additional variables generally lead to more complex circuit realizations.

In many situations it is possible to examine the status variables sequentially rather than simultaneously. This can dramatically reduce the complexity of the discrete logic or the size of the ROM used to implement the logic. In the extreme, all decisions based on status variables may be done sequentially using a single **multiplexed status variable**.

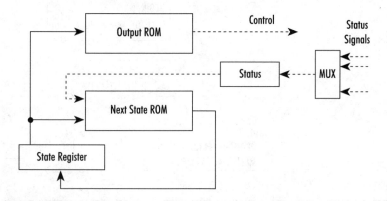

Figure 4.31   Multiplexing Status Variables to Reduce the Size of the Control ROM

Figure 4.31 shows an added status register to assure continuity of the status variables through the clock cycle. Loading the status register must be incorporated into the microoperation diagram. This may require additional processing blocks and increase the number of states in the controller. The number of state variables may likewise increase. Additional control signals are also necessary to load the DECISION register; this may increase the width of the ROM. In most cases, the decrease in logic complexity or ROM size greatly outweighs a slight increase in the number of control signals.

In the controller for the repeated addition multiplier, an extra 1-bit register, DECISION, holds the selected status signal. Contents of DECISION will either be the variable START or the variable (MPLIER=0), which is obtained from logic associated with the MPLIER register.

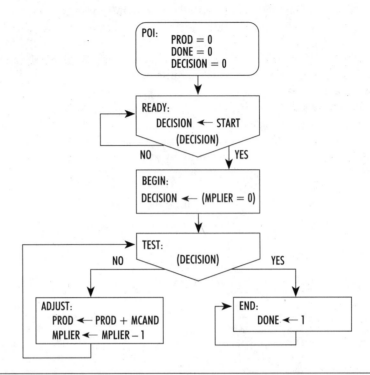

Figure 4.32   Microoperation Diagram for Multiplier Using a Single Status Variable for the Controller

Figure 4.32 indicates some subtle changes needed to accommodate the DECISION register in the microoperation diagram of Figure 4.18. The DECISION register is initialized in the POI block. DECISION is loaded with the value of START in block READY. This operation synchronizes the START signal if the signal is initially asynchronous. Note that the first time block READY is encountered, the value of DECISION must be 0, the value it was initialized to in POI. This will force a second clock period in READY. At the end of the clock cycle in READY, the value of START is loaded into DECISION.

An extra block, BEGIN, allows DECISION to be loaded with the logic value (MPLIER=0). The value must be loaded before it can be tested in TEST. If a single ROM implements both the next-state and output logic, the width must be increased by two bits to accommodate the two additional control signals needed: one to load the DECISION register and one to control the multiplexer. One additional state is needed in the processing algorithm. This does not increase the number of states beyond the threshold that three state variables can accommodate.

 Without multiplexing status variables (microoperation diagram of Figure 4.18) there are three state variables, two status variables, and five control signals. The ROM size is 32 (2^{2+3}) words, each of five bits.

With multiplexing status variables (microoperation diagram of Figure 4.32) there are three state variables, one status variable, and seven control signals. The ROM size is 16 (2^{1+3}) words, each of 7 bits. The savings is far more dramatic when more decision variables are involved. Except for a custom application, it is unlikely that ROMs of the exact desired size would be available.

4.7   Supervisory Operations

In addition to START, many controllers must respond to other supervisory input variables such as **PAUSE**, **RESUME**, ABORT, and **RESET**. These variables may be manually generated or originate in external systems. It is assumed that the signals are synchronized with the system clock.

PAUSE/RESUME

PAUSE causes the datapath to stop processing so that it can easily be resumed later. During a pause condition, no changes in register contents are performed, no control signals are generated, and all devices using buses shared with other systems are tri-stated. RESUME causes the processing action to continue from the condition in which it was stopped.

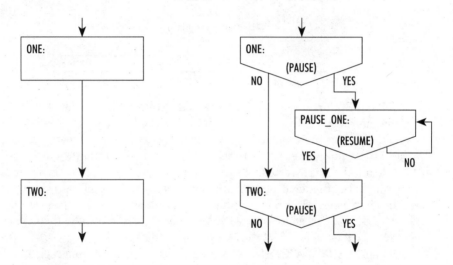

Figure 4.33   Possible Implementation of PAUSE and RESUME Controller Functions

Figure 4.33 shows one possible implementation of PAUSE and RESUME. Each block of the microoperation diagram has a shadow block that is entered synchronously when PAUSE is asserted. No control signals are asserted and no operations are performed in the shadow block. RESUME causes the successor state to be entered and the processing algorithm to continue. Since the PAUSE condition must be tested in every processing block, the number of required states doubles. If PAUSE is used as an input into a multiplexer and decision register, additional complexity and lower performance are likely.

It is tempting to create PAUSE using logic to inhibit the system clock. This method is not preferred since it introduces potential problems such as clock skew and clock glitches.

An alternative method of implementing PAUSE/RESUME involves creating a PAUSE state in which no control actions are performed. The controller enters the PAUSE state when PAUSE is asserted. It remains in state PAUSE until RESUME is asserted. At this time the controller returns to the state it would have entered had PAUSE not been asserted. Figure 4.34 shows a hardware modification to the controller that accomplishes this.

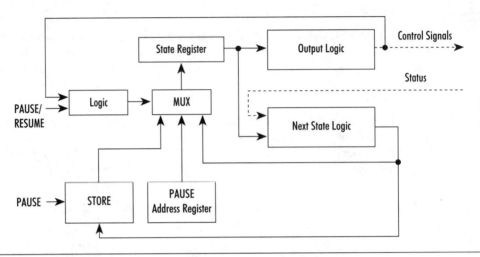

Figure 4.34   Implementation of a PAUSE/RESUME Capability

A PAUSE address register, PAR, stores the state value associated with the PAUSE state. PAR is either hardwired to this value or initialized to the value during power-up. A multiplexer selects the proper source of information for the state register. When supervisory control signals are not asserted, the multiplexer provides next-state information from the next-state logic.

When PAUSE is asserted, the value from the next-state logic (the value of the next state that would have been entered had PAUSE had not been asserted) is placed in register STORE. It remains in register STORE for the duration of PAUSE. The next state for the state register is provided by the PAUSE address register. When in the PAUSE state, the next-state logic causes the controller to remain in this state. No control signals are asserted.

When RESUME is asserted, the state register is loaded from register STORE via the multiplexer. This returns the previously stored value of the next state. Since the contents of datapath registers are not changed, the system is able to proceed with its processing actions. The only effect is the time delay introduced by the duration of the PAUSE/RESUME supervisory signals. Appropriate logic to control the multiplexer and the STORE register must be derived from the PAUSE and RESUME signals.

This combination of actions may suggest a subroutine in a programming language. This similarity is not surprising, since a digital computer similarly implements its subroutine operations. The role of the STORE register is similar to the role of a stack in implementing a computer subroutine. Both save the location of the address (state) to be entered following the completion of the subroutine (pause).

ABORT

The **ABORT** supervisory control signal is usually reserved for situations in which further processing of the algorithm is not needed or for situations having safety implications. The purpose of ABORT is to initiate the execution of an algorithm to safely terminate the current processing action. Processing operations can resume only with manual intervention or through a RESET command. It is not necessary to be able to resume the processing algorithm from where it was aborted.

ABORT initiates an unconditional jump to the shut-down algorithm from anywhere in the microoperation diagram. As with the case of PAUSE, the inclusion of ABORT in the processing algorithm specified by the microoperation diagram requires additional hardware and is unwieldy.

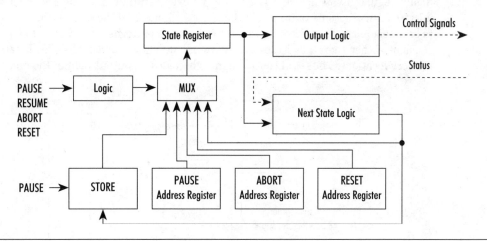

Figure 4.35   Implementing the ABORT and RESET Functions

The ABORT action may be implemented using a technique similar to that used in PAUSE and RESUME, as shown in Figure 4.35. The ABORT signal causes the value of the ABORT address register to be placed in the system state register. This diverts the processing algorithm to state ABORT which is the beginning of the safe shut-down procedure. Because of the way it is implemented, ABORT can be initiated from any state in the processing algorithm. Since there is no need to resume the algorithm after an ABORT, it is not necessary to store a return address in the STORE register.

RESET

A manual reset has the same effect as a power-on reset; it causes the system to return to the power-on initialize (POI) state. RESET can be asserted in any state. The RESET operation can be implemented similar to ABORT, as shown in Figure 4.35. When RESET is asserted a new value is loaded into the state register. The reset algorithm terminates with the device in POI.

4.8   Microprogrammed Control

A distinction may be drawn between implementing a finite state machine control unit using hardware logic gates and the implementation using a highly structured ROM-based approach. The former is called **hardwired control** and the latter is **microprogrammed control**.

Hardwired control has usually been the method chosen for implementing small systems using relatively simple algorithms. Hardwired systems are able to operate at very high clock rates and are used where system processing speed is of great importance.

Microprogrammed control, first suggested by Maurice Wilkes of the University of Manchester in 1951, provides a much more versatile environment by replacing discrete logic components with highly regular ROM devices. The memory devices, rather than simply replacing logic, contain microcode instructions that specify the control signals and the manner in which the next state is to be determined. The microprogramming approach patterns its activities to a programming environment rather than a logic design environment. The sequence of states becomes a sequence of memory locations. The set of control signals becomes an output field in the microcode.

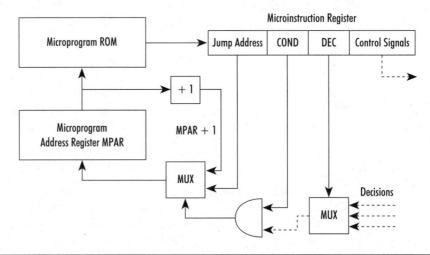

Figure 4.36   A Microprogrammed Control Unit

Figure 4.36 shows an organization for a typical microprogrammed control unit. A microprogram address register (MPAR) provides the address for the current microcoded instruction. Its role is equivalent to that of the state register in a state machine controller. The contents of the ROM are the microcoded instructions that may be used directly or latched in a microinstruction register (MIR). The MIR has the same role as an output register in hardwired control.

A microprogrammed device also contains a method of altering the flow of the program using the status/decision variables from the datapath. In Figure 4.36 this is accomplished by allowing the microprogram to execute sequentially (MPAR←MPAR + 1) or to jump to a new location (MPAR←JUMP address) under the influence of a decision variable. The JUMP address is one of the fields in the microcode. Another field of the microcode, COND, participates in the selection of the next address. A fourth field, DEC, chooses the decision variable to be used in determining the next address. If one of the decision variables is "1," an unconditional jump is performed.

An implementation of the next address feature is shown in Figure 4.36. The 1-bit conditional field, COND, is ANDed with the selected decision variable and used as a select input to the multiplexer. When the result of the AND is "0," MPAR + 1 is selected; with a result of "1," the address in the JUMP address field is chosen.

Microcode Instruction Formats

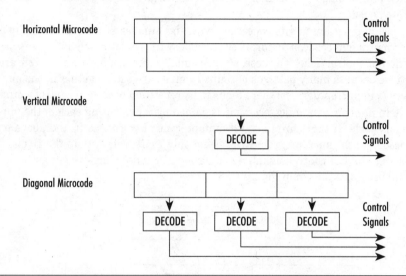

Figure 4.37   Horizontal, Vertical, and Diagonal Microcode Structures Used in Microprogramming

The control field of the microcoded instruction generates the control signals for the datapath. The control field format may be chosen to use one bit position for each possible control bit. This is known as **horizontal microcode**, as shown in Figure 4.37. Versatility is the main advantage of horizontal microcode. Any combination of control signals may be simultaneously asserted; this allows a maximum of parallel activities. For complex systems involving many control signals, the horizontal structure may lead to microinstructions that are quite wide.

For situations not requiring simultaneously asserted control signals, a **vertical** structure for the microcode may be used. The designation of the control signal is encoded in binary and stored in the control field. Upon retrieving the microinstruction, external decode logic recovers the control bit to be asserted. The vertical format offers decreased width of the control field and may lower its cost. The major disadvantage of the vertical structure is its inability to assert more than one control signal simultaneously. This may seriously limit performance.

A third situation, sometimes known as a **diagonal** structure may be used. In the diagonal structure, vertical fields encode control signal selection. Multiple vertical fields allow simultaneous assertion of control signals provided they originate from different vertical fields. Choosing a method of encoding the microinstructions using the minimum memory storage is known as **microcode compaction**.

4.9   Nanoprogramming

Nanoprogramming, as the name suggests, is a technique for reducing the control word structure to even finer levels than those used in microprogramming. The goal is to implement a microcontroller using less memory than is required for microprogramming. The control field width for a microprogrammed application is often quite large and the total ROM size required may also be quite large.

Nanoprogramming is concerned with the number of unique control fields used in the fully horizontal microcode structure. For example, with 10 control signals, there are $2^{10} = 1024$ possible different control fields. Of these, a much smaller number, for example 128, are actually used. This is true because many patterns of control signals are used together in the microcode. Also, even in very long microprograms, there tends to be repetition of certain control words.

In nanoprogramming, economy is gained by representing each of the unique control words in binary code. If there are only 128 unique control words, as in the above example, they can be encoded in the microprogram ROM using only 7 bits, rather than the 10 bits initially required. To recover the full control signal, decoding logic or a decoding nanoprogram ROM is required. The total process is shown in Figure 4.38.

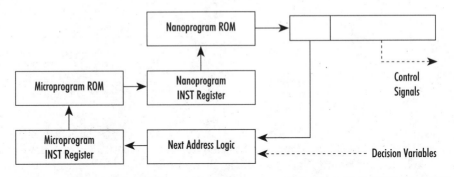

Figure 4.38   A Nanoprogrammed Control Unit

Nanoprogramming is effective when the size of the combined microprogram ROM and nanoprogram ROM is smaller than the size of a single microprogram ROM from a conventional microprogrammed operation. It is slower than microprogramming since information must be obtained from two ROM memories.

Nanoprogramming is used where it is necessary to minimize ROM memory size. This is often the case in VLSI designs for complex microprocessors. A reduced memory size reduces the need for costly chip area to implement the memory. Motorola has used nanoprogramming to fabricate many of the 680x0 series of microprocessors. They have claimed a reduction of about 40% in storage requirements.

Summary

This chapter investigates the structure and implementation of controllers for digital systems. The controller provides the asserted control signals to sequence the datapath to perform digital data processing. Controllers are especially important for achieving high performance in digital computers.

There are many ways to implement controllers. Using conventional logic design techniques, controllers can be designed using flip-flops, shift registers, and counters. They can be designed using classical state machine design procedures or heuristic techniques such as the one-hot method. Programmable logic devices, PLDs, offer the opportunity to fabricate controller structures using relatively few highly integrated components. Devices are available to provide a complete multivariable state machine on a single chip. Corresponding devices to support computer aided design facilities are available in the component libraries.

Controllers, in addition to responding to status and decision variables from the datapath, also must respond to supervisory signals such as PAUSE, RESUME, RESET, and ABORT. These signals must be synchronized to the system clock to allow reliable detection.

Microprogramming provides a highly structured method of implementing complex processing algorithms with a minimum of hardware uniqueness and complexity. Microprogramming, despite some performance disadvantages, is widely used to implement controller structures for digital computers.

For additional information and references on the topics presented in this chapter, access the World Wide Web page provided at: http://www.ece.orst.edu/~herzog/docs.html.

Key Terms

ABORT

asynchronous inputs

clock cycle

controller state diagram

diagonal microcode

hardwired control

horizontal microcode

microcode compaction

microinstruction

microprogrammed control

multiplexed status variable

nanoprogramming

next-state logic

output logic

PAUSE

registered inputs

registered outputs

RESET

RESUME

sequencer

vertical microcode

Exercises

4.1 For the controller state diagram and datapath structure in Figure 4.39, determine the microoperation diagram (without control signals). All registers are 4 bits in length. The device is to be initialized with values $A = 0001_2$, $B = 0010_2$, $C = 0011_2$, $D = 0100_2$.

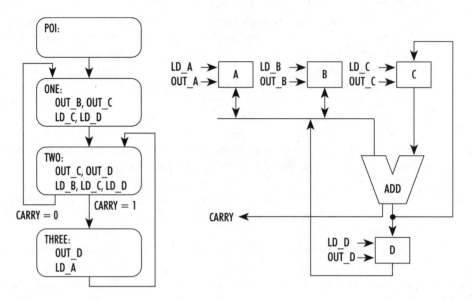

Figure 4.39

4.2 Figure 4.40 shows a datapath structure and several microoperation diagrams. Find the corresponding controller state diagram for each.

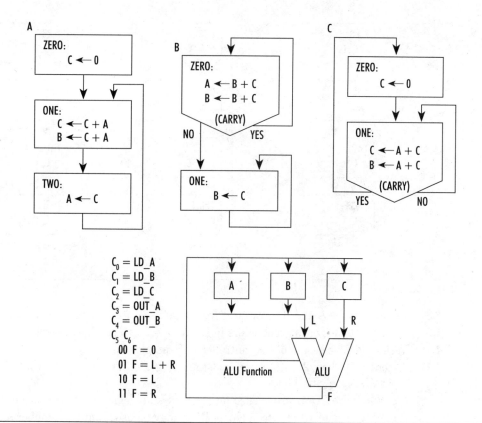

Figure 4.40

4.3 Give an example in which a glitch in a control signal could cause an erroneous operation in a digital system.

4.4 A digital system simultaneously examines multiple external signals. There may be some skew in the values. Show a technique to synchronize the signals with the system clock. Assume that any skew is less than one clock period.

4.5 A controller uses two asynchronous input lines to encode four operations.

```
00 = CONTINUE
01 = ABORT
10 = PAUSE
11 = START
```

Is synchronization necessary? How could it be accomplished?

4.6 Which of the microoperation diagrams from Exercise 4.2 can be implemented with a sequencer controller? Why?

4.7 Implement the A and B controllers in Figure 4.41 with a shift register sequencer.

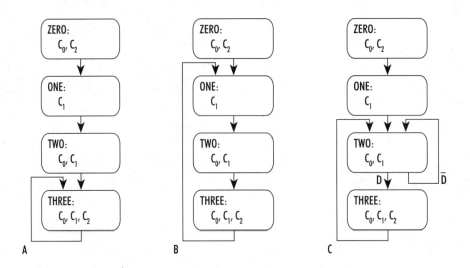

Figure 4.41

4.8 Implement the A and B controllers from Exercise 4.7 using a reloadable counter.

4.9 Implement each of the controllers in Exercise 4.7 as a one-hot controller.

4.10 Verify the logic design of the controller in Example 4.5.

4.11 A controller has 25 states in its state diagram. There are 10 control signals and 8 status/decision variables.

    (a) What size ROMs are needed for the next-state logic and for the output logic?

    (b) With registered inputs, a single ROM may be used, but the number of states expands to 31. What size ROM is needed?

    (c) If the status variables are MUXed to one (still using a registered input), the number of states expands to 36. What size ROM is needed?

4.12 Four-bit registers are used in Figure 4.42. Initial values are $A = 0000_2$, $B = 0001_2$, $C = 0100_2$. The controller, which is constructed from type D flip-flops, is initialized to state 00. The generated control signals are applied to the appropriate register. Register C is permanently connected to one input of the adder.

    (a) Find the microoperation diagram for the device.

    (b) Find the controller state diagram for the device.

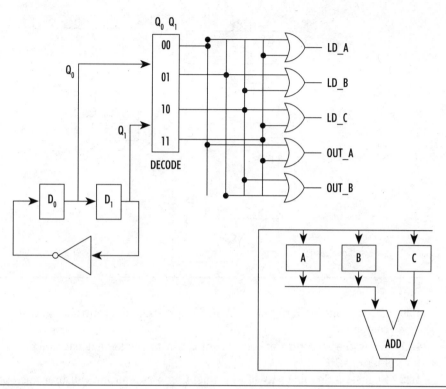

Figure 4.42

4.13 Design a controller for the state diagram of Figure 4.19 using flip-flops and AND/OR/NOT logic.

4.14 In Example 4.5, find the contents of the next-state ROM and output ROM used in the logic implementation.

4.15 If a finite state machine controller is designed using a single ROM with no status register, show how oscillation might be possible.

4.16 What is the purpose of multiplexing decision variables? How does this affect performance?

4.17 Why are control outputs registered? How does this affect performance?

4.18 Why are input variables registered? How does this affect performance?

4.19 The microoperation diagram for a digital device is shown in Figure 4.43. For simplicity, only the control signals (C_0, C_1, C_2) are shown. Design a controller for this device using a ROM memory as shown. Determine the size and contents of the ROM.

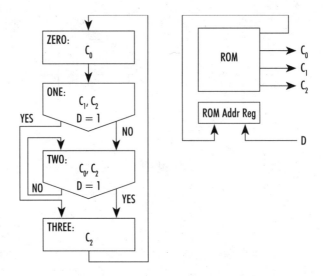

Figure 4.43

4.20 Discuss the similarity of the ABORT operation in a controller and an interrupt in a digital computer.

4.21 Show the structure of an RTL program to implement each of the control structures in Figure 4.16.

4.22 Suggest a possible way of implementing a subroutine call in a microprogrammed control unit. In what way would this be useful?

4.23 Show how the microprogrammed controller of Figure 4.36 could be used to implement each of the controller state diagrams of Exercise 4.7. Show the structure of the microcode and the encoding of the fields used.

4.24 Is a microinstruction register needed in a microprogrammed controller? Why?

4.25 What is the reason for encoding the control field in a microinstruction? Does encoding affect performance?

4.26 When designing digital devices with microprogrammed control units, it is often desirable to achieve the design goals with the minimum width of control ROM. A horizontal microcode structure represents a starting point associated with the widest field for the control signals. A diagonal or vertical structure may allow minimization of the control field width, an optimization process known as compaction. Minimum width of the control field for a microprogrammed controller is desired. Seven control signals (a, b, c, d, e, f, g) must be asserted at various times to perform the required algorithm. Four different microinstructions (I_1, I_2, I_3, I_4) must be available. Each of these microinstructions involves control signals as follows:

I_1 (a, d, e)
I_2 (b, d, e)
I_3 (c, d, g)
I_4 (a, f)

Find the minimum total width encoding of the microcode fields.

4.27 Microprogramming techniques have sometimes been described as a computer within a computer. What does this mean?

4.28 What is the purpose of nanoprogramming? Will it always succeed? Explain.

4.29 Bit bashing involves generating clock and control signals for a datapath using single bits of the output port of a microprocessor. Each of the control signals is generated by toggling bits under program control of the microprocessor. Discuss in quantitative terms the likely performance penalty of using bit bashing.

4.30 Under what circumstances is bit bashing (as described in the previous problem) a reasonable method of developing a controller?

5
Data Representation and Arithmetic

Blaise Pascal

A brilliant young Frenchman named Blaise
Designed calculators in his early days
Oh what a rascal
was that lad Pascal
But he sure changed our programming ways

Prologue

Blaise Pascal, born in France in 1623, was one of those geniuses able to contribute to almost every field he entered. His life had many contradictions: He never attended a day of school in his life, yet made outstanding contributions to the fields of mathematics, science, religion, and literature.

Blaise was drawn into a self-study of geometry at an early age. Due to his lack of formal education, he was able to impart fresh insights and even unique names to many elements of geometry. His mind was also attracted to the study of mathematics. He is credited with developing the first mathematically based theory of probability.

At the age of 19 Blaise became aware, through his father's employment, of the tedious and repetitious calculations required as part of the French tax system. For the next three years he combined his skill in mathematics and geometry to construct a small mechanical calculator capable of addition, subtraction, multiplication, and division.

In its final form, Pascal's calculator was housed in a polished brass box $14 \times 5 \times 3$ inches. It contained eight movable dials. Three of the lower dials were unique to French coinage, which worked in multiples of 12 and 20. The five most significant dials were decimal. After zeroing the contents of all display windows, the operator used a stylus to sequentially dial numbers to be added into the calculator. The calculator automatically produced carries and presented the accumulated sums in display windows. Multiplication was performed by a series of shifts and adds. Despite its impressive performance, Pascal's calculator was not a financial success.

Pascal's other contributions are equally impressive. A brilliant experimenter, he devised a series of demonstrations to describe the behavior of gasses and liquids as they interrelate in vacuum systems. The mathematical formulation, known as Pascal's Law, is a topic of most university physics curricula.

At the age of 31, as the result of a miraculous escape from death as his carriage plunged over the parapet of a bridge, Pascal became convinced that this was a direct warning from God to change the direction of his life. Pascal entered a period of extreme religious introspection and self-criticism. He died at the age of 39.

5.1   Introduction

Blaise Pascal recognized the importance of automating arithmetic operations. His calculator, in 1642, greatly affected calculations in science and commerce.

This chapter will examine methods for representing data and the implementation of algorithms to perform arithmetic operations on data. Integer and floating-point data representations, as well as algorithms will be considered. Many methods exist for performing digital arithmetic. This chapter will present techniques to map these methods onto the datapath structures studied in previous chapters.

The primary interest in this chapter will be arithmetic algorithms for **addition**, **subtraction**, **multiplication** and **division**. Algorithms and hardware implementation of both integer and floating-point operations will be considered.

5.2   Number Systems

Most methods of representing numbers use a weighted number system. A numeric value, N, can be represented as:

$$N = d_{n-1}\, d_{n-2} \dots d_1\, d_0 \, . \, d_{-1}\, d_{-2} \dots d_{-m}$$

Each of the integer digits, d_i, is a multiplier for a power of the radix of the system. Numerically, N can be evaluated as:

$$N = d_{n-1}\, r^{n-1} + d_{n-2}\, r^{n-2} + \dots + d_1\, r^1 + d_0\, r^0 + d_{-1}\, r^{-1} + d_{-2}\, r^{-2} + \dots + d_{-m}\, r^{-m}$$

$$0 \le d_i < r$$

or:

$$N = \sum_{i=-m}^{n-1} d_i\, r^i$$

In the above equations, d is a digit, r is the base or radix of the number system, n is the number of integer digits, and m is the number of fractional digits. The radix point (.) separates the integer and fractional part of the number. If the position of the radix point is fixed, the number system is a **fixed point** number system. A position to the right of the least significant bit results in all numbers being represented as integers. A position to the left of the most significant bit results in all numbers being stored as fractions. Other locations are also possible but seldom used. Both methods have advantages and disadvantages when used with arithmetic operations. Most of the arithmetic discussion in this chapter will assume an integer representation.

In electronic digital systems, the radix, r, is 2. The digits are then restricted to $\{0,1\}$, a binary number system. Because so many binary bits are required to represent even modest sized numbers, it is common to group the bits into sets of three or four. Each of the unique collections of bits is then referred to by a single designation. When the grouping is by sets of three, the result is an octal representation. The digits from 0 through 7 represent the possible groupings.

When groups of four bits are used, the result is a hexadecimal number. There are 16 possible representations. The 10 decimal digits (0–9) are used to name the first 10; the alphabetic letters A, B, C, D, E, and F are used for the other six.

Example 5.1:   *Binary to Octal, Binary to Hexadecimal Conversion*

Convert the 12-bit binary number 010100001101 to its octal and hexadecimal representation.

Solution:

$010/100/001/101_2 = 2415_8$
$0101/0000/1101_2 = 50\text{D}_{16}$

Octal and hexadecimal notation is used only to aid humans. Internal to the hardware, digital systems use only two-valued logic to represent binary numbers.

By weighting the radices, conversions between various number systems is possible. For computing devices, conversions between decimal- and binary-based systems are of the most interest.

Example 5.2:   Binary to Decimal Conversion

Convert the binary number 1001.0101 to decimal.

Solution:

The conversion is done using weights associated with each bit position. For this situation $n = 4$, $m = 4$.

$$1001.0101_2 = 1 * 2^3 + 0 * 2^2 + 0 * 2^1 + 1 * 2^0 + 0 * 2^{-1} + 1 * 2^{-2} + 0 * 2^{-3} + 1 * 2^{-4}$$
$$= 8 + 0 + 0 + 1 + 0 + 0.25 + 0 + 0.0625$$
$$= 9.3125_{10}$$

Example 5.3:   Decimal to Binary Conversion

Convert the number 25.38 to binary.

Solution:

The integer (25) and fractional (.38) parts of the original number are converted separately. In converting the integer portion, we can make the following association:

$$N = 25 = d_{n-1} 2^{n-1} + d_{n-2} 2^{n-2} + \dots + d_1 2^1 + d_0 2^0$$

If both the left and right side of this equation are divided by 2, the remainders after the division must be the same. The remainder from the right side is $d_0 2^0 = d_0$. After determining d_0, the remainders can be discarded. The integer portion can again be divided by two, the remainders compared, and d_1 determined. The process is shown Table 5.1.

n	$n/2$	Remainder	Binary Bit
25	12	1	d_0
12	6	0	d_1
6	3	0	d_2
3	1	1	d_3
1	0	1	d_4

Table 5.1

$25_{10} = 11001_2$ (Remember d_0 is the least significant bit)

Similarly, the fractional part of the representation is determined by examining the binary representation.

$$0.38 = d_{-1} 2^{-1} + d_{-2} 2^{-2} + \dots + d_{-m} 2^{-m}$$

If both sides of the equation are multiplied by two, the integer portion of the right side is d_{-1}. The integer resulting from the multiplication is deleted and the process is repeated to determine the remaining bits. Table 5.2 shows the conversion process for finding the binary representation for 0.38. The algorithm has been terminated after 8 bits.

n	$2 * n$	Int. Value	Binary bit
0.38	0.76	0	d_{-1}
0.76	1.52	1	d_{-2}
0.52	1.04	1	d_{-3}
0.04	0.08	0	d_{-4}
0.08	0.16	0	d_{-5}
0.16	0.32	0	d_{-6}
0.32	0.64	0	d_{-7}
0.64	1.28	1	d_{-8}

Table 5.2

$0.38_{10} = 0.01100001_2$

Putting both parts together gives:

$25.38 = 11001.01100001$

When converting between different radices, there is no assurance that fractional portions can be fully represented in a finite number of bits.

5.3   Representation of Signed Integers

There are three common methods for binary representation of negative numbers. The method chosen depends on how the resulting numeric values will be used. The following examples assume that numbers are to be represented in 8 bits. **Size extension** is expanding a binary integer to be represented by a larger number of bits. This is often required to perform arithmetic operations on data represented by different bit lengths.

Sign-magnitude Representation

Sign magnitude is similar to the methods humans use to indicate negative numbers. A minus sign is represented by a 1; a plus sign is represented by a 0. The most significant bit location is allocated for the sign bit. For $X(7:0)$, an 8-bit integer, $X(7)$ provides the sign information and $X(6:0)$ provides the magnitude information.

$+ X = 0, X(6:0)_{\text{sign mag}}$
$- X = 1, X(6:0)_{\text{sign mag}}$

$+ 5_{10} = 00000101_{\text{sign mag}}$
$- 5_{10} = 10000101_{\text{sign mag}}$

Size extension, for example conversion from an 8-bit integer to a 16-bit integer, involves relocating the sign bit and filling expansion bits with 0s.

Decimal Representation	8-Bit Sign Magnitude	16-Bit Sign Magnitude
+5	00000101	00000000 00000101
−5	10000101	10000000 00000101

Table 5.3

Sign magnitude is often used in calculators. It is not a convenient representation for arithmetic processing hardware since it requires complex rules to perform arithmetic operations.

Two's Complement

Two's complement is based on a special characteristic of a negative number; when a negative number is added to its positive counterpart, the result is 0. For fixed-length registers, two numbers may be added to produce an overflow carry and leave the remaining bits with a value of 0. A number having this property is the two's complement of the corresponding positive number. For an 8-bit integer number, the two's complement is formed by subtracting the number from 2^8.

$+ X = 0, X(6:0)_{\text{two's comp}}$
$- X = 2^8 - X(6:0)_{\text{two's comp}}$

Then

$$X + (-X) = X + (2^8 - X) = 2^8$$

Since the register can store only 8 bits and the 2^8 term is in the ninth bit position, it is lost to the register.

It is easy to verify that the sum of the two's complement representations for +5 and −5 yields 100000000. The ninth bit is an overflow carry bit and is lost if representations are limited to 8 bits. Two algorithms exist to find the two's complement of a number.

- **Algorithm A:** Take the number and complement all of its bits; $1 \leftarrow 0, 0 \leftarrow 1$. Add 1. The algorithm is the same whether converting from a positive to a negative or from a negative to a positive number. For example, to convert +5 to its negative:

  $00000101 = +5_{10}$
  11111010 complement all of the bits
  $+ \underline{00000001}$ add 1
  $11111011 = -5$ (in two's complement binary)

- **Algorithm B:** Starting from the right-most bit, scan the number from right to left to find the first appearance of a 1. Complement all of the bits to the left of the first 1.

$00101100 = +44_{10}$ (in two's complement binary)
$\underline{00101}100$ Bits to the left of first 1 are identified
$\underline{11010}100$ Bits to the left of first 1 are complemented
$11010100 = -44_{10}$ (in two's complement binary)

As with the sign-magnitude representation, the most significant bit position indicates if the number is positive or negative; 0 identifies positive, 1 identifies negative. When adding two's complement numbers, verify that the sum has not exceeded the allowable range of values and affected the sign position. This would be a magnitude overflow.

Expanding a two's complement number from 8 to 16 bits requires replication of the most significant bit.

$+ X = 0, X(6{:}0)_{\text{8-bit}} = 0, 0, 0, 0, 0, 0, 0, 0, 0, X(6{:}0)_{\text{16-bit}}$
$- X = 1, X(6{:}0)_{\text{8-bit}} = 1, 1, 1, 1, 1, 1, 1, 1, 1, X(6{:}0)_{\text{16-bit}}$

Implementing this algorithm in hardware is not difficult; the most significant bit is replicated throughout the expanded range.

Decimal Representation	8-Bit Two's Complement	16-Bit Two's Complement
+5	00000101	00000000 00000101
−5	11111011	11111111 11111011

Table 5.4

Biased Number Systems

A biased number is based on number representations from the most negative number to the most positive number. The most negative number representation is 00–0. Each number is generated in sequence from most negative to most positive by incrementing its predecessor in a binary counting sequence. This characteristic is not true with sign magnitude or two's complement. Both have discontinuities in the counting sequence as they pass through zero.

In an 8-bit representation of integers, bias values of 128 and 127 are usually used. In an excess 128 representation, the most negative number is −128 with a representation of 00000000. The representation of all integers is formed by first adding 128 to the integer and then taking its binary representation.

$X = (X + 128)_{\text{excess 128}}$
$+5 = (5 + 128)_{\text{excess 128}} = (133)_{\text{excess 128}} = 10000101_{\text{excess 128}}$
$-5 = (-5 + 128)_{\text{excess 128}} = (123)_{\text{excess 128}} = 01111011_{\text{excess 128}}$

Table 5.5 shows a range of values using excess 128 and excess 127 representation. While the excess 128 representation is easier for human interpretation, excess 127 is used as a component of the IEEE floating-point number system presented later in this chapter.

Decimal Value	Excess 128	Excess 127
128	No Representation	11111111
127	11111111	11111110
126	11111110	11111101
...	...	...
2	10000010	10000001
1	10000001	10000000
0	10000000	01111111
−1	01111111	01111110
−2	01111110	01111101
...	...	...
−127	00000001	00000000
−128	00000000	No Representation

Table 5.5

The following discussion and examples, while using excess 128 representation, can be easily modified to cover excess 127 or any other bias value.

In excess 128, positive numbers have a 1 in the most significant position, indicating that the number is positive. A 0 indicates a negative number. In both excess 128 and excess 127 representations, the transitions from the most negative number to the most positive number follow a binary counting sequence. This is an advantage when determining the difference in the magnitudes of the numbers. This is required for comparing exponents in floating-point representations.

Size extension is more complex with a **biased number system**. Extending an 8-bit representation to a 16-bit representation usually involves changing from an excess value of 128 (2^7) to an excess value of 32,768 (2^{15}). Converting a positive 8-bit excess 128 integer [$X(7)=1$] to a 16-bit excess 128 integer involves moving bit 7 to bit 15 and placing 0s in all other expanded bit positions.

$$X = 1, X(6:0)_{8\text{-bit}} = 1, 00000000, X(6:0)_{16\text{-bit}}$$

For negative numbers bit 7 is moved and bit 6 is replicated.

$-X = 0, X(6:0)_{8\text{-bit}} = 0, X(6), X(6), \ldots, X(6), X(6:0)_{16\text{-bit}}$

Decimal Representation	8-Bit Excess 128	16-Bit Excess 128
+5	10000101	10000000 00000101
−5	01101011	01111111 01101011

Table 5.6

Table 5.7 compares the various number systems for an 8-bit numerical representation.

Numeric Value	Unsigned Binary	Sign Magnitude	Two's Complement	Excess 128
255	11111111			
254	11111110	None	None	None
...	...			
128	10000000			
127	01111111	01111111	01111111	11111111
126	01111110	01111110	01111110	11111110
...	...	...	...	...
2	00000010	00000010	00000010	10000010
1	00000001	00000001	00000001	10000001
0	00000000	00000000	00000000	10000000
−1		10000001	11111111	01111111
−2		10000010	11111110	01111110
...	None	...	...	...
−127		11111111	10000001	00000001
−128		None	10000000	00000000

Table 5.7

Table 5.7 shows that each number system uses all 256 combinations of the 8 bits. The range of representation of the systems is different. Both the sign magnitude and two's complement have discontinuities in the counting sequence at the value zero, while the biased system has a smooth transition through zero.

5.4   Floating-point Numbers

Floating-point numbers can represent a much greater range of values than integers. To accommodate these very large (or small) numbers, a representation is used to specify a magnitude (known as the mantissa, M) and an exponent value, E. This information is encoded into two fields, or groups of bits. When used in digital computers, the full representation often requires several data words.

The magnitude is often scaled so its value is greater than 1 but less than two. This assures that the first bit location to the left of the binary point has a value of 1. The location of the binary point is implicit; no special hardware or symbol is used to mark its location.

The mantissa can have either a positive or negative value. The sign of the mantissa is the sign of the overall numeric value. The mantissa is usually expressed in sign magnitude format with a sign bit preceding the exponent field. Exponents can also be either positive or negative. They are usually represented using a biased number system.

When comparing magnitudes of two numbers, the one with the larger exponent is greater. If the exponents are equal, the one with the greater mantissa is larger. The use of biased exponents allows the magnitudes of two floating-point numbers to be compared using the same algorithm for comparing the magnitudes of two integers.

IEEE Floating-point Representation

Many floating-point representations have been defined. Variations include the number of bits for each field, location of the binary point, and the representation of negative exponents and mantissas. Indeed, many manufacturers historically have defined a representation and designed hardware to perform arithmetic on their selected representation. With the arrival of the personal computer in the early 1980s, a standard representation was sought for manufacturers to create coprocessors for floating-point mathematics. The IEEE 754 standard was developed to address these needs. The IEEE standard also considered many possible computational conditions such as magnitude overflow.

The IEEE standard includes two formats for **floating-point** numbers. The short real (single precision) format is based on a 32-bit representation; a long real (double precision) is based on a 64 bit representation. Both use three fields within the data bits. Both are shown in Figure 5.1.

A) 32-bit, Short Real Floating-point Format

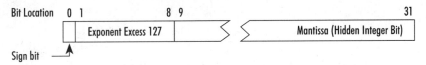

B) 64-bit, Long Real Floating-point Format

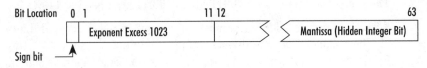

Figure 5.1   *IEEE Floating Point Data Formats*

Bit 0, the left-most bit in the IEEE representation, is a sign bit for the overall numerical value. In a sign-magnitude representation, 0 indicates a positive number; 1 indicates a negative number. The next 8 bits indicate the exponent value in an excess 127 code. Thus an exponent value of 0 would be represented by the value 01111111.

The remaining 23 bits represent the mantissa of the number. All floating-point numbers are assumed to be normalized in the format 1.M, where M is the mantissa. In a binary number system, the normalization will always result in a 1 bit immediately to the left of the binary point. Because the bit is known, there is no need to actually store it. Thus, the 23-bit mantissa field is sufficient to store 24 significant bits of the magnitude. The leading 1 is often referred to as the hidden or phantom bit.

Example 5.1:   *Interpret the value of the following short real number.*

   0 10000010 11000000000000000000000 (the spaces are for interpretation purposes)

Solution:

   The leading bit is 0; therefore the number is positive.
   The exponent is $130 - 127 = 3$.
   The magnitude field is 0.11. By adding in the phantom bit, this becomes 1.11.
   The value is $1.11 * 2^3 = 1110_2 = 14_{10}$

Example 5.2:   *Encode N = –87$_{10}$ as a short real.*

Solution:

   $87 = 1010111 = 1.010111 * 2^6$
   The leading bit is 1; number is negative.
   The exponent field is $127 + 6 = 133_{10} = 10000101_2$.
   The magnitude field is 01011100000000000000000 (the phantom bit has been removed).
   The representation is 1 10000101 01011100000000000000000.

Example 5.3:   Encode N = 21.3 as a short real.

Solution:

The leading bit is 0; the number is positive.

The integer portion of the binary number is $21_{10} = 10101_2$.

The fraction portion of the binary number is $0.3_{10} = 0100\ 1100\ 1100\ 1100\ 1100$.

$21.3 = 10101.01001100110011001100 = 1.0101\ 0100\ 1100\ 1100\ 1100\ 1100 * 2^4$.

The magnitude field is 01010100110011001100 (the phantom bit has been removed).

The exponent (+4) is represented as 10000011 in excess 127.

The representation is 0 10000011 0101 0100 1100 1100 1100 110.

With the short real format, approximately six to seven decimal digits of magnitude can be stored. The range of representations is from a minimum of 8.43×10^{-37} through $3.37 \times 10^{+38}$.

In the long real format, a 64 bit representation is used. A single sign bit is followed by eleven exponent bits and 52 mantissa bits. The exponent field uses excess 1023 representation. The phantom bit expands the mantissa to 53 bits.

The IEEE standard also allows representation of exception conditions. These conditions are identified by a 11111111 or 00000000 in the exponent field. For example:

Exponent	Magnitude	Numeric Value
$E = 11111111$	$M \neq 0$	Not a valid number.
$E = 11111111$	$M = 0$	N = Infinite, Sign bit determines +/–
$00000000 < E < 11111111$	M = Any value	$N = (-1)^8\ 2^{E-127}\ (1.M)$
00000000	$M \neq 0$	$N = (-1)^8\ 2^{-126}\ (0.M)$
00000000	$M = 0$	$N = 0$

Table 5.8

An exponent field of 255 with a non-zero mantissa field indicates an invalid number. Such a number might result from an overflow or underflow of an arithmetic operation. An exponent value of 255 with a zero mantissa and a 0 sign bit indicates a value of infinity. This might result from a division by 0. Negative infinity has the same exponent and mantissa representation with a 1 sign bit. When the exponent field is 0, the denormalized format is used to allow a linear decrement to 0. The value 0 is, by definition, represented by an exponent field of 0 and a mantissa field of 0.

5.5   Other Representations for Numeric Information

While standard binary representations are the most commonly encountered numeric representations, several other forms are useful in specific applications.

Binary-coded Decimal

Binary-coded decimal (BCD) is used when there is extensive interaction or display with a human operator. In BCD each of the 10 decimal integers is assigned a 4-bit binary code. In BCD representation, binary bits are logically grouped into fields of 4-bits. For example:

569 (decimal) = 0101 0110 1001 (BCD)

When determining the decimal value of a BCD number, each of the four bit fields is weighted by a power of 10. For example:

100001110101 (BCD) = $(8 \times 100) + (7 \times 10) + (5 \times 1) = 875$

When performing arithmetic on BCD numbers, special algorithms must be used to produce a result in BCD. It is common to perform BCD arithmetic using binary adders. The sum is then corrected using a processing algorithm to return the result to BCD form. Calculators often use BCD representation with a sign bit.

Residue Number Representations

Not all number representations are based on a weighted number system. A number may also be represented as a set of remainders obtained through integer division by a set of base numbers. This set of remainders is defined as a **residue**. The residue is generated by taking the original number modulo n_i, where n_i is one of the residue bases. Arithmetic performed on a residue representation has the potential for very high speed addition and multiplication.

For ease of interpretation, the following example is provided in decimal. Each of the decimal numbers could be encoded in binary for a purely binary system. The base numbers for residues are selected as 2, 3, and 5.

Addition using residue numbers is accomplished by performing the column-by-column add operation. The column sum is then reduced by taking its modulus with respect to the residue base for the column. There is no carry between digits. The sum's decimal value is obtained by performing a table-look-up operation.

For example, to add 3 and 4, first obtain the residue representation of each of the operands. The residue representation for 3 is obtained as $3_{\text{mod } 2} = 1$; $3_{\text{mod } 3} = 0$; $3_{\text{mod } 5} = 3$. The representation for 3 is 103. In a similar manner the representation of 4 is 014. The representation of integers from 0 through 14 is shown in Table 5.9.

Numeric Value	Residue Modulo 2	Residue Modulo 3	Residue Modulo 5	Residue Representation
0	0	0	0	000
1	1	1	1	111
2	0	2	2	022
3	1	0	3	103
4	0	1	4	014
5	1	2	0	120
6	0	0	1	001
7	1	1	2	112
8	0	2	3	023
9	1	0	4	104
10	0	1	0	010
11	1	2	1	121
12	0	0	2	002
13	1	1	3	113
14	0	2	4	024

Table 5.9

To add 3 and 4, the digit-by-digit sum of the individual residue positions is then performed. Each of the sum digits is also expressed in residue form. The addition of digits 3 and 4 in the mod 5 position produces $7_{\text{mod } 5} = 2$.

$$
\begin{array}{cl}
3 & 103 \\
+\ 4 & \underline{014} \\
\hline
7 & 112
\end{array}
$$

By examination of Table 5.9, 112 is the residue representation for the number 7.

Multiplication is performed in a manner similar to addition. A column-by-column multiplication of the digits is performed. Each digit-by-digit product is reduced by taking its modulus with respect to the base number for the column. There is no carry or interaction of digits in one column with those from another.

For example, to multiply 3 and 4 the digit-by-digit product of the residues is expressed in the residue form ($0 * 1 = 0_{\text{mod } 2} = 0$; $1 * 0 = 0_{\text{mod } 3} = 0$; $3 * 4 = 12_{\text{mod } 5} = 2$).

$$
\begin{array}{cl}
3 & 103 \\
\times\ 4 & \times\ \underline{014} \\
\hline
12 & 002
\end{array}
$$

By examination of Table 5.9, 002 is the residue representation for the number 12.

The residue system has some computational advantages. The absence of carry signals allows very high speed operations, and the algorithm for integer multiplication is especially fast. It also has significant disadvantages. The foremost problem is the lack of a good algorithm to convert between binary and residue representation of a numeric value. The use of table-look-up ROM can be useful in some situations.

Residue representations have been revived recently for applications in digital signal processing. The ability to process large amounts of data in a short period of time is especially useful in implementing digital filters.

5.6   Representation of Text—ASCII Coding

In addition to representing numerical values, computing devices must also reliably exchange and store text. American Standard Code for Information Interchange, ASCII, is a code used for representing alphabetic and numeric text information. ASCII is the logical descendant of the 5-bit Baudot code that was originally used in printing telegraphic terminals.

00	16	SOH	ETB	2D	43	59	6F	0	F
01	17	STX	S0	2E	44	5A	70	1	G
02	18	ETX	S1	2F	45	5B	71	2	H
03	19	EOT	S2	30	46	5C	72	3	I
04	1A	ENQ	ESC	31	47	5D	73	4	J
05	1B	ACK	S4	32	48	5E	74	5	K
06	1C	BELL	S5	33	49	5F	75	6	L
07	1D	BKSP	S6	34	4A	60	76	7	M
08	1E	HT	S7	35	4B	61	77	8	N
09	1F	LF	*	36	4C	62	78	9	O
0A	20	VT	!	37	4D	63	79	:	P
0B	21	FF	"	38	4E	64	7A	;	Q
0C	22	CR	#	39	4F	65	7B	<	R
0D	23	SO	$	3A	50	66	7C	=	S
0E	24	SI	%	3B	51	67	7D	>	T
0F	25	DLE	&	3C	52	68	7E	?	U
10	26	DC1	'	3D	53	69	7F	@	V
11	27	DC2	(	3E	54	6A	+	A	W

12	28	DC3	)	3F	55	6B	,	B	X
13	29	DC4	2A	40	56	6C	-	C	Y
14	SP	NAK	2B	41	57	6D	.	D	Z
15	NULL	SYNC	2C	42	58	6E	/	E	[
\	a	f	k	p	u	z	DEL		
]	b	g	l	q	v	{			
^	c	h	m	r	w	|			
_	d	i	n	s	x	}			
`	e	j	o	t	y	~			

Table 5.10

ASCII, an 8-bit code, represents all letters, numbers, and special characters most often used in written English communication. Additional code words are used to denote a carriage return, a line feed, and an audible bell. Other codes, such as SOH (start of header) and EOT (end of transmission) represent operations useful in controlling communication activities. Table 5.10 lists the character representations used in 7-bit ASCII representation. The eighth bit is available as a parity bit or to designate an alternate set of 128 characters.

5.7   Arithmetic Operations—Integer Addition

In a digital computer, integer addition is the most important arithmetic operation. Two n-bit operands are combined to produce an n-bit sum. An overflow carry may also be generated. Addition is an important component of other operations such as multiplication. Some rules for addition algorithms are related to the method of encoding the data. Unsigned binary, two's complement, and sign magnitude will require slightly different processing algorithms. For arithmetic purposes, two's complement is preferred and is assumed in the following sections.

Adding Two Positive, n-bit Integers

If two integers, X and Y, are both positive, then both have a sign bit value of 0. Adding two positive numbers in a binary adder will give a correct result, including the sign bit, if the magnitude of the sum requires fewer than n bits for its representation. Any carry into the sign bit indicates that the sum is too large to be represented with the available bits.

Adding Two Negative, n-bit Integers

In two's complement

$$-X = 2^n - X$$
$$-Y = 2^n - Y$$
$$(-X) + (-Y) = (2^n - X) + (2^n - Y) = 2^{n+1} - (X + Y) = 2^n + [2^n - (X + Y)]$$

The portion of the result within square brackets is recognized as the two's complement representation of the sum of X and Y. The leading 2^n term represents a carry overflow from the most significant bit position, and is ignored. As long as the sum of X and Y allows representation in $n - 1$ bits, the result will be correct.

Adding Two n-bit Integers with Opposite Signs

It can be verified that adding numbers with mixed signs will always produce a correct result. Magnitude overflow is not possible.

Parallel Addition

Addition is distinguished from most logic operations by the interaction of the bits as the sum is formed. Because of the carry process, each bit of the sum is a function of all of the less significant input bits. The logic to perform a fully parallel addition, in which only two levels of AND/OR logic are used, grows very rapidly with the number of bits in the data words.

Adding two n-bit numbers, A and B, would involve the operations shown in Equation 5.1. No carry is explicitly generated by logic circuitry since the sum bits are determined directly as a function of the input operands.

$$
\begin{array}{cccccc}
& A_{n-1} & A_{n-2} & & A_1 & A_0 \\
+ & B_{n-1} & B_{n-2} & & B_1 & B_0 \\
\hline
S_n & S_{n-1} & S_{n-2} & & S_1 & S_0 \\
\end{array}
$$

$$
\begin{aligned}
S_0 &= f(A_0, B_0) \\
S_1 &= f(A_1, B_1, A_0, B_0) \\
S_{n-1} &= f(A_{n-1}, B_{n-1}, A_{n-2}, B_{n-2}, \ldots, A_0, B_0) \\
S_n &= f(A_{n-1}, B_{n-1}, A_{n-2}, B_{n-2}, \ldots, A_0, B_0)
\end{aligned}
$$

Equation 5.1

For modest bit widths, this circuitry may be realized and packaged as a single component. For example, the 74LS181 integrated circuit contains circuitry to produce a 4-bit summation. It accepts a carry input and produces a carry output to allow cascading to process larger data widths. Using a table-look-up ROM may also be appropriate in some applications.

Ripple Adders

A cascade or ripple processor, as shown in Figure 5.2, can decrease the circuitry required for addition. Each of the identical single-stage adders accepts two operand bits and a carry bit from the previous adder in the cascade. Each produces a sum bit and a carry bit to pass to the next unit. Each sum bit and carry bit is a function of all of the lower order data bits.

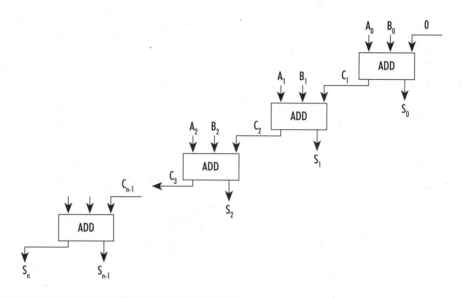

Figure 5.2   A Ripple (Cascade) Adder

If a two-level logic circuit requires t time units to produce a result, the carry from adder unit 0 is available as input to adder unit 1 after t time units. Unit 1 can then produce its correct sum and carry after a total of $2t$ time units: t time units waiting for the correct carry and t additional time units to compute its outputs. By this analysis, the sum and carry bits for the last unit, unit $n - 1$, requires nt time units.

A cascade processor requires adequate time to complete the processing before storing results. This can be accomplished either by having a clock cycle long enough for the processing to finish, or by allowing multiple clock cycles for each ADD operation.

Carry Completion Detection

When using ripple adders, the time delay associated with ripple propagation can be considerable. In the worst-case situation, a carry generated in adder stage 0 propagates through the entire range of adders and affects the sum or carry bit in stage $n - 1$. A carry chain of this duration does not always occur. In most cases the carry propagations will terminate long before nt time units.

Carry completion detection attempts to hasten completion of carries by determining in advance the delay required for a particular addition operation. The delay associated with carry propagation is data dependent. Each adder section has three inputs: two are operands, one is an input carry. To produce an output carry, at least two of the inputs must be 1. Therefore, if at least two of the inputs are 0, there can be no carry output from the section. Although the condition of the carry input is not known, the simultaneous appearance of 0 in both operand bits assures that no carry will be present at the output of the stage. The longest carry chain is equal to the longest string of bit positions to propagate a carry. This calculation is easier than precisely determining the length of a carry propagation.

For example, consider the following 16-bit addition.

A(15:0) Operand 1:   1 0 1 1 $\underline{0}$ 1 0 $\underline{0}$ 1 0 0 1 1 1 $\underline{0}$ 1
B(15:0) Operand 2:   0 1 1 0 $\underline{0}$ 0 1 $\underline{0}$ 1 1 1 1 0 0 $\underline{0}$ 1

The underlined entries indicate positions in which both operand bits are 0. The longest string of bit pairs in which one or both of the bits are 1 is from position 2 through position 7, a total of 6 adder sections. This means that if a carry were generated in position 2, it would propagate no further than position 7, requiring no more than $6t$ time units. It should be noted in this example that a carry is *not* generated in position 2; therefore the actual carry propagation time is less than $6t$ time units. If a worst-case analysis were used, $16t$ time units would be required.

The price paid for using the carry completion detection technique is the logic necessary to determine the length of the possible carry string and the complexity of using a variable length of time for completing the processing algorithm.

Carry Look-ahead Addition

Carry propagation through multiple adder stages is the main time-consuming activity in cascade addition. The carry look-ahead technique, as shown in Figure 5.3, attempts to simultaneously generate all the carries for each stage of the adder in a single AND/OR logic circuit.

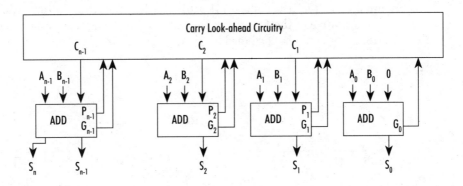

Figure 5.3   A Carry Look-ahead Adder

Adder stage 0 will generate a carry, G_0, for adder stage 1 if both A_0 and B_0 have a value of 1. Adder stage 1 will receive a carry, C_1, if it was generated in stage 0. This can be indicated logically with AND/OR operations as:

$G_0 = A_0 B_0$
$C_1 = G_0$

Adder stage 1 will generate a carry, G_1, for adder stage 1 if both A_1 and B_1 have a value of 1 OR if stage 0 generates a carry, G_0, which is propagated, P_1, through stage 1 to stage 2. A carry will be propagated if at least one of the operand bits for the stage has a value of 1. Logically ("+" is an OR operation), this becomes:

$G_1 = A_1 B_1$
$P_1 = A_1 + B_1$
$C_2 = G_1 + G_0 P_1$

By similar reasoning, the carry input for any adder stage, n, can be calculated if the Gs and Ps from stages 0 through $n-1$ are known.

$$C_n = G_{n-1} + G_{n-2}P_{n-1} + G_{n-3}P_{n-2}P_{n-1} + G_0P_1P_2P_3 \ldots P_{n-1}$$

Since each of the Gs and Ps are functions of a single pair of operand bits, they may be computed using AND/OR logic in each of the adder stages. With access to all of the P_i and G_i, logic within the carry look-ahead device can then compute the carry signals for each stage of the adder. Each adder stage, provided with its carry signal, can then compute its sum bit. The entire addition process can be completed in $3t$ time units: t units to compute all G_i and P_i terms, a second t units to compute all C_i terms, a third t units to compute the S_i terms.

The carry look-ahead technique can be extended to additional bits. The logic required to compute the carries gets more complex as the number of bits grows. Carry look-ahead circuitry that can work with four adder stages is available in integrated circuit form. Carry look-ahead devices are designed to be cascaded for working with longer data lengths with only slight time delay increases.

The 74LS181 is a commercially available integrated circuit with a 4-bits data width. It has internal circuitry capable of combining 4-bit operands and an input carry to produce a 4-bit sum and an output carry. It also produces output signals indicating when the entire unit will generate a carry, G, or propagate a carry, P. These signals are used by a companion chip, the 74LS182, to produce input carries for each of the 74LS181s. Each of the '182s, in turn, can produce G and P signals representative of the four carry generate and four carry propagate signals they receive from the '181s. These can be combined by higher order '182s.

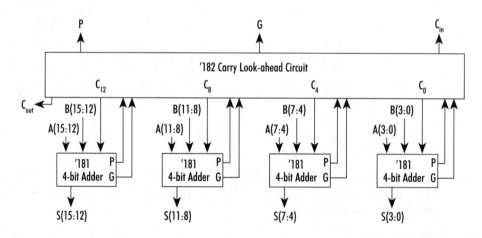

Figure 5.4   The '181 and '182 Carry look-ahead Adder Devices to Perform a 16-bit Addition

Figure 5.4 shows how '181 and '182 integrated circuits can be used to create a 16-bit adder. Since each of the 4-bit adders within the '181 requires only a single time delay to produce the G_i and P_i terms, the entire addition can be completed in $3t$ time units. With an additional level of '182s, a 64-bit addition could be performed using only one additional time unit.

Pipeline Adders

In Figure 5.5 a pipeline datapath allows propagation of carries with each succeeding clock pulse. This example involves operands of only 2 bits. Operands A(1:0) and B(1:0) are presented to an adder. The sum and carry bits from the adders are captured in registers. The second stage of the pipeline allows the carries to propagate one stage to the left, where they are introduced to the second-level adders. Again the sum and carry bits are saved in registers. Eventually all carries finish their propagation and a sum is available.

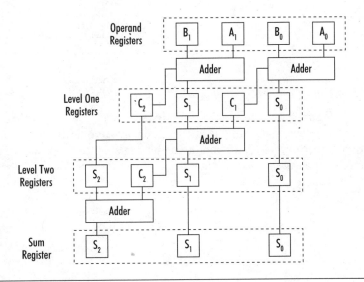

Figure 5.5   A Pipeline Adder

The pipeline adder is advantageous when adding multiple pairs of operands. Once the pipeline is filled, every clock cycle produces a new sum. No carry look-ahead circuitry is required. Although the addition of 2-bit operands is shown, the pipeline technique can be expanded to handle arbitrary data widths. The length of the pipeline grows proportionately. As an alternative, each stage of the pipeline may process a multibit addition, for example four.

5.8   Multiple Operand Addition

Some situations require adding three or more operands. These include matrix operations, processing data from data acquisition systems, and multiplication. The following methods are designed to provide datapaths suitable for these operations.

Accumulating Adders

Accumulating adders are useful when a set of operands are presented sequentially to a datapath structure, and the sum of the entire sequence is required. This can be readily accomplished using an accumulator to retain the accumulated sum of the operands, as shown in Figure 5.6. The adder receives one of its input operands from the input device and the other from the accumulator.

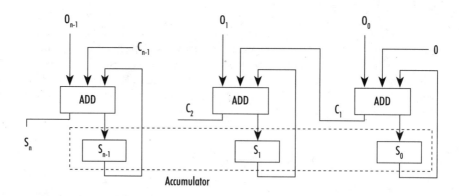

Figure 5.6   An Accumulating Adder Used with Multiple Operands

In Figure 5.6, performance is limited by carry-propagation time delays. When multiple operands are involved, a technique known as **carry-save addition** can be used. This is shown in Figure 5.7. In carry-save addition, a one-bit register captures the carry from each adder stage to prevent ripple propagation. A new operand is introduced during every clock cycle. The overall process is similar to a pipeline processor. This is considerably faster than is possible with techniques that propagate carries.

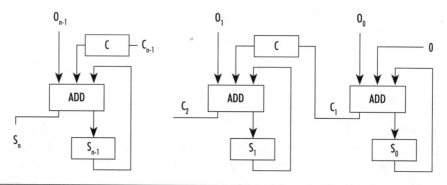

Figure 5.7   A Carry-save Adder

Each new operand combines with the current sum and carry registers to produce new values for the sum and carry register. The carries propagate one stage during each clock cycle. The carries in the system are a complex function of current and past operands. After the last operand is presented, an additional $(n - 1)t$ time units are required for the carries to finish propagating.

By using carry-save techniques, a fast adder can be obtained without the added complexity and expense of using carry look-ahead components. The efficiency of the carry-save technique is related to the number of operands to be combined. The technique is particularly well suited to high-speed data acquisition systems that may produce a new data value every clock cycle.

Tree Adders

Tree adders combine multiple operands to produce a single sum. While carry-save adders are useful for sequentially generated operands, tree structures are best used where a large array of data is simultaneously available. Tree structures emphasize parallelism in the data path to achieve high performance.

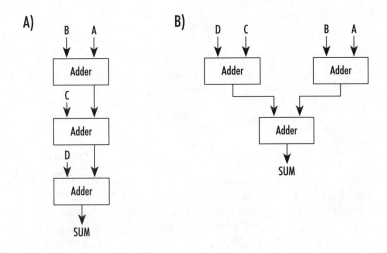

Figure 5.8   Alternate Methods of Combining Four Operands

Figure 5.8 shows two methods of combining four operands. The adders are assumed to be multibit adder segments (for example the 4-bit 74LS181) which may involve carry look-ahead to speed up their internal operation. The first method, sometimes called a linear addition, introduces the additional operands to combine with previously produced sum terms. This is a form of cascade processing. In the second method the four operands are combined as pairs in a first level of operations. The resulting sums are combined in a second level. This first level of processing is performed in parallel.

In addition to the possible ripple propagation time required for each adder segment, other time delays are required for vertically cascading the adder segments. By performing multiple additions in parallel, the effective time for the vertical data propagation is decreased.

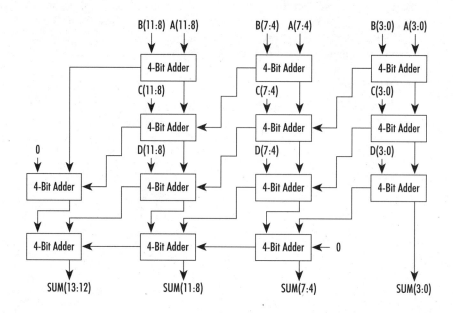

Figure 5.9   Use of Carry-save Techniques for a Multiword Adder

Figure 5.9 shows a variation on the carry-save technique as used for linear addition. In this figure four 12-bit numbers are combined using 4-bit adder segments. Carries generated within each adder segment are not propagated horizontally but instead are saved and propagated diagonally to the next group of adders. In this way all valid information arrives at the second level at the same time and can be processed further. The level at which the carry is introduced will not affect the final summation. There is no horizontal propagation of carries until the final adder. A carry look-ahead technique may be used at the final level to increase performance; it is of no benefit in earlier segments.

The technique shown in Figure 5.9 could also be modified and used with a pipeline structure by introducing registers to capture the sum and carry bits after each stage of the addition.

The Wallace Tree Adder

The **Wallace tree adder** is designed to combine many operands to generate a sum as rapidly as possible. The method uses both parallelism and carry-save techniques.

A single-bit adder stage may be viewed as a logic device with three inputs (two from operands and one from a carry) and two outputs (sum and carry). The three inputs are logically equivalent; it makes no difference which of the possible inputs is presented to which input pin. In a Wallace tree adder, the three inputs are connected to bits from three different operands; carries, since they are not immediately available, are introduced at lower levels of the cascade.

As an example, consider the addition of four 3-bit numbers: W, X, Y, and Z. The operands are first divided into groups of three. Three operands are applied to the three inputs of the first level of adders. Note that an operand is applied to the "carry" input as well as the two operand inputs, as shown in Figure 5.10. The first level of adders reduces the three inputs to two outputs, sum and carry, which are then combined with the sum and carry outputs of other devices at a second level of logic.

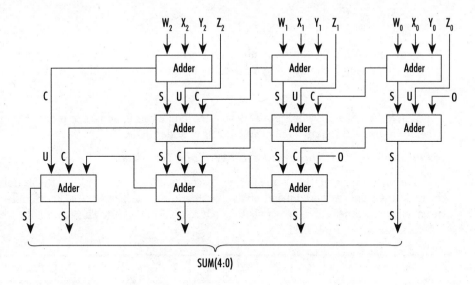

Figure 5.10   A Wallace Tree Adder

Each level of adders reduces the number of variables from 3 to 2, a factor of 1.5. This is accomplished at each level without waiting for carry propagation. When the reduction reaches a value of two, a single fast adder completes the operation.

The operation of the Wallace tree is shown in the following tables. Each level of the table corresponds to a level of logic used to implement the sum. Os represent bits of the original operands. Ss represent bits of the sums obtained from the adder stages. Cs represent carry bits and are introduced one bit position higher than the level at which they were produced. U represents bits that have not been processed in the preceding level. Every time three bits are combined in an adder, an S is created with the same binary weight as the operands that produced it. A C is also created with a weight one unit greater than that of the operands.

Original Operands

$$
\begin{array}{ccccc}
O & O & O & & Z_2 \ Z_1 \ Z_0 \\
O & O & O & & Y_2 \ Y_1 \ Y_0 \\
O & O & O & = & X_2 \ X_1 \ X_0 \\
O & O & O & & W_2 \ W_1 \ W_0
\end{array}
$$

Three single-stage adders combine bits from W, X, and Y to each produce an S (sum) and C (carry) bit. The single remaining bits of Z are not combined and appear as Us (unprocessed) in the following representation. After the first level of addition:

$$
\begin{array}{cccc}
 & C & C & \\
 & U & U & U \\
C & S & S & S
\end{array}
$$

The combination of operands is repeated in the second layer of processing. After the second level of addition:

```
C   C   C
U   S   S   S
```

Now conventional addition combines the remaining bits. Carries may propagate internally or a carry look-ahead technique may be used. After the third level of addition:

```
S   S   S   S
```

Conventional adders in Figure 5.9 require four levels of adders to produce a sum. The Wallace tree in Figure 5.10 requires only three levels of adder logic.

Example 5.4:   12-operand 8-bit Wallace Tree Adder

The advantage of the Wallace tree becomes more apparent as the number of operands increases. Consider the addition of twelve operands, each of 8 bits. Four first-level adders are used; each combines three operands to produce an S and C bit. The combination and distribution of bits is shown in Table 5.11.

Processing Level	2^{11}	2^{10}	2^9	2^8	2^7	2^6	2^5	2^4	2^3	2^2	2^1	2^0
Original Operands					O	O	O	O	O	O	O	O
					O	O	O	O	O	O	O	O
					O	O	O	O	O	O	O	O
					O	O	O	O	O	O	O	O
					O	O	O	O	O	O	O	O
					O	O	O	O	O	O	O	O
					O	O	O	O	O	O	O	O
					O	O	O	O	O	O	O	O
					O	O	O	O	O	O	O	O
					O	O	O	O	O	O	O	O
					O	O	O	O	O	O	O	O
					O	O	O	O	O	O	O	O
After First Level of ADD Processing					C	C	C	C	C	C	C	
					C	C	C	C	C	C	C	
					C	C	C	C	C	C	C	
					C	C	C	C	C	C	C	
				C	S	S	S	S	S	S	S	S
				C	S	S	S	S	S	S	S	S
				C	S	S	S	S	S	S	S	S
				C	S	S	S	S	S	S	S	S

Processing Level	2^{11}	2^{10}	2^9	2^8	2^7	2^6	2^5	2^4	2^3	2^2	2^1	2^0
After Second Level of ADD Processing					C	C	C	C	C	C		
				U	C	C	C	C	C	C		
			C	C	C	C	C	C	C	C	C	
				C	S	S	S	S	S	S	S	U
			C	S	S	S	S	S	S	S	S	S
After Third Level of ADD Processing			U	C	C	C	C	C	C	C	U	
			C	C	S	S	S	S	S	S	C	
			C	S	S	S	S	S	S	S	S	S
After Fourth Level of ADD Processing				U	U	U	U	U	U			
			C	C	C	C	C	C	C	C		
			S	S	S	S	S	S	S	S	S	S
After Fifth Level of ADD Processing				C	C	C	C	C	C			
		C	S	S	S	S	S	S	S	S	S	S
Final Sum	S	S	S	S	S	S	S	S	S	S	S	S

Table 5.11

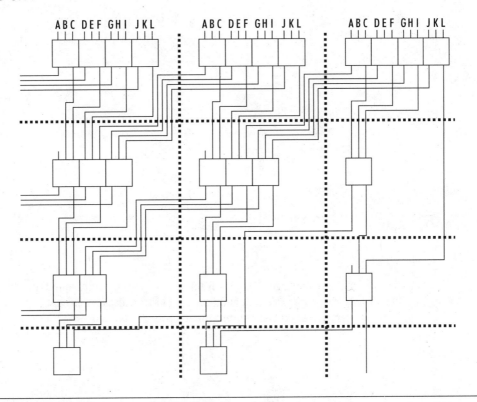

Figure 5.11 Interconnection for Wallace Tree Adder

In this example the twelve operands A through L are combined in six levels of adders to produce the sum. Figure 5.11 show the interconnection of processors for the lowest three bit positions. Because of the way carries are delayed for introduction in a lower-level adder, five of the levels do not require carry look-ahead to improve performance. In the sixth level, where each stage has two or fewer inputs, a high-speed adder section could be used for those sections still propagating carries.

5.9   Subtraction

Techniques for addition have their counterparts in subtraction. Two techniques can be used. The first is to directly implement the logic of subtracting the subtrahend (lower number) from the minuend (upper number). The second technique negates the subtrahend and performs an addition.

Direct Subtraction

Subtraction, like addition, can be performed bit-by-bit. For each bit location, i, there is a minuend bit, M_i, and a subtrahend bit, S_i. A borrow input, B_i, is provided from the previous bit position. The subtracter logic produces a difference bit, D_i, and a borrow output, B_{i+1}, which is sent to the next bit position. Table 5.12 shows the truth table for the difference and borrow bits.

Inputs			Outputs	
M_i	S_i	B_i	D_i	B_{i+1}
0	0	0	0	0
0	0	1	1	1
0	1	0	1	1
0	1	1	0	1
1	0	0	1	0
1	0	1	0	0
1	1	0	0	0
1	1	1	1	1

Table 5.12

The logic for a single stage of subtraction can be directly implemented. The borrow, and the possibility of borrow propagation, are similar to carry and carry propagation in addition. Techniques such as borrow look-ahead can minimize propagation delays.

Subtraction Using Addition Circuitry

Subtraction can also be performed by taking the negative of the subtrahend and performing an addition.

$$A - B = A + (-B)$$

With two's complement representation, negation is performed using algorithms from previous sections. This approach is an alternative to developing a full set of subtraction procedures. By taking the two's complement negation of the subtrahend, subtraction can be performed by an adder. Figure 5.12 shows a way to implement a combined adder/subtractor. The adder may be any of the types discussed in previous sections.

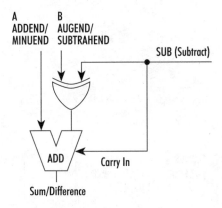

Figure 5.12   A Combined Adder/Subtractor

When the SUB control signal is asserted at logic 1, the exclusive OR functions as an inverter to the input from B, the subtrahend (number to be subtracted). SUB also provides a 1 to the carry input to the adder. The combination of the two operations provides an increment and an inversion of the subtrahend. This converts the subtrahend to its negative two's complement representation during subtraction. The adder then adds the minuend and the negative representation of the subtrahend. When SUB is not asserted, the exclusive OR circuit allows the operand to pass through the circuitry without change. A zero is provided to the carry input. An ADD operation is then performed.

5.10   Multiplication

Multiplication is one of the primary arithmetic operations. Multipliers are used in data processing activities and are also important components in devices for digital filtering and digital control. Because of its importance, many processing algorithms and datapath structures, with widely varying performance, exist for its implementation.

Multiplication by Repeated Addition

Multiplication may be viewed as a form of multiple addition. As such, its circuitry and techniques are similar to those discussed in previous sections. By definition, multiplication means adding the multiplicand to itself a number of times specified by the multiplier to produce a product. In integer binary multiplication, the number of bits of the product may equal, but not exceed, the sum of the number of bits in the multiplier and multiplicand. The design of such a multiplier has been considered as an example in previous chapters.

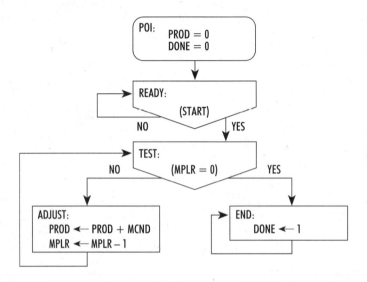

Figure 5.13   A Microoperation Diagram for a Repeated Addition Multiplier

The microoperation diagram for a multiple-addition multiplier is repeated as Figure 5.13. In the microoperation diagram, register PROD accumulates multiple additions of MCND. The operation is repeated until MPLR, the original multiplier, is decremented to 0.

Shift/Add Multiplication

As an alternative to multiple addition techniques, shift/add methods reduce the total number of adds required by generating and adding partial product terms.

Multiplication can be written as an operation involving a **multiplicand** and a **multiplier** to produce a **product**.

$$MCND(n - 1:0) * MPLR(n - 1:0) = PROD(2n - 1:0)$$

The multiplier may be expanded using the binary weight of each bit position.

$$MPLR(n - 1:0) = MPLR(n - 1) * 2^{n-1} + MPLR(N - 2) * 2^{n-2} + ...$$
$$+ MPLR(2) * 2^2 + MPLR(1) * 2^1 + MPLR(0) \cdot 2^0$$

The multiplication process can be viewed more systematically by defining a **partial product** of a n-bit $\times$ n-bit multiplication as the bit-by-bit product of a multiplier bit and the multiplicand. The partial product is weighted by the bit position of the multiplier bit.

$$PP_0 = MCND * MPLR(0) * 2^0$$
$$PP_1 = MCND * MPLR(1) * 2^1$$
$$PP_{n-1} = MCND * MPLR(n-1) * 2^{n-1}$$

The full product can then be obtained by adding together the partial products.

Figure 5.14 shows a possible datapath for shift/add multiplication. MPLR is a n-bit register containing the multiplier. MCND is an $2n$-bit register to allow shifting of the n-bit multiplicand. PROD, the product register, is $2n$ bits. A $2n$-bit adder is required. MPLR requires a shift control to allow the controller to sequentially examine the multiplier bits. MCND requires a shift control to perform the left shift.

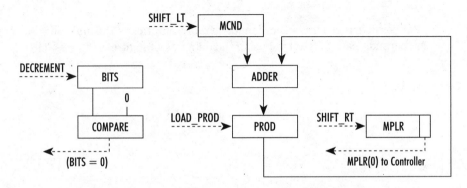

Figure 5.14   Datapath Structure for a Shift/Add Multiplier

Multiplying MCND by a single bit from MPLR and then multiplying by 2^i, where i is the bit position of the MPLR bit, produces the partial products. The datapath of Figure 5.14 performs the whole process. The multiplication by 2^i is accomplished by shifting MCND one bit position to the left for each partial product term. MCND is multiplied by the appropriate bit in MPLR and added to PROD. If the MPLR bit is 1, PROD is loaded with the current PROD plus the partial product. If the MPLR bit is 0, no addition is performed. Register BITS counts the number of shift operations.

The datapath configuration in Figure 5.14 can be economized by noting that instead of shifting MCND to the left after every addition, PROD can be shifted to the right. Since addition affects only the most significant bits of PROD, the adder does not need to be capable of double-length addition; an adder sufficient to work with the width of MCND is adequate. Low-order bits of MPLR can be discarded after use. Since the number of significant bits of PROD increases by one with every operation and the number of unused bits of MPLR decreases by one, economy can be achieved by having both share a common hardware register, as shown in Figure 5.15.

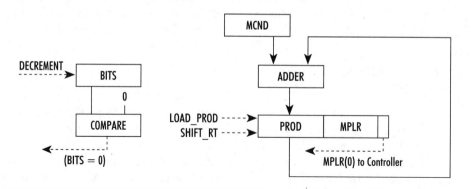

Figure 5.15   Datapath Structure for a Shift / Add Multiplier with a Shared Product and Multiplier Register

A possible register transfer language representation of the multiplication of two positive integer operands, each of four bits, using the datapath of Figure 5.15, is as follows:

```
NAME:        MULTIPLIER

REGISTERS:   PROD[4],          /*Shift register for partial product.     */
             BITS[2],          /*Down counter for bit operations.        */
             MPLR[4],          /*Shift register to store multiplier.     */
             MCND[4];          /*Register to store multiplicand.         */

LOGIC:       ADDER[4];         /*Four-bit adder.                         */

ALGORITHM:

POI:         PROD = 0,         /*Initialize PROD.                        */
             BITS = 4;         /*Initialize BITS.                        */

START:       IF (MPLR(0) = 1) then /*Test MPLR bit.                      */
               GOTO ADD        /*MPLR bit is 1.                          */
             else
               GOTO SKIP       /*MPLR bit is 0.                          */
             end IF;

ADD:         PROD←PROD + MCND, /*Perform the add.                        */
             BITS←BITS − 1,    /*Decrement the count.                    */
             GOTO TEST;

SKIP:        BITS←BITS − 1;    /*Decrement the count.                    */

TEST:        PROD←SHR(PROD),   /*Shift the combined PROD                 */
             MPLR←SHR(MPLR),   /*and MPLR registers.                     */
```

```
              IF (BITS = 0) then      /*Test if finished.                    */
                 GOTO DONE
              else
                 GOTO START
              end if;

DONE:         NOP,                    /*No operation.                        */
              GOTO DONE;              /*All done.                            */
```

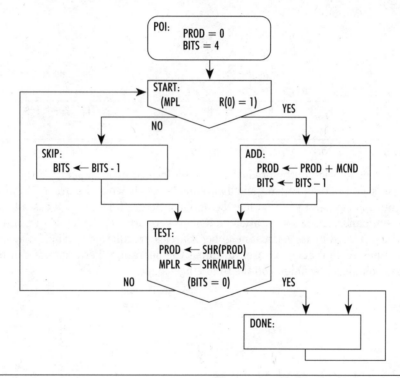

Figure 5.16   Microoperation Diagram for a Shift/Adder Multiplier

Figure 5.16 shows a microoperation diagram for the shift/add multiplication algorithm for the datapath of Figure 5.15.

Array Multiplication

The multiplication operation can also be viewed as an array process. The multiplier and multiplicand generate an array of elements called **summands**, which are then combined in an algorithm involving row and column operations.

A summand is the one-bit product produced by a bit of the multiplier and a bit of the multiplicand. One-bit multiplication is conveniently performed using an AND gate.

$S_{00} = MCND(0) * MPLR(0)$
$S_{01} = MCND(0) * MPLR(1)$
$S_{10} = MCND(1) * MPLR(0)$

The partial products can then be defined in terms of the summands.

$$PP_0 = (S_{n-1,0} * 2^{n-1} + S_{n-2,0} * 2^{n-2} + \ldots + S_{10} * 2^1 + S_{00}$$
$$PP_1 = (S_{n-1,1} * 2^{n} + S_{n-2,1} * 2^{n-1} + \ldots + S_{11} * 2^2 + S_{01} * 2^1$$
$$PP_i = (S_{n-1,i} * 2^{n-1+i} + S_{n-2,i} * 2^{n-2+i} + \ldots + S_{1i} * 2^{1+i} + S_{0i} * 2^i$$
$$PP_n = (S_{n-1,n} * 2^{2n-1} + S_{n-2,n} * 2^{2n-2} + \ldots + S_{1n} * 2^{n+1} + S_{0n} * 2^n$$

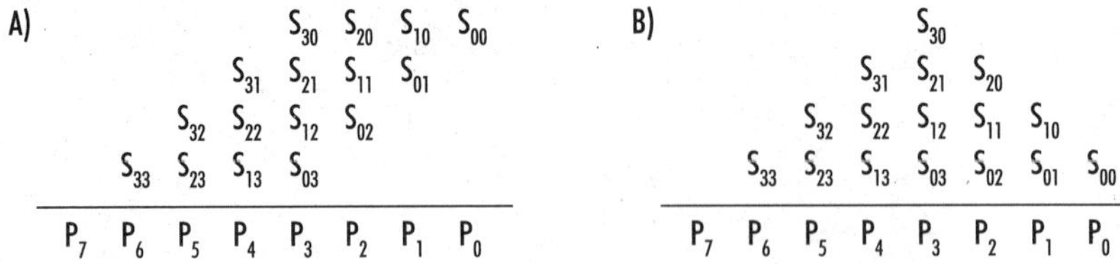

A)
S30 S20 S10 S00
S31 S21 S11 S01
S32 S22 S12 S02
S33 S23 S13 S03
P7 P6 P5 P4 P3 P2 P1 P0
B)
S30
S31 S21 S20
S32 S22 S12 S11 S10
S33 S23 S13 S03 S02 S01 S00
P7 P6 P5 P4 P3 P2 P1 P0

Figure 5.17 Addition of Summands to Produce a Product

The generation of the product can be performed by manipulating the individual summands. To generate the product, summands are added as shown in Figure 5.17A and 5.17B. Techniques for multiple operand addition may be applied. Figure 5.18 shows a cascade technique for adding the summands. Each computational element generates the value of the summand and adds it to values produced by its vertical neighbor. Carries are allowed to ripple diagonally down the array technique. A final adder is used to finish the operation. Performance can be improved using a carry-look-ahead technique on the final stage of addition.

0
S30
S20
S10
S00
S31
S21
S11
S01
S32
S22
S12
S02
S33
S23
S13
S03
Adder
P7 P6 P5 P4 P3 P2 P1 P0

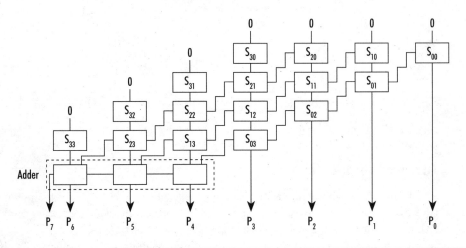

Figure 5.18 An Array Multiplier Using Adder Segments to Add Summands

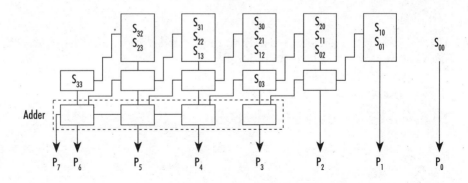

Figure 5.19 A Wallace Tree Multiplier

Figure 5.19 shows how a Wallace tree can be applied to the 4-bit multiplication. Adder sections combine summands in groups of three. Sum bits, carries, and single summand entries not combined in earlier stages are then combined in groups of three. The process continues until a single adder can combine the remaining bits to produce the product. The advantages of the Wallace tree become more apparent as the number of bits in the operands increases.

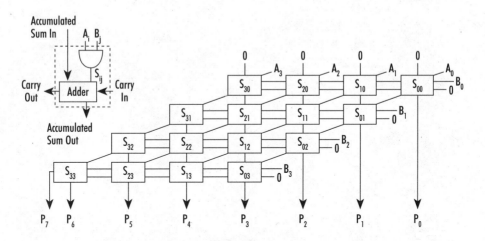

Figure 5.20 A Multiplier Using a Systolic Array Structure

The very regular structure of a multiplier, particularly as shown in Figure 5.18, can be an advantage. In Figure 5.20, an array structure computes and combines the summands. Each bit of the multiplier is introduced to all cells in each row of the device. Bits of the multiplicand are introduced to all elements on a diagonal. Accumulated sums are propagated downward from cell to cell. Carries are propagated horizontally from right to left.

The multiplier structure shown in Figure 5.20 is an example of a **systolic array**. All cells are identical. Each communicates only with its neighbors. Such devices are important because they can readily be synthesized in large repetitive structures using integrated circuit fabrication technology.

5.11   Division

Dividing binary integers involves a **dividend**, D, and a **divisor**, V. The result of the division produces a **quotient**, Q, and a **remainder**, R.

Consider the example of dividing the number 11000110 (198_{10}) by 1101 (13_{10}). Several techniques are available.

Division by Repeated Subtraction

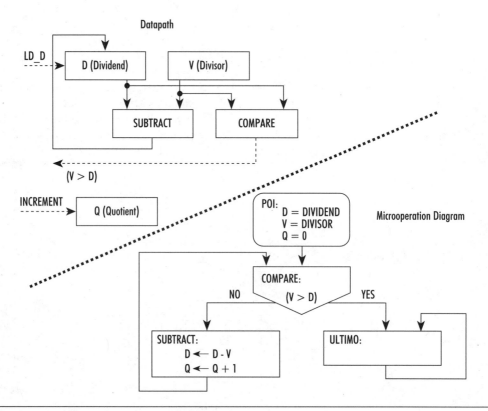

Figure 5.21   Division by Repeated Subtraction

Figure 5.21 shows a datapath structure in which the dividend (D) and divisor (V) initially reside in registers. Repeatedly subtracting the divisor from the dividend until the accumulated subtraction result is smaller than the dividend produces the quotient. With each subtraction the quotient register is incremented.

The following table shows the register contents when dividing 198_{10} (11000110_2) by 13_{10} (1101_2).

n	D (Dividend)	V (Divisor)	Q (Quotient)	D – V	V > D
0	11000110 (198)	1101 (13)	0000 (0)	10111001 (185)	N
1	10111001 (185)	1101	0001	10101100 (172)	N
...	...	...	...	...	...
13	00011111 (31)	1101	1101 (13)	00010000 (16)	N
14	00010010 (18)	1101	1110 (14)	00000011 (3)	N
15	00000011 (3)	1101	1111 (15)	—	Y

Table 5.13

After performing the iterative looping operation 15 times, the remaining dividend is less than the divisor and the algorithm stops. The quotient is in the Q register. The remainder, R, is the remaining dividend in the D register.

Direct Division

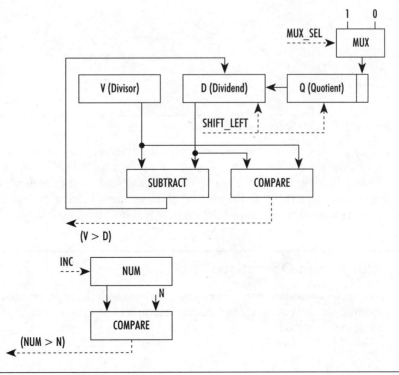

Figure 5.22   Datapath for Division

Because of the large number of subtract operations, the repeated subtraction technique is not an attractive method for division. An alternative is to use the compare, subtract, and shift algorithm commonly used for base 10 division. Figure 5.22 shows a datapath to support the algorithm.

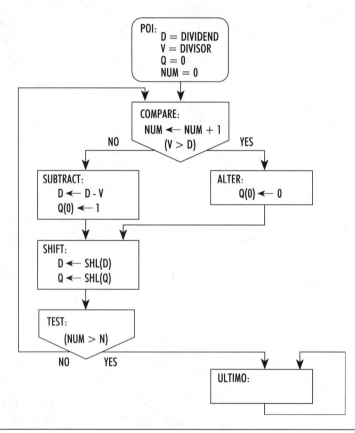

Figure 5.23   Microoperation Diagram for a Divider

Figure 5.23 shows a possible Microoperation diagram to perform a division. Initially, the dividend occupies the combined D and Q register. In block COMPARE of the microoperation diagram, the divisor, V, is compared with the most significant digits of the dividend, D. If $V > D$, the next bit of the quotient is 0. If $V \leq D$, the next bit of the quotient is 1 and the dividend subtracts the divisor. The contents of the combined DQ register is shifted to the left to create room for the new quotient bit.

5.12   Floating-point Operations

Floating-point operations are more complex because both the exponent and magnitude are used in the calculation. Only addition and multiplication will be considered. An IEEE floating-point representation is assumed.

Floating-point Addition

Two floating-point numbers, A and B, have magnitude fields MA, MB and exponent fields EA, EB. Their addition requires several steps.

- **Restoration.** In IEEE 754 format, one bit of the magnitude (the phantom bit) has been suppressed. It must be restored to create MA and MB, the magnitude fields of the operands.

- **Magnitude Comparison.** The larger (in magnitude) of the two exponents, EA and EB, must be identified. This can be done by comparing the exponent fields. The larger operand and the smaller operand are subjected to different processing.

- **Exponent Comparison/Magnitude Shifting.** Before addition can be performed, the exponent fields must agree. The magnitude field of the smaller number is shifted to the right a number of positions equal to the difference in the magnitude of the exponent fields. After the shifting operations, both exponents are the same.

- **Add/Subtract.** Depending on the sign bits of the two numbers, an add or subtract is performed on the magnitude fields. Sign-magnitude representation is used for the mantissa fields. The proper sign for the result is determined.

- **Exponent Adjust.** The addition or subtraction will affect the magnitude field of the sum. An adjustment in the exponent field may be necessary to maintain the correct floating-point format.

- **Recoding.** The phantom bit must be removed and the magnitude field may require truncation or expansion to agree with the IEEE data format.

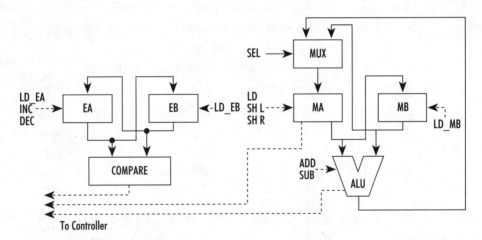

Figure 5.24   Datapath Structure for Floating-Point Addition

Figure 5.24, shows a possible datapath structure for floating-point addition. It is assumed that any phantom bits have been restored prior to placement in the registers. EA, MA and EB, MB are the exponent registers and magnitude registers for floating-point numbers A and B. The registers associated with the A operand can perform the shifts, increments, and decrements necessary for adjustment of the exponent field.

Using the datapath of Figure 5.24, a floating-point algorithm might proceed as follows:

1. EA and EB, the exponent fields of A and B, are compared. If necessary they are switched, along with MA and MB, so that EA contains the smaller exponent.

2. EA and EB are again compared and MA, the magnitude of A, is shifted to the right while EA is incremented. Data in MA requires rounding-off or truncation. This continues until the exponents are the same.

3. The magnitudes, MA and MB, are then added with the result returning to register MA.

4. By examining the carry bit from the adder and the most significant bits of MA, the contents of MA are shifted and EA incremented or decremented to properly normalize the sum.

Example: Add $10_{10} + 3_{10}$ using the floating-point processor structure of Figure 5.24. EA and EB, the exponent registers, are 8 bits long and use excess 127 representation of the exponent. MA and MB, the magnitude registers, are 8-bits long. Initially, the magnitude is expressed as 1.XXXXXXX.

Solution: In floating-point representation:

$A = 10_{10} = 1010_2 = 1.01000000 * 2^3$

$B = 3_{10} = 11_2 = 1.10000000 * 2^1$

$EA = 3_{10} + 127_{10} = 130_{10} = 10000010_2$ (in excess 127)

$EB = 1_{10} + 127_{10} = 128_{10} = 10000000_2$ (in excess 127)

Step	EA	EB	MA	MB	Comments
0	10000010	10000000	1.01000000	1.10000000	Original values placed in EA, EB, MA, MB.
1	10000000	10000010	1.10000000	1.01000000	Since A > B, swap (EA, EB), (MA, MB).
2	10000010	10000010	0.01100000	1.01000000	To make exponents agree, shift MA to the right two positions.
3	10000010	10000010	1.10100000	1.01000000	Perform the addition of mantissas, placing sum in MA.
4	10000010	10000010	1.10100000	1.01000000	Magnitude adjustment; none necessary.

Table 5.14

The result is $1.101 * 2^3 = 1101_2 = 13_{10}$

Floating-point Multiplication

Floating-point multiplication is similar to integer multiplication. The magnitudes are multiplied using an integer multiplication datapath structure; the exponents are added. The addition of exponents is performed by using a biased number system for their representation.

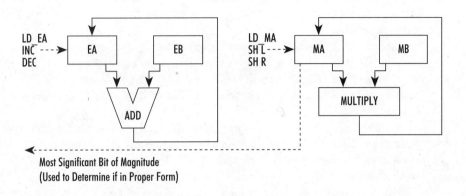

Figure 5.25 Datapath for Floating-point Multiplication

Figure 5.25 shows a possible datapath structure for floating-point multiplication. Registers EA, MA and EB, MB hold the initial values of the respective magnitudes and exponents. The addition circuitry for the exponents must operate with a biased number system. After adding exponents and multiplying magnitudes, registers EA and MA are adjusted to put the result in the proper floating-point form.

Example: Using the same register structure as in the previous example, perform a floating-point multiplication of $10_{10} * 3_{10}$.

Solution: The floating-point representation of the two numbers has already been determined.

Step	EA	EB	MA	MB	Comments
1	10000010	10000000	1.01000000	1.10000000	Original number placed in EA, EB, MA, MB.
2	10000011	10000000	1.11100000	1.10000000	Add exponents (in excess of 127); multiply mantissas.
3	10000011	10000000	1.11100000	1.10000000	Adjust magnitude if needed.

Table 5.15

The result is $1.11100000 * 2^4 = 11110_2 = 30_{10}$

Summary

This chapter studies data representation and the implementation of datapaths to perform arithmetic operations on data. Integer digital data can be represented as an unsigned integer, sign magnitude, or two's complement form. Binary-coded decimal is useful for decimal representation. Residue techniques offer advantages for a limited class of applications. Text is usually represented in 8-bit ASCII data format.

Integer addition, because of its widespread use in data processing, is an important processing function. Techniques such as carry-save and carry look-ahead minimize the time delays caused by carry propagation. Highly parallel techniques, such as Wallace tree adders, attempt to minimize the time required to add multiple operands. Algorithms for subtraction are similar to those used for addition.

Integer multiplication is also an important fundamental digital operation. Techniques ranging from multiple addition to high-speed techniques involving Wallace tree adders are used. Systolic arrays provide a method for implementing multiplication using cellular arrays on integrated circuit chips.

Floating-point techniques can represent a much wider range of values than is possible using integers. Floating-point data representation uses two fields: one for the exponent and one for the mantissa. The IEEE floating-point standard is the dominant method of representing floating-point numbers. The algorithms for adding and multiplying floating-point numbers are moderately complex and require a supporting datapath.

For additional information and references on the topics presented in this chapter, access the World Wide Web page provided at: http://www.ece.orst.edu/~herzog/docs.html.

Key Terms

carry-save addition	quotient
dividend	remainder
divisor	residue number system
fixed point	size extension
floating point	summands
multiplicand	systolic array
multiplier	tree adders
partial product	Wallace tree adders
product	Wallace tree multiplier

Readings and References

Pollard, L. Howard. *Computer Design and Architecture*. Englewood Cliffs, N.J.: Prentice Hall, 1990.

This textbook contains a good summary of historical computing devices, as well as well-written material on information representation and arithmetic units.

Stallings, William. *Computer Organization and Architecture*. New York: Macmillan Publishing Company, 1987.

This book has an especially interesting introductory chapter on the history of computing devices.

Exercises

5.1   Find the binary representation of the following. Truncate your answer to three positions beyond the binary point.

(a) 133

(b) 0.175

(c) 25.31

(d) 13.2_4

(e) 21.2_3

5.2   Convert the following to decimal. Truncate your answer to three positions beyond the decimal point.

(a) 11011.11_2

(b) 2101.22_3

(c) $A2_{16}$

(d) 713.11_8

5.3   Represent the following in the specified format:

(a) $+3_{10}, -12_{10}$ in sign-magnitude binary

(b) $+25_{10}, -13_{10}$ in two's complement binary

(c) $+1_{10}, -120_{10}$ in excess 128 binary

5.4   Find the 8-bit two's complement binary negation of the following numbers. Express the resulting negation in both binary and decimal.

(a) 01000001_2

(b) 27_{10}

(c) 11111111_2

(d) 00000000_2

5.5   Perform the following operations in 8-bit binary. Use two's complement to represent negative numbers.

(a) $25_{16} + 17_{16}$

(b) $10_{16} + (-7_{16})$

(c) $50_{16} + (-7F_{16})$

(d) $(-3F_{16}) + (-3F_{16})$

(e) $40_{16} + 50_{16}$

5.6   Design a circuit to detect overflow when adding 8-bit signed binary integers using 2's complement.

5.7   Represent the following numbers in binary-coded decimal.
   (a) 128_{10}
   (b) 100_8
   (c) 11001100_2
   (d) 1111.1111_2

5.8   Use residue representation from Table 5.9 to perform the following decimal numerical operations. Verify the result.
   (a) $3 + 7$
   (b) $12 + 1$
   (c) $4 * 2$
   (d) $1 * 14$

5.9   Extend the table of residues to include the values from 15 through 35. How is the next modulus selected?

5.10 Although residue arithmetic offers some advantages in processing speed, the problem of converting between binary and binary-coded residue representation is not trivial. Discuss at least one way of performing the conversions.

5.11 Find the short real IEEE floating-point representation for the following numbers.
   (a) 375.41
   (b) 0.595
   (c) -2.78×10^{-2}
   (d) 9.3622×10^2
   (e) -101111.01_2

5.12 Find the decimal value of the following short real IEEE floating point numbers.
   (a) 1 10000000 00000000000000000000000
   (b) 0 10000100 11110000000000000000000
   (c) 1 01111110 10000000000000000000000

5.13 Decode the following message in 7-bit ASCII.
   (a) 1001000 1100101 1101100 1101100 1101111 0100000 1010111 1101111 1110010 1101100 1100100 0100001
   (b) 0110110 0101011 0111000 0111101 0110001 0110100
   (c) 0111001 0001101 0001010 0101101 0110001 0001101 0001010 01011111 0001101 0001010 0111000

5.14 A 1024×8 RAM contains the ASCII representation of some text material. Design a system to read the text and convert all lowercase letters (a, b, c, etc.) to uppercase letters (A, B, C, etc.). Numbers and punctuation are to remain unchanged. The text is to be returned to the RAM.

5.15 Design a 1-bit binary adder to add operands A_i and B_i, with a carry input, C_{in}. There is to be a sum output, S_i, and carry out, C_{out}. Use two-level AND/OR logic.

5.16 Determine the number of logic gates required to perform a fully parallel addition for operands of 1 bit, 2 bits, and 3 bits. All the sum bits are to be determined with no more than two levels of logic. (NOT gates are not considered a level of logic.) What conclusions can be drawn?

5.17 Show that adding two numbers, with differing signs, in two's complement will always produce a valid representation of the sum.

5.18 In a system using 16-bit arithmetic with positive integers, use carry completion detection techniques to determine the maximum number of ripple carries in each of the following situations.

  (a) 1010101010101010 + 1100110011001100

  (b) 1100011000110001 + 0011100111001111

5.19 An arithmetic circuit uses ripple addition and operates on 8-bit positive integers. If each adder stage requires 10 nanoseconds and load time for a register is 10 nanoseconds, how much time is required for the following operations if a carry completion detector adder is used? The clock period is 10 nanoseconds.

  (a) 01111111 + 00000001

  (b) 01010101 + 10101010

  (c) 10010100 + 00100100

5.20 Discuss a possible algorithm that can be implemented by logic components and will determine the number of clock cycles that must be used in a carry completion adder.

5.21 Which is faster, a pipeline adder or a carry-save adder for each of the following situations? Which uses more hardware components?

  (a) For adding a single pair of operands.

  (b) For 25 pairs of operands which must be added as pairs.

  (c) For 25 operands which must be added to produce a single sum.

5.22 Consider the addition of four 4-bit numbers using '181 devices. If each two-level logic segment requires a time delay of t, how much delay is required for the following situations? Assume that when carry look-ahead is used, the 4-bit addition can be performed in $2t$ time units.

  (a) The numbers are combined sequentially as in Figure 5.8A. No carry look-ahead is used.

  (b) same as *(a)* but carry look-ahead is used.

  (c) combine as in Figure 5.8B with carry look-ahead.

5.23 Using the following adder array, show a Wallace tree adder configuration to add 5 numbers V, W, X, Y, and Z each of 2 bits.

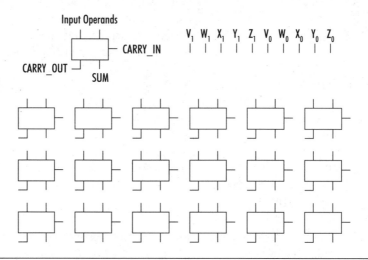

Figure 5.26

5.24 Why is a Wallace tree adder faster than a carry-save adder for adding multiple numbers?

5.25 Show how a Wallace tree adder could be pipelined. Compare its processing rate to a pipelined carry-save adder.

5.26 In Figure 5.10, which shows a Wallace tree adder, what is the effect on performance of routing carry signals horizontally instead of diagonally?

5.27 Design a 1-bit binary subtracter to subtract operand B_i from operand A_i. Consider a borrow input, R_{in}. There is to be a difference output, D_i, and borrow out R_{out}. Use two-level AND/OR logic.

5.28 Figure 5.15 shows a possible datapath structure for performing shift/add binary multiplication. Show the contents of registers MCND, PROD, MPLR, and BITS during each clock cycle of the multiplication of multiplier 1010 and multiplicand 0111.

5.29 Determine the summands for the multiplication of 1010 and 0111. Verify that addition of the summands produces the product.

5.30 Determine the structure of a Wallace tree adder to add the summands for the product of 1010 and 0111.

5.31 Using the floating-point addition algorithm of Section 5.12, show the register contents during each clock cycle for the following additions.
   (a) $25_{10} + 3_{10}$
   (b) $8.5_{10} + 0.05_{10}$

5.32 Using the floating-point multiplication algorithm of Section 5.12, show the register contents during each clock cycle for the following multiplications.
   (a) $12_{10} * 3_{10}$
   (b) $2.5_{10} + 0.08_{10}$

5.33 Develop a microoperation diagram for the floating-point adder described in Section 5.12.

5.34 Develop a microoperation diagram for the shift/add multiplier of Figure 5.14.

5.35 Develop a microoperation diagram for the floating-point multiplier described in Section 5.12.

6
The Digital Computer

Charles Babbage

Charles Babbage was a genius from Britain
Who with an idea was smitten
With axle and gear
He made numbers appear
Using programs he'd cleverly written

Prologue

Charles Babbage was born in England near the beginning of the 19th century. He was driven by a desire to design and fabricate computing devices.

Before mechanical computational devices, major computation projects such as developing tide tables, trigonometric tables, or tables of logarithms were performed by large groups of clerks. Each of these computations involve evaluating polynomials. Using the method of **finite differences**, polynomials could be evaluated using only addition and subtraction. The tasks were tedious and repetitious, and the results were often erroneous.

Babbage proposed the **Difference Engine** to mechanically implement finite difference computations and began developing it with £17,000 provided by the British government. Babbage based the addition and subtraction operations on mechanical techniques developed earlier by Pascal. Geared wheels were used to mechanically provide a memory for 15-digit numbers. The machine was to be capable of computing third-degree polynomials.

Because of numerous mechanical, financial, and personnel problems, the Difference Engine was never completed. The British government refused to provide additional funds. Babbage's mechanical designer abandoned the project and took many of the specialized tools with him.

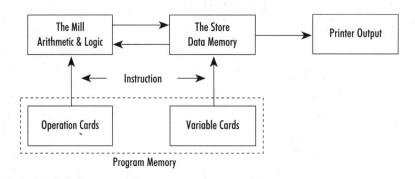

Figure 6.1   The Babbage Analytical Engine

Babbage himself lost interest in the Difference Engine as the idea of an **Analytical Engine** began to evolve. In contrast to previous computational devices, the Analytical Engine was to be programmable. Babbage envisioned a device, shown in Figure 6.1, which included:

- **Store.** This was the memory for the device. Using mechanical storage devices, Babbage proposed a capacity of 1000 numbers each of 50 digits.
- **Mill.** The mill was the arithmetic unit. It was to be capable of performing the four basic arithmetic operations and branch operations based on the sign of an arithmetic result.

- **Output.** Like the Difference Engine, the Analytical Engine had a provision for providing a printed record of computational results.
- **Operation Cards.** These cards, using a technology developed earlier by Jacquard for controlling the automated weaving of fabrics, provided the instruction codes for the machine.
- **Variable Cards.** These cards specified the memory locations for operands and results.

Although the Analytical Engine was never built, the concepts presented are clearly the groundwork for the modern digital computer. Babbage is usually given credit as the intellectual father of the digital computer. Babbage's major problems were related to the technology of the day. It wasn't until almost 100 years later that electrical relays and electronic vacuum tubes made the digital computer feasible.

6.1   Introduction

Before Charles Babbage, computing devices could perform arithmetic and logical operations. There was no provision for automatically sequencing such operations to perform a more complex algorithm. Babbage is credited with proposing the main concepts for a stored-program digital computer. This included a mechanism for storing instructions that could be executed sequentially. His machine also envisioned using conditional instructions that could alter the program execution sequence depending on the condition of a data element.

Previous chapters developed techniques for designing digital systems capable of executing processing algorithms specified at a hardware-dependent level. The sequence of operations is specified by the sequential behavior of a finite state machine controller. The processing algorithm is fixed at the time the machine is fabricated.

In this chapter, a digital processing device's activity will be specified at a higher level: the instruction level. A program memory will present coded instructions to the system. The control unit will interpret each instruction and initiate a sequence of register transfer operations to perform higher-level operations specified by the instruction. By changing the sequence of instructions, a diverse range of processing algorithms can be performed. If the instruction set is useful for a wide range of applications, the resulting device is a general-purpose digital computer.

6.2   Programmable Systems

In the design model presented in Chapter 1, digital systems are partitioned into a controller and a datapath. The controller has an internal model of the algorithm to be performed; it generates control signals and monitors status information produced by the datapath.

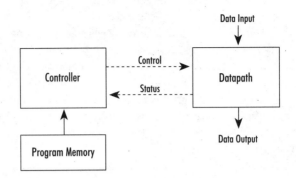

Figure 6.2   A Programmable System

It is often useful to have a method to readily change the processing algorithm. In a **programmable system**, such as a digital computer, the algorithm is changed by additional input to the controller from a program memory. This is shown in Figure 6.2. An instruction from the program memory specifies the operation to be performed.

In a programmable system, the controller fetches an instruction from the program memory. After interpretation, the controller generates control signals that allow the datapath to execute the sequence of specific microoperations associated with that instruction. When this is completed, another instruction is fetched from the program memory. The succession of fetch and execute cycles continues until the program is terminated. The overall processing algorithm can be altered by changing the contents of the program memory; no modification of the controller is required.

Programmable systems are important in the evolution of computing structures. They have several advantages over nonprogrammable devices:

- Programmable systems generally use a datapath capable of many simple processing operations.
- The system can be programmed to perform sequential operations.
- A general-purpose system can economically be mass-produced.
- Highly specialized high-performance programs (or software) can control the general purpose-hardware.
- The same general-purpose machine can be adapted for many different uses by changing the program rather than the machine.

6.3   The Digital Computer

The digital computer, propelled by the inspiration of Charles Babbage in 1856, came to fruition in the mid 1940s. At this time a workable blend of electronic, logical, and memory devices simultaneously became available. The resulting devices, ENIAC (Eastern Numerical Integrator And Computer) at The University of Pennsylvania, The IAS (Institute of Advanced Studies) computer at Princeton University, and others formed a technological base from which further advances were rapidly made.

The stored-program digital computer represents the current state of progress in programmable systems that are optimized to manipulate information. The same basic machine can be highly specialized with application-dependent programs. Modern digital computers come in an extensive assortment of sizes and capabilities, yet their internal organization remains similar. Most computers can be divided into five functional parts, as shown in Figure 6.3: the program memory, the data memory, the datapath, the input/output, and the control unit (controller).

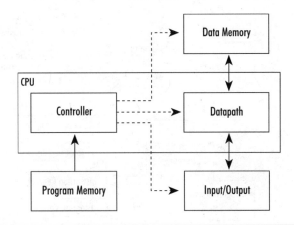

Figure 6.3   Organization of a Digital Computer

The **program memory** is the initial repository for the user program. The **data memory** is a bulk storage unit for storing data. In some situations both may share the same storage media, for example a solid state read/write memory. In some instances, particularly when microprocessors are used, the program memory utilizes permanent ROM storage since there is no need to alter the program after the device has been designed.

The **CPU (Central Processing Unit)** is the main processing element of the computer. Using storage registers and electronic logic devices, the datapath processes data to produce useful results. The CPU is comprised of a datapath and a controller.

The **datapath** contains registers, interconnection paths, and the arithmetic and logic circuitry to perform processing actions. In some older references, the term ALU (arithmetic and logic unit) along with registers, is used instead of datapath but the term *datapath* is used extensively in the architecture of modern computers. Both the datapath and controller have the functions presented earlier in this text.

The **controller** is responsible for transforming the computer instructions, which are present in the form of binary code in the memory, into sequences of control signals to initiate a sequential series of microoperations within the datapath.

The **input/output** section of the computer provides an interface point for controlled transfer of information to and from the logical and physical confines of the computer. It may be extensive, as in the case of some large computers, or almost non-existent, as in the case of some microprocessors.

The Harvard Computer Architecture

The major components of a digital computer using the **Harvard architecture** are shown in Figure 6.4. The datapath contains the main data manipulation elements. Only registers affecting the memory and controller have been shown. The interconnection structure between registers is not shown. Signals produced by the controller perform the sequencing of processing actions. The status register, shown in the datapath, is a collection of all of decision information affecting the controller.

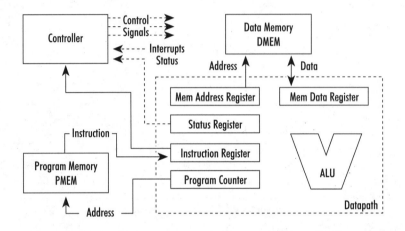

Figure 6.4   A Digital Computer Using a Harvard Architecture

In the Harvard computer architecture, the program memory, PMEM, and data memory, DMEM, are implemented separately. This allows simultaneous memory fetches for instructions in the program memory and for data in the data memory. In high-performance computers this capability is exploited to achieve high instruction-processing rates.

The von Neumann Computer Architecture

John von Neumann proposed a computer organization, now known as the **von Neumann architecture**, in which the data memory and program memory are physically stored in the same media. This architecture is shown in Figure 6.5. A single memory address register and memory data register are used to access both data for datapath operations and instructions for the instruction register. Only the usage determines if the content of a memory cell are data or an instruction. Since the same memory is used for two different purposes, some extra complexity is involved to route the retrieved information to the proper register.

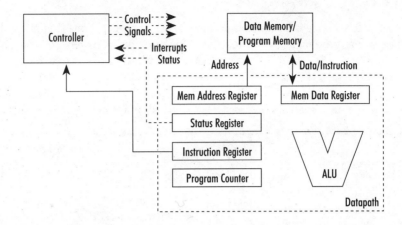

Figure 6.5   A Programmable System Using a von Neumann Architecture

The von Neumann structure offers a great deal of flexibility since it accommodates large programs using only a small amount of data storage, or small programs requiring a large amount of data storage. This architecture also offers the economy associated with the need to implement only one memory structure. When memory components were relatively expensive, dual-purpose memory was an important consideration.

The von Neumann architecture allows the computer to move and modify its own program. This allows portions of the program to be swapped from other storage media such as disks. It is also possible for the computer to modify its own instructions while executing a program. Von Neumann thought this would be useful to allow looping and indexing operations. Modifying a program during execution is now considered an inappropriate programming technique.

6.4   Instruction Set Processing

In a programmable system, the processing algorithm is not fixed, but is specified in high-level system directives called instructions. The term **macroinstruction** distinguishes, if necessary, between instructions provided by the programmer (for example an assembly code instruction) and the microinstructions used by the designer to implement microoperations. The macroinstructions are stored in the program memory. Microinstructions are a function of the control unit.

Sequencing and executing instructions requires two actions. First, a **fetch cycle** must access the instruction from the program memory. Second, an **execute cycle** must perform the unique sequence of microinstructions specified by the macroinstruction. At the end of the execute cycle, another fetch cycle is entered to retrieve the next instruction. The sequence of fetch and execute cycles then continues until the processing action is halted.

The instructions for a programmable device initially reside in the program memory. The starting location of the program is at a fixed location, typically address 0000. The program counter, PC, maintains the location of the current instruction address.

The Fetch Cycle

The fetch cycle is a sequence of register transfer operations that bring the instruction from its place in program memory to the instruction register, IR, where the control unit can interpret it. The fetch cycle affects only the program memory and its associated address and data registers. At the conclusion of the fetch cycle, an instruction is placed in the instruction register.

The initial clock cycle will be referred to as t_0. The clock cycle also becomes a convenient name for use in a microinstruction program. Table 6.1 lists abbreviations to be used extensively.

Register		Description
Memory Address Register	MAR	Provides an address to the memory.
Memory Data Register	MDR	Destination register for a memory read; source register for a memory write.
Program Counter	PC	Stores location of next instruction to be processed.
Instruction Register	IR	Stores the current instruction.
Status Register	SR	Maintains information concerning the status of present and past arithmetic and logic operations.

Table 6.1

When data is accessed from memory, the register providing the address is indicated. For example, IR←MEM(PC) means that memory contents from the location specified by the program counter is transferred to the instruction register.

The following sections show a fetch cycle for a single word instruction as it might appear for the Harvard and von Neumann machines.

Fetch Cycle (Harvard)

```
t₀:     IR←PMEM(PC),     /* Get the instruction from the program memory, put it   */
        PC←PC + 1;       /* in the instruction register and increment the PC to   */
                         /* point to the next instruction.                        */
```

The fetch cycle for the Harvard architecture needs only one clock cycle. The address is provided directly by the PC; the resulting data (an instruction) is loaded directly into the instruction register. This example assumes that the program counter has provisions for incrementing its value. A parallel load counter is appropriate for its implementation. Note that the memory value retrieved is specified by the current value of the program counter, not the new value (PC + 1) loaded at the end of the clock cycle.

Fetch Cycle (von Neumann)

```
t₀:        MAR←PC,              /* Load the memory address register with the program  */
           PC←PC + 1;           /* counter, increment the program counter.            */
t₁:        MDR←M(MAR);          /* Get the instruction from memory.                   */
t₂:        IR←MDR;              /* Put the instruction in the instruction register.   */
```

The fetch cycle for the von Neumann architecture requires two more clock cycles than that of the Harvard architecture. This is because a single MAR provides addresses for both instructions and data; a single MDR retrieves both instructions and data.

In both the Harvard and von Neumann representations, the program counter is incremented in t_0 in preparation for the next fetch operation. The next fetch is anticipated to be from the next sequential location in the program memory. The program counter may be affected by jump and skip instructions.

Since the controller does not know in advance which instruction it will read, the fetch operation is the same for all instructions. If the instruction is longer than one memory word, two or more read operations must be performed to obtain the complete instruction. If some instructions are longer than others, the first word of the instruction must be examined to determine if additional memory read operations are required.

The Execute Cycle

The sequence of microoperations performed in the execute cycle depends on the instruction presented during the fetch cycle. Two general types of instructions exist: arithmetic/logic and flow control. Arithmetic/logic instructions cause data in the datapath architecture to be modified. Examples include adding operands stored in registers and moving information from one register to another. Flow control instructions affect the sequencing of the instructions. Examples include conditional and unconditional jumps, skips, and subroutine calls.

6.5   BART, a Stored-Program Digital Computer

In the next several sections a small programmable system, BART (Boolean Algorithmic Register Trainer), will be described and designed. BART uses a von Neumann architecture with a single memory shared for instructions and data. BART is not a practical computing structure for most applications and is doomed for a life of perpetual underachievement. BART does illustrate many of the techniques and decisions involved in implementing a programmable system. A datapath structure will be selected. An instruction set will be chosen. BART's operations will be expressed in terms of microoperation statements. A microoperation diagram will be developed to show the sequential behavior of the device. A control unit will be designed to accommodate the instruction set.

Designing a complete digital computer is challenging because of the interdependence of the composite parts. There are multiple decisions to be made. A datapath hardware structure must be selected. A set of macroinstructions must be chosen, along with an encoding format. A set of microinstructions for implementing each computer macroinstruction must be formulated. A control unit must be designed. The design process requires consideration of the conflicting requirements of its various aspects. The choices selected for BART are workable, but certainly not unique.

Datapath

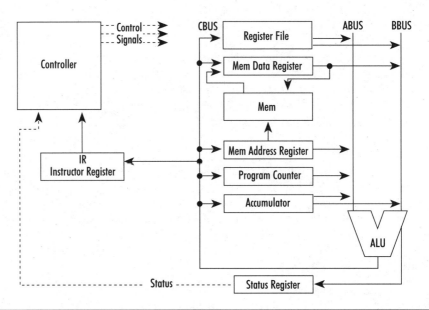

Figure 6.6   BART, A Simple Programmable Digital System

Figure 6.6 shows the structure of BART. The datapath involves the buses ABUS, BBUS, and CBUS. This is an example of a three bus architecture considered in Chapter 3. All buses are 8 bits wide. There are six 8-bit registers attached to the buses. Another four registers are available through a register file.

- **AC—Accumulator.** This register provides a source and destination register for most processing activities. It uses three control signals: LD_AC loads the accumulator from CBUS; OUT_A places the accumulator on the ABUS; OUT_B places the accumulator on the BBUS.

- **PC—Program Counter.** This register holds a numeric value that maintains the location of the instruction being executed. It uses two control signals; LD_PC and OUT_PC. When necessary, the ALU increments the PC.

- **MAR—Memory Address Register.** The MAR provides an address for all data transfers to and from the memory. Control signal LD_MAR is used.

- **MDR—Memory Data Register.** The MDR is the source of data for all memory data writes; it is the destination for all data read from the memory. The MDR uses control signals OUT_MDR to place information on BBUS. LD_MDR causes information to be loaded. A multiplexer selects either the memory or the CBUS as the source of information. During a memory read, the RD_MEM control signal selects the MUX to accept information from the memory; at all other times it selects information from the CBUS.

- **SR—Status Register.** The status register holds status information originating in the ALU. In this implementation, Z indicates that the last ALU operation produced a result of 0. A second bit, C, indicates that an overflow carry resulted from the last ALU operation. Expanding the status register will allow input/output flags to also be included.
- **IR—Instruction Register.** This register holds macroinstructions fetched from memory and presents them to the controller for interpretation. Control signal LD_IR loads the register.
- **RF—Register File.** Four registers are included in the file; each is 8-bits. The registers are selected by providing a 2-bit code, A_ADDR, to specify the information to be placed on ABUS. B_ADDR specifies information to be placed on BBUS and also indicates the destination register to receive information provided on CBUS. Additional control signals OUT_RFA, OUT_RFB, and LD_RF provide output control of a specified register to ABUS, BBUS and input control from CBUS.

Memory

BART has an 8-bit × 256 word read/write memory. While not large, it conveniently allows all registers to use the same 8-bit width. BART utilizes a von Neumann structure; the memory contains both data and instructions. Modifications necessary for a 16-bit address will be considered as an exercise.

ALU

BART's 8-bit ALU provides an 8-bit output (and 1-bit carry) for the eight data processing functions that can be specified with the 3-bit code $F_2F_1F_0$. Its data processing operations are shown in Table 6.2.

CONTROL INPUTS			ALU FUNCTION	
F_2	F_1	F_0	Function	F_{ALU} Symbolic Name
0	0	0	0000	ZERO
0	0	1	ABUS	ABUS
0	1	0	BBUS	BBUS
0	1	1	ABUS + 1	INC
1	0	0	$\overline{\text{ABUS}}$ (Complement of ABUS)	NOT
1	0	1	ABUS OR BBUS	OR
1	1	0	ABUS AND BBUS	AND
1	1	1	ABUS + BBUS	ADD

Table 6.2

One operand for the ALU is provided by a register via ABUS; the other is from BBUS. The results from all ALU operations are returned to the appropriate register through CBUS. Data moves are accomplished by putting the source register on one of the buses, selecting an ALU function that passes either ABUS or BBUS through the ALU without modifying it, and returning the results to the destination register via CBUS.

The ALU also performs status check operations. The output from the ALU is examined by zero detect logic to determine if the result was zero. This information, provided as an output from the ALU, is stored in the Z bit of the status register. An overflow carry is stored in the C bit of the status register. Later in this chapter selected bits of the status register will be tested in executing conditional instructions.

6.6   BART Instruction Set

BART uses a small instruction set emphasizing operations with the accumulator. Instructions requiring addressing capability to access an operand or specify a memory location use a 4-bit operation code, or OpCode. The remaining 4 bits specify an addressing mode. Instructions affecting only a single operand in the accumulator use the full 8 bits to specify the instruction. The instruction set along with assigned operation codes is shown in Table 6.3.

Operation Code	Symbolic Name	Operation Performed	Description
0000	ADD	AC←AC + Opnd, SR←status result	ADD the specified operand to the contents of the accumulator; adjust the status register.
0001	AND	AC←AC AND Opnd, SR←status result;	AND the specified operand to the contents of the accumulator; adjust the status register.
0010	OR	AC←AC OR Opnd, SR←status result;	OR the specified operand to the contents of the accumulator, adjust the status register.
0011	STA—Store AC	MEM(ADDR)←AC;	Store the accumulator in the specified address.
0100	LDA—Load AC	AC←MEM(ADDR);	Load the accumulator with the contents of the specified memory location.
0101	—	—	Spare (for future expansion).
0110	—	—	Spare (for future expansion).
0111	—	—	Spare (for future expansion).
1000	JMP—Jump	PC←ADDR;	Jump to the specified address.

Operation Code	Symbolic Name	Operation Performed	Description
1001	CALL—Subroutine Call	STACK←PC, PC←ADDR;	Jump to subroutine; save return address on the stack (for future expansion).
1010	RET—Subroutine Return	PC←STACK	Return from subroutine; load PC with return address from the stack (for future expansion).
1011	—	—	Spare (for future expansion).
1100	BOS—Branch On Status	If Status Bit = 1 then take Branch;	If the status bit selected is asserted, then the PC is loaded with the branch address.
1101	—	—	Spare (for future expansion).
1110	—	—	Spare (for future expansion).
11110000	NOP—No Operation	AC←AC	No Operation.
11110001	CLA—Clear AC	AC←0, Z←0;	Clear the accumulator and status register.
11110010	INC—Increment AC	AC←AC + 1, SR←Status Result	Increment the accumulator, adjust the status register.
11110011	CMP—Complement AC	AC←$\overline{AC}$	Complement the accumulator.
11110100	PUSH—Push AC to stack	Top of Stack←AC	Put the accumulator on the stack (for future expansion).
11110101	POP—Pop AC from stack	AC←Top of Stack	Put the top of the stack in the accumulator (for future expansion).

Table 6.3

Three instructions (ADD, AND, OR) perform arithmetic and logic operations using two operands. The location of one operand is specified by the instruction; the other is in the accumulator. Two instructions (BOS, JMP) are used for conditional and unconditional branching. Three instructions (INC, CLA, CMP) use a single operand located in the accumulator. Two instructions (LDA and STA) exchange information between the accumulator and the data memory. CALL and RET are used for subroutines. Their implementation requires an additional register, a stack pointer (SP), to be considered later in this chapter. Several of the possible operation codes are not currently used and will allow future expansion.

Instruction Formats

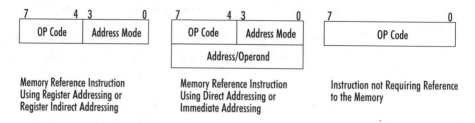

Figure 6.7   BART Instruction Formats

All BART instructions, as shown in Figure 6.7, are either one or two bytes long. Instructions requiring the specification of an address have a 4-bit operation code followed by 4 bits to specify the address mode. If necessary, a second byte contains additional addressing information. Instructions not requiring an address use all 8 bits to specify the operation code. This increases the number of available operation codes. The use of the bits in the instruction varies with the type of instruction. The set of instructions may be divided into groups to make similar use of their bit patterns. A regular and consistent pattern of bit usage allows easier implementation by the control unit.

Arithmetic and Logic Instructions (ADD, AND, OR)

Arithmetic and logic instructions are characterized by their need to access two operands. One operand is assumed to be in the accumulator. The second operand must be obtained from another register or the data memory, or perhaps the instruction itself (immediate addressing).

The operation codes for most BART instructions, including ADD, AND, and OR, are 4 bits long. The remaining 4 bits of the instruction indicate the location of the second operand. BART uses four methods, called **address modes**, to indicate the location of the second operand. Figure 6.8 illustrates each of the modes used by BART.

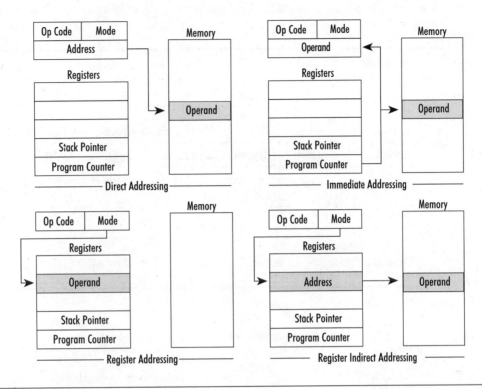

Figure 6.8   BART Addressing Modes

- **Direct Addressing.** The second operand is located in memory, as indicated by the second byte of the instruction. To obtain the operand, the second byte of the instruction (which contains the address of the operand) must be accessed and placed in the MAR.
- **Immediate Addressing.** The second operand is located in the second byte of the instruction. Since the PC is incremented during the fetch cycle, it is pointing to the operand. A "#" is used in the macrocode to specify immediate addressing.
- **Register Addressing.** The second operand is located in one of the registers in the datapath. The participating register is specified by the macroinstruction.
- **Register Indirect Addressing.** One of the registers in the datapath contains the memory address for the second operand. A "@" is used in the macroinstruction to specify register indirect addressing.

In BART, the four addressing modes are specified by bits I(3:0) of the first byte of the instruction. Table 6.4 shows the encoding formats for addressing modes. An "xx" in the table indicates that the bits are not used. "RR" is a 2-bit binary code to identify one of the four registers in the register file.

Address Mode	I(3:0)	Comment
Direct	00xx	The second byte of the instruction specifies the data memory location of the instruction.
Immediate	01xx	The second byte of the instruction contains the operand.
Register	10RR	Register RR in the register file contains the operand.
Register Indirect	11RR	Register RR in the register file contains the address of the operand in data memory.

Table 6.4

Table 6.5 illustrates some encoding examples for BART's ADD instruction.

Instruction		Instruction (using mnemonics)	Comment
Binary	Hex		
00000000 01010010	00 52	ADD 52H	Direct Addressing: Add location 52_{16} in data memory to the accumulator.
00000100 01010010	04 52	ADD #52H	Immediate Addressing: Add the number 52_{16} to the contents of the accumulator.
00001011	0B	ADD R3	Register Addressing: Add the number in R_3 of the register file to the accumulator.
00001111	0F	ADD @R3	Register Indirect Addressing: Get the required operand by using the address contained in R_3. Add it to the accumulator.

Table 6.5

Data Transfer Instructions—LDA, STA

The LDA (load accumulator) and STA (store accumulator) instructions move information between the accumulator and data registers or memory. In BART they use the same addressing modes as the arithmetic and logic instructions. LDA uses the addressing modes to determine the source of information to be moved to the accumulator. STA uses the addressing modes to determine the destination of information moved from the accumulator; this may be a register or the data memory. Immediate addressing is not meaningful for the STA instruction.

LDA and STA may be used as a pair to initialize registers in the register file; they may also initialize storage locations in the data memory. Performing LDA with immediate addressing to the accumulator, followed by an STA to a register, loads a register in the register file with specified contents. A data memory location can be changed by performing an LDA immediately to the accumulator followed by an STA to the data memory.

Table 6.6 shows examples of the addressing modes for the LDA and STA instructions.

Instruction		Instruction (using mnemonics)	Comment
Binary	Hex		
01000000 01010010	40 52	LDA 52H	Direct Addressing: Move location 52_{16} in data memory to the accumulator.
01000100 01010010	44 52	LDA #52H	Immediate Addressing: Load the number 52_{16} into the accumulator.
01001011	4B	LDA R3	Register Addressing: Load the contents from Register 3 to the accumulator.
01001111	4F	LDA @R3	Register Indirect Addressing: Get the required operand by using the address in R_3.
00110000 01010010	30 52	STA 52H	Direct Addressing: Store the contents of the accumulator in location 52_{16} of data memory.
00111011	3B	STA R3	Register Addressing: Store the contents of the accumulator in R_3.
00011111	3F	STA @R3	Register Indirect Addressing: Store the accumulator at the address contained in Register R_3.

Table 6.6

Branch Instructions (JMP)

Branch instructions can modify the flow of program instructions. Branches may be unconditional, the branch is always taken; or they may be conditional, the branch is taken only if a condition is met. BART's JMP instruction is unconditional.

The JMP instruction also requires an effective address. It uses the same format as the arithmetic and logic instructions. Since the jump location must be in program memory, only direct and register-indirect addressing may be used. Examples are shown in Table 6.7.

| Instruction | | Instruction (using mnemonics) | Comment |
Binary	Hex		
10000000 10000010	80 82	JMP 52H	Direct Addressing: Jump to location 52_{16} in program memory.
10001111	8F	JMP @R3	Register Indirect Addressing: Jump to the address contained in Register R_3.

Table 6.7

Conditional Instructions (BOS)

The BART instruction BOS (branch on status) tests the condition of a status bit and alters the program counter if the specified status condition is met. Bits $I(2:0)$ in the first byte of the instruction specify the status condition to be tested. Bit $I(3)$ is not assigned a use. The second byte contains the branch address. This address is loaded into the program counter if the tested condition is true. For BART, the ALU generates only two status bits. This leaves several spare designations for future expansion such as input/output flags. Table 6.8 shows the code assignment for the status bits. Table 6.9 gives some examples of macroinstructions using the available status bits.

$I(2:0)$	Status	Comment
000	Z, zero bit	Branch to the branch address if the last ALU operation produced a result of 0.
001	C, carry bit	Branch to the branch address if the last ALU operation produced an overflow carry.
010–111	Spares	Spare locations for future status selection.

Table 6.8

| Instruction | | Instruction (using mnemonics) | Comment |
Binary	Hex		
11000000 01010010	C0 52	BOS Z, 524	Branch to location 524 if the Z bit of the status register is set.
11000001 01010010	C1 52	BOS C, 524	Branch to location 524 if the C bit of the status register is set.

Table 6.9

Accumulator Instructions

Instructions such as CLA (clear accumulator), INC (increment), CMP (complement accumulator), and NOP (no operation) do not require addressing since the operand is in the accumulator. All accumulator instructions, as shown in Table 6.10, use the same first 4 bits of the operation code; the instructions are distinguished by the remaining 4 bits.

Instruction		Instruction (using mnemonics)	Comment
Binary	Hex		
11111000	F8	CLA	Clear accumulator.
11110100	F4	INC	Increment the accumulator.
11110010	F2	CMP	Complement the accumulator.
11110000	F0	NOP	No operation.

Table 6.10

Example 6.1:   Programming BART

Required: Write a BART program that starts at location 00_{16}. The program adds the contents of locations 10_{16} and 11_{16}. The value $1A_{16}$ is to be added to the result. The sum of the three numbers is to be stored in location 13_{16}. Ignore any overflow resulting from the addition. Show the code using mnemonics (ADD, etc.). Show the actual binary contents to be placed in the memory.

Solution:

Address	Contents	Instruction	Comments
00	1111 1000	CLA	Initialize by clearing the accumulator.
01 02	0100 0000 0001 0000	LDA 10H	Load accumulator, direct addressing, with contents of location 10_{16}.
03 04	0000 0000 0001 0001	ADD 11H	Add, direct addressing, location 11_{16} to the accumulator.
05 06	0000 0100 0001 1010	ADD #1AH	Add immediate addressing, the value $1A_{16}$ to the accumulator.
07 08	0011 0000 0001 0011	STA 13H	Store accumulator, direct addressing, in memory location 13_{16}.

Table 6.11

In the above example, "xx" values for portions of the address field not used have been replaced by "00_2".

6.7   BART Instruction Processing

Processing for BART is divided into two cycles: S_f, the fetch cycle, and S_e, the execute cycle. Each of the cycles is further divided into four clock cycles, t_0 through t_3. The designation $S_f t_0$ specifies time slot t_0 of the fetch cycle.

Fetch Cycle

The purpose of the fetch cycle is to retrieve an instruction from the program memory and place it in the instruction register. The fetch cycle also increments the program counter in preparation for the next memory operation. The microinstructions for the Fetch Cycle are the same for all BART instructions and are shown in Table 6.12.

Clock Cycle	Microoperation	Description
$S_f t_0$	MAR←PC;	The program counter is moved to the memory address register.
$S_f t_1$	MDR←MEM(MAR), PC←PC + 1, MAR←PC + 1;	The instruction from memory is placed in the memory data register. The program counter is incremented. The incremented value of the PC is then placed in MAR.
$S_f t_2$	IR←MDR;	The instruction from memory is placed in the instruction register.
$S_f t_3$	No Operation	Spare. Used to simplify design of the controller.

Table 6.12

The MAR is loaded with the contents of the PC during $S_f t_0$. The PC contains the address of the next instruction. The instruction is retrieved and the program counter is incremented during $S_f t_1$. After it is incremented the address in the PC is either the address of the next instruction or the address of the second byte of the current instruction. These two microoperations can be performed simultaneously without conflict. In anticipation that the instruction might use direct or immediate addressing, the MAR also receives the incremented value of the PC in $S_f t_1$. Incrementing the MAR at this time saves a processing step during the execute cycle of instructions that use direct or immediate addressing. It causes no problem if other forms of addressing are used.

The instruction is moved to the instruction register by passing it through the ALU in $S_f t_2$. No operations are performed during $S_f t_3$. This wasted time slot, while slowing down BART, eliminates complexity in the design of the control unit. Using a microprogrammed control unit, discussed later in this chapter, will eliminate the wasted time slots.

Execute Cycle

For the execute cycle, BART requires a separate set of microinstructions for implementing each of the instructions. The BART instructions will be examined in groups.

Arithmetic/Logic Instructions (AND, ADD, OR)

	Direct $I(3{:}2) = 00$	Immediate $I(3{:}2) = 00$	Register $I(3{:}2) = 00$	Register Indirect $I(3{:}2) = 11$
	ADD: I_0 ADD (AND and OR are similar)			
$S_eI_0t_0$	MDR←MEM(MAR), PC←PC + 1;	MDR←MEM(MAR), PC←PC + 1;	No operation.	No operation.
$S_eI_0t_1$	MAR←MDR;	No operation.	No operation.	MAR←R_i;
$S_eI_0t_2$	MDR←MEM(MAR);	No operation.	No operation.	MDR←MEM(MAR);
$S_eI_0t_3$	AC←AC + MDR, SR←STATUS;	AC←AC + MDR SR←STATUS;	AC←AC + R_i, SR←STATUS;	AC←AC + MDR, SR←STATUS;

Table 6.13

Since AND, ADD, and OR are identical in their addressing modes and quite similar in their processing, only ADD will be considered. Table 6.13 indicates the microinstructions to be performed in each clock cycle of the four address modes. For brevity, the operation codes for each of the instructions have been decoded. I_0 corresponds to the instruction with a 4 bit operation code of 0000, or ADD. Bits $I(3{:}2)$ of the instruction specify the addressing mode.

Note that with the ADD instruction, only the direct-addressing mode uses all of the available clock cycles. ADD with immediate addressing, for example, requires only two clock cycles to perform necessary microinstructions. Uniform processing activity among the four address modes is attained by performing microinstructions in only the S_et_0 and S_et_3 clock cycles. This leads to less complex logic to implement the hardware of the control unit. For example, the accumulator is modified in clock cycle S_et_3 for all addressing modes of ADD.

Data Transfer Instructions (STA, LDA)

Both STA and LDA use addressing modes similar to those of the arithmetic and logic instructions of the previous section. Microinstructions for the various addressing modes are shown in

Tables 6.14 and 6.15. Specific inclusion of the control signals has been eliminated. Immediate addressing for STA is not meaningful.

STA: I_3 Store Accumulator				
	Direct $I(3:2) = 00$	Immediate Not Used	Register $I(3:2) = 10$	Register Indirect $I(3:2) = 11$
$S_e I_3 t_0$	MDR←MEM(MAR), PC←PC + 1;	—	No operation.	No operation.
$S_e I_3 t_1$	MAR←MDR;	—	No operation.	MAR←R_i;
$S_e I_3 t_2$	MDR←AC;	—	No operation.	MDR←AC;
$S_e I_3 t_3$	MEM(MAR)←MDR;	—	R_i←AC;	MEM(MAR)←MDR;

Table 6.14

LDA: I_4 Load Accumulator				
	Direct $I(3:2) = 00$	Immediate $I(3:2) = 01$	Register $I(3:2) = 10$	Register Indirect $I(3:2) = 11$
$S_e I_4 t_0$	MDR←MEM(MAR), PC←PC + 1;	MDR←MEM(MAR), PC←PC + 1;	No operation.	No operation.
$S_e I_4 t_1$	MAR←MDR;	No operation.	No operation.	MAR←R_i;
$S_e I_4 t_2$	MDR←MEM(MAR);	No operation.	No operation.	MDR←MEM(MAR);
$S_e I_4 t_3$	AC←MDR, SR←STATUS;	AC←MDR, SR←STATUS;	AC←R_i, SR←STATUS;	AC←MDR, SR←STATUS;

Table 6.15

Branch Instructions (JMP)

	Direct $I(3:2) = 00$	Immediate Not Used	Register Not Used	Register Indirect $I(3:2) = 11$
		JMP: I_8 Jump		
$S_e I_8 t_0$	MDR←MEM(MAR);	—	—	No operation.
$S_e I_8 t_1$	PC←MDR;	—	—	PC←R_i;
$S_e I_8 t_2$	No operation.	—	—	No operation.
$S_e I_8 t_3$	No operation.	—	—	No operation.

Table 6.16

The JMP instruction, shown in Table 6.16, changes the content of the PC. Only the direct and register indirect addressing modes are reasonable. In the direct jump, the next byte of the instruction, which contains the destination address, is accessed in t_1. In t_2 this address is placed in the program counter. This causes the next fetch to occur from the specified jump location. In the register indirect addressing mode, the destination address is stored in an internal register and moved to the program counter in t_1.

Branch Instructions (BOS)

BOS: I_c Branch On Status $I(2:0)$ Specifies status condition	
$S_e I_C t_0$	MDR←MEM(MAR), PC←PC + 1;
$S_e I_C t_1$ TEST	PC←MDR;
$S_e I_C t_1$ $\overline{\text{TEST}}$	No operation.
$S_e I_C t_2$	No operation.
$S_e I_C t_3$	No operation.

Table 6.17

The BOS (branch on status) instruction selects one of two different microinstructions in clock cycle t_1, depending on the contents of the selected test bit of the status register. The selected bit becomes TEST, which is presented to the control unit. If TEST is asserted, the PC is loaded with the branch address found in the second byte of the instruction. This overwrites the value of PC, which was modified in t_0. If TEST is not asserted, the PC is not changed and BART fetches the next instruction in sequence. Table 6.17 shows the microinstructions for the BOS instruction.

Accumulator Instructions (CLA, CMP, INC, NOP)

	CLA I(3:0) = 1000	INC I(3:0) = 0100	CMP I(3:0) = 0011	NOP I(3:0) = 0000
	Accumulator Instructions: I_F			
$S_e t_0$	No operation.	No operation.	No operation.	No operation.
$S_e t_1$	No operation.	No operation.	No operation.	No operation.
$S_e t_2$	No operation.	No operation.	No operation.	No operation.
$S_e t_3$	AC←0, SR←0;	AC←AC + 1, SR← STATUS;	AC←$\overline{AC}$;	No operation.

Table 6.18

Accumulator instructions require no addressing. They all use 1111 as the first four bits of the operation code. The remaining bits, since they are not needed for address purposes, are used to fully identify the instruction. This effectively expands the operation code field from 4 to 8 bits. Table 6.18 shows the microinstructions for CLA, INC, CMP, and NOP.

Other possible BART instructions involving only the accumulator include those shown in Table 6.19.

SHL	Shift the accumulator to the left; insert "0" in vacated position.
SHR	Shift the accumulator to the right; insert "0" in the vacated position.
DEC	Decrement the accumulator.
NEG	Take the two's complement of the number.

Table 6.19

6.8   Additional Addressing Modes for BART

BART is restricted to four addressing modes by limitations in the allocation of bits in the instruction. BART was kept simple to ease design of a control unit. This section describes some possible additional addressing modes and some additional instructions found on most modern computers. R_2 is dedicated as an index register, XR; R_3 is dedicated as a stack pointer, SP. Figure 6.9 shows several addressing modes. The following sections indicate how to implement these modes.

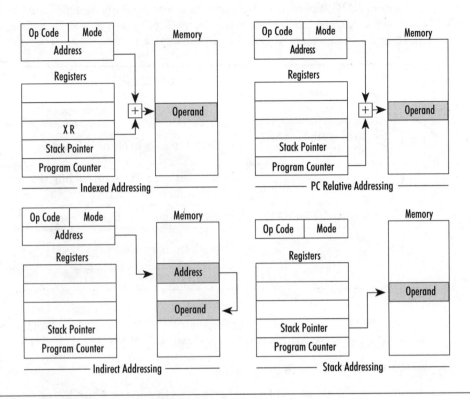

Figure 6.9   Additional Address Modes for BART

Indexed Addressing

All modern computers contain an **index register** that allows indexed addressing modes. Indexed addressing is useful for accessing information stored in tables and array structures in memory. By using register R_2 as an index register, XR, BART can locate an item in memory by adding the address provided by the instruction to the value in the index register. The index register acts as a pointer to a section in memory. The instruction address provides an offset to the pointer. The offset is expressed in two's complement, which allows both positive and negative offsets.

Effective Address = XR + Offset

Within BART, the register file provides a convenient place for the index register. R_2 will be redefined as XR, the index register for the machine. Later, R_3 will be defined as a stack pointer, SP. R_0 and R_1 will remain as general purpose registers.

ADD: I_0 Indexed Addressing		
	Microoperations	Comments
$S_eI_0t_0$	MDR←MEM(MAR), PC←PC + 1;	Get the address byte of the instruction from memory.
$S_eI_0t_1$	MAR←MDR +XR;	Get the effective address by adding the address field of the instruction to the index register.
$S_eI_0t_2$	MDR←MEM(MAR);	Get the operand from data memory.
$S_eI_0t_3$	AC←AC + MDR, SR←STATUS;	Complete the addition.

Table 6.20

Table 6.20 shows a sequence of microinstructions that provide indexed addressing. For example purposes, only the ADD instruction is shown.

Relative Addressing

Relative addressing is a form of indexed addressing in which the program counter is used as the index register. All addresses are relative to the PC. This is especially useful if the program is to be executed from different locations in program memory. In direct addressing, if the program is relocated, all code for memory locations must be changed. In relative addressing, if the program is relocated, program branches and jumps are relative to the location of the instruction in the program and are unaffected.

ADD: I_0 Program Counter Relative Addressing		
Clock Cycle	Microoperations	Comments
S_et_0	MDR←MEM(MAR), PC←PC +1;	Get the address byte of the instruction from memory.
S_et_1	MAR←MDR + PC;	Get the effective address by adding the address field of the instruction to the program counter.
S_et_2	MDR←MEM(MAR);	Get the operand from data memory.
S_et_3	AC←AC + MDR, SR←STATUS;	Complete the addition.

Table 6.21

Table 6.21 shows an ADD instruction using program counter relative addressing.

Indirect Addressing

With indirect addressing, the address of the operand is located in memory. The instruction requires three memory cycles: one for the instruction (fetch cycle), one for the address of the operand, and one for the operand. As shown in Figure 6.9, the address included in the instruction is the address of the operand in memory. Implementing indirect addressing will be left for an exercise.

PUSH and POP Instructions for BART

Two additional useful instructions make use of a **stack pointer**, SP, as a means of addressing a section of data memory called the **stack**. Within BART, R_3 is the stack pointer. The PUSH instruction moves data from the accumulator to the stack and places it in the next available location. POP retrieves the last information placed on the stack and returns it to the accumulator. Programming sequences involving PUSH and POP provide a convenient method of storing and retrieving data without program management of the storage space.

PUSH: I_{F4} PUSH		
	Microoperations	Comments
$S_e I_{F4} t_0$	MAR←SP;	Move the stack pointer to the memory address register.
$S_e I_{F4} t_1$	MDR←AC;	Put the information to be "pushed" in the MDR.
$S_e I_{F4} t_2$	MEM(MAR)←MDR, SP←SP + 1;	Store the information and increment the stack pointer.

Table 6.22

The stack pointer provides a reference pointer for PUSH and POP operations. It stores the address of the next available memory location for a PUSH operation. As data is stored, the SP increments to maintain its pointer function. When used in a POP operation, the SP must be first decremented to point to the last piece of information stored in the stack. Table 6.22 shows possible microinstructions for implementing the PUSH instruction. The stack can be managed in two ways; it can fill upward from its lowest address or downward from its highest address. For BART, the stack will fill upward. The PUSH and POP instructions do not require address modes. They are assigned operation codes 11110100 (F4) and 11110101 (F5).

The PUSH instruction requires three clock cycles. In both PUSH and POP, no attempt has been made to make the microoperations compatible with the BART controller to be designed later in this chapter. In clock cycle $S_e I_{F4} t_0$, the stack pointer, which contains the location of the next free location, is placed in the MAR. During the next clock cycle the content of the accumulator is moved to MDR. In $S_eI_{F4}t_3$ the value of the accumulator is transferred to the memory and the stack pointer is incremented.

POP: I_{F5}POP		
	Microoperations	Comments
$S_eI_{F5}t_0$	SP←SP − 1, MAR←SP − 1;	Decrement the stack pointer; put the decremented value in the MAR.
$S_eI_{F5}t_1$	MDR←MEM(MAR);	Access the information from the top of the stack.
$S_eI_{F5}t_2$	AC←MDR, SR← STATUS;	Move the information to the accumulator and update the status register.

Table 6.23

BART's POP instruction of Table 6.23 is more difficult to implement than PUSH because the ALU, as originally specified, does not support a decrement operation. Despite this limitation, the microoperations for POP are shown as if the decrement were available from the ALU.

6.9   A Subroutine Feature for BART

Subroutine calls, an important programming feature, are available on all modern digital computers. The CALL instruction causes the current contents of the program counter to be pushed on the stack. The address of the subroutine is placed in the program counter in preparation for the next fetch cycle. The subroutine executes with a fetch of the first instruction in the subroutine.

Within a subroutine, further subroutines can be called. This is called **nesting** subroutines. During nesting, additional return addresses are pushed sequentially on the stack.

A subroutine ends with a return (RET) instruction. The address saved on the stack is popped and returned to the program counter. The next instruction is fetched from the location following the subroutine call. Using the stack assures an orderly procedure for calling subroutines and eventually returning, using the RET instruction, to the main program. Addresses for nested subroutines are returned in the proper order.

Table 6.24 indicates how CALL, a subroutine call, and RET, a subroutine return can be implemented with BART.

CALL Operation Code = 1001

CALL has some features in common with the JMP instruction. Both result in a change of the value of the program counter. With a CALL, however, the current value of the PC is saved on the stack. The CALL and RET instructions are always used as a pair.

CALL: I_5 Subroutine Call		
	Microoperations	Comments
$S_eI_5t_0$	MDR←MEM(MAR), MAR←PC + 1;	Get the direct address for the subroutine location from the memory. Temporarily store the value of the incremented program counter in the MAR. This points to the next sequential instruction.
$S_eI_5t_1$	PC←MDR;	Put the address of the subroutine in the program counter.
$S_eI_5t_2$	MDR←MAR;	Move the current value of the PC from the MAR to the MDR in preparation for storage on the stack.
$S_eI_5t_3$	MAR←SP;	Put the stack pointer address in the memory address register.
$S_eI_5t_4$	MEM(MAR)←MDR, SP←SP + 1;	Store the return address on the stack; increment the stack pointer.

Table 6.24

Table 6.24 shows a possible implementation of a subroutine CALL using the features of BART. The first clock cycle acquires the subroutine address from its memory location. The current value of the program counter is incremented and then stored temporarily in the MAR; it will eventually be placed on the stack. Clock cycle $S_eI_5t_1$ puts the address of the subroutine in the program counter. Next, the old value of the program counter is moved from the MAR to the MDR. $S_eI_5t_3$ gets the storage location from the SP. The final clock cycle stores the return address on the stack and increments the SP. The next fetch will then occur from the location specified in the subroutine call.

Since CALL uses an address, multiple addressing modes are possible. In Table 6.24, only direct addressing is considered. Register addressing will be left as an exercise.

In the sequence of microinstructions for the CALL, it is necessary to use the MAR as a temporary storage location for the original program counter before it is stored on the stack. The careful observer will notice that in the original specifications for BART, there were no provisions for a data transfer from the MAR to the ALU. Provisions for such a data move would be required to implement the algorithm as indicated. Inside a digital computer, temporary storage registers are usually provided for the designer's use. While not available to the programmer, they provide a temporary repository for information used to process microinstructions.

RET Operation Code = 1010

	Microoperations	Comments
$S_e t_0$	SP←SP – 1, MAR←SP – 1;	Decrement the SP. Return the decremented value to both the SP and MAR.
$S_e t_1$	MDR←MEM(MAR);	Access the old value of the program counter from the top of the stack.
$S_e t_2$	PC←MDR;	Restore the address to the program counter.
$S_e t_3$	No operation.	

Table 6.25

Table 6.25 shows the microoperations for the return instruction. In implementing RET, the return address must be obtained from the stack and placed in the program counter. This is the value of the PC that existed prior to the subroutine call. Its restoration allows the computer to resume operations from the location following the subroutine call. The stack pointer must be decremented prior to retrieving the address from the stack. The microoperations shown in Table 6.25 assume that the ALU can provide a decrement operation. Although awkward, the decrement can be performed using the operations of the existing ALU.

6.10   Input and Output for BART

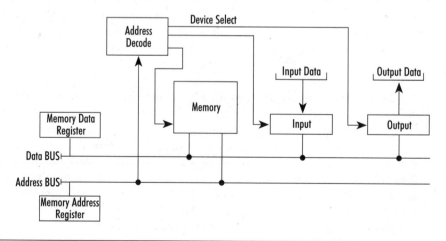

Figure 6.10   A Memory Mapped Input/Output Structure for BART

For input/output in BART, data links from the memory address register and the memory data register are extended as an address bus and a data bus, as shown in Figure 6.10. Data provided by the memory data register may be transferred either to the memory or to the output register. The memory data register may receive information from either the memory or the input register. The address decode logic examines the address provided on the address bus and selects either a memory element or a register as the desired source/destination.

In this structure, memory and input/output registers share the address space. Some addresses reference memory; others reference the input and output registers. Both respond to the same instructions. There are no separate instructions for performing input/output operations. Output is accomplished using the STA (store accumulator) with an instruction address to select the output register. Input is performed by an LDA (load accumulator) instruction in which the address selects the input register. This structure is known as **memory mapped input/output** and provides a method of allowing system expansion without adding instructions.

BART performs input/output operations using an input register and output register on the memory bus. Flag bits may be added to the status register to allow testing of the condition of the input and output registers. The input flag is set when an external device transfers information to the input register. The flag may then be tested using the BOS (branch on status) instruction. Reading the Input register clears the input flag. An input flag prevents the computer from reading a register that has not been updated with new information.

In a similar manner, moving information to the output register sets the output flag. When an external device reads the register, the output flag is cleared. This simple flag system prevents the computer from sending additional data to the output register before the previous information has been read. I/O structures and techniques will be considered in greater detail in Chapter 10, "System Input/Output and Information Exchange."

6.11   A Control Unit for BART

Figure 6.11 shows an overview of the structure of a finite state machine control unit for BART. The structure is similar to the controllers studied in Chapter 4. The state register is formed from three registers: a one-bit cycle register (CR), a 2-bit timer register (TR), and the 8-bit instruction register. The state register has a total of 11 bits.

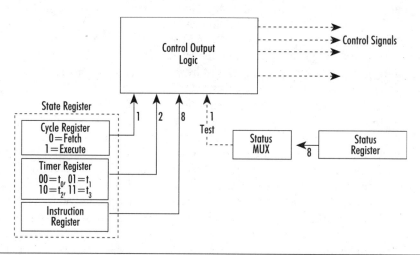

Figure 6.11   The Control Unit for BART

The cycle register distinguishes between the fetch cycle and the execute cycle. The timer register delineates the four clock cycles within both the fetch and execute cycle. Overall, the combination of the cycle register and timer register functions as a 3-bit binary counter that is advanced with each system clock pulse. Four clock periods of the fetch cycle are followed by four clock periods of the execute cycle. The instruction register completely specifies the instruction being processed.

The control output logic receives input from the 11 state variables and the single status input, TEST. Based on the twelve input variables, the logic generates the set of control signals necessary to execute the required microinstructions. The next state for the state register is obtained by advancing the 3-bit counter formed by CR(0) and TR(1:0) and by updating the contents of the instruction register when a new instruction is fetched.

Figure 6.12 shows a portion of the structure of the microoperation diagram for BART. Four processing blocks are used for the fetch cycle, and four are used for the execute cycle of each instruction. Since the fetch cycle is the same for all instructions, state variables provided by the instruction register are ignored during fetch operations.

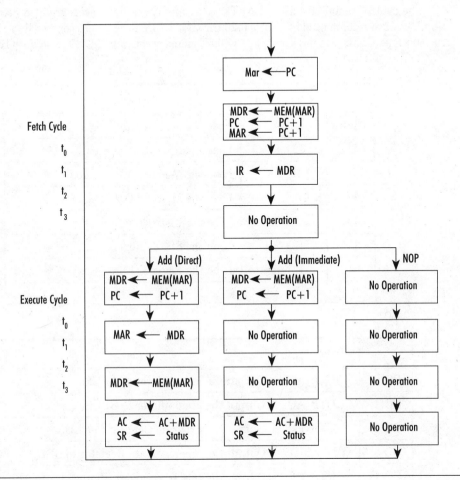

Figure 6.12   A Microoperation Diagram for BART

At the beginning of the execute cycle, the instruction register contains the operation code, addressing mode, and other information provided by the instruction. The sequence of microinstructions is unique for each instruction. The processing blocks for three of the BART instructions are shown. The contents of the others can be inferred from the cycle-by-cycle description of the computer operation included in previous sections. At the conclusion of the execute cycle, the controller loops back to the common fetch cycle.

The control signals for all of BART's microoperations have been grouped according to function. They are summarized in Tables 6.26 through 6.29. There are 23 control signals referenced as register load CL(0:6), register output CO(0:5), memory control CM(0:1), and data routing CD(0:7).

	Control Signal	Comment
CL(0)	LD_AC	Load the accumulator.
CL(1)	LD_PC	Load the Program Counter.
CL(2)	LD_MAR	Load the Memory Address Register.
CL(3)	LD_MDR	Load the Memory Data Register.
CL(4)	LD_IR	Load the Instruction Register.
CL(5)	LD_SR	Load the Status Register.
CL(6)	LD_RF	Load the Register File.

Table 6.26

	Control Signal	Comment
CO(0)	OUTA_AC	Output the accumulator to ABUS.
CO(1)	OUTB_AC	Output the accumulator to BBUS.
CO(2)	OUTA_PC	Output the Program Counter to ABUS.
CO(3)	OUTB_MDR	Output the Memory Data Register to BBUS.
CO(4)	OUTA_RF	Output the selected register file register to ABUS.
CO(5)	OUTB_RF	Output the selected register file register to BBUS.

Table 6.27

	Control Signals	Comment
CM(0)	RD_MEM	Read the system memory.
CM(1)	WR_MEM	Write to the system memory.

Table 6.28

	Control Signals	Comment
CR(0:1)	ADDR_RFA	Address of register in register file to be put on ABUS.
CR(2:3)	ADDR_RFB	Address of register in register file to be put on BBUS; also provides destination address for information provided to register file.
CR(4)	MMUX	Multiplexer signal to select source of information for memory data register.
CR(5:7)	SMUX	Multiplexer signal to select status signal to be used as TEST input to controller.

Table 6.29

Hardwired Controller for BART

Each of BART's control signals is asserted under certain conditions of the cycle register, timer register, instruction register, and the TEST input. If each of the major cycles (S_f, S_e), time slots (t_0, t_1, t_2, t_3), instruction OpCodes (I_0, I_1, ... , I_F), and address modes (A_0, A_1, A_2, A_3) are decoded, the control signals can be expressed as logic functions of these variables.

The logic circuitry required to generate the control signals may be derived from the above tables. Some of the control signal logic is not complicated. For example:

$CL(4) = LD\_IR = S_f t_2$—The instruction register is loaded during clock cycle t_2 of an instruction fetch cycle.

Logic conditions for two of the control signals, $CL(0)$ and $CL(1)$ are more complex.

$CL(0) = LD\_AC = S_e t_3 (I_0 + I_1 + I_2 + I_4 + I_F)$—Load the accumulator during clock cycle t_3 of the ADD, AND, OR, LDA, NOP, CLA, INC, and CMP instructions.

$CL(1) = LD\_PC = S_f t_1 + S_e t_0 (I_0 A_0 + I_0 A_1 + I_1 A_0 + I_1 A_1 + I_1 A_0 + I_0 A_1 + I_2 A_0 + I_2 A_1 + I_3 A_0 + I_4 A_0 + I_4 A_1) + S_e I_C \text{ TEST } (t_0 + t_1)$

Logic expressions for the remaining control signals can be determined in a similar fashion. The logic can be reduced to an AND/OR expression and then realized with appropriate logic components.

The above tables indicate that 23 control signals are necessary to implement BART. The output control logic is well suited for programmable logic devices or ROM. The 12 inputs from the state register and TEST can be applied to the address lines of a 4K byte ROM. The content of the location specified is the bit pattern of the control signals. Three byte-wide memory units would be required to obtain a word width sufficient to provide all of the control signals. Such an implementation would not require discrete logic and could easily be changed in response to new design requirements.

When the state machine controller is implemented using AND/OR logic (or an equivalent PLD device), the result is known as a **hardwired controller**. Hardwired control has usually been the method of choice for implementing small systems using relatively simple algorithms.

Hardwired systems are characterized by the ability to operate at very high clock rates and are used when system processing speed is of great importance. Control structures using ROM are key elements of the **microprogrammed controller** for BART to be designed in the next section.

Microprogrammed Controller for BART

Figure 6.13 shows a method to use microprogramming to implement the control structure of a digital computer. The structure is similar to microprogrammed control units studied in Chapter 4. A microprogram ROM holds the microcode. Included as a field in the microcode are control signals associated with each block of the microoperation diagram. In this implementation, a microinstruction output register is not used.

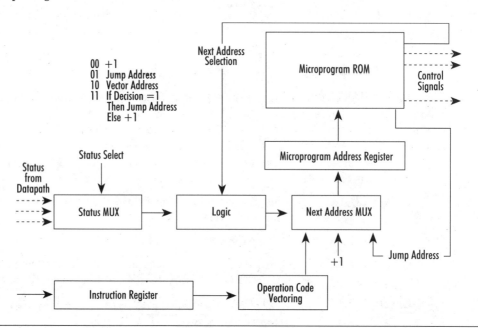

Figure 6.13   A Microprogram Hardware Structure for a Digital Computer

To perform the fetch cycle and execute cycle it is necessary to sequence through the microcode addresses in the proper order. The next address MUX provides several address choices for the microprogram address register.

At the conclusion of the fetch cycle the controller branches to the portion of the control algorithm responsible for the execute cycle of the current instruction. This process is called **vectoring**. For example, the fetch cycle may reside at locations 0 through 2 in the microprogram ROM. The execute cycle for ADD with direct addressing may begin at location 3; for ADD with immediate addressing at location 7, etc. Based on the contents of the instruction register, a logic block generates the address of the next microinstruction for the controller. One way of generating the vector address is using a look up table. For each of the operation codes and address modes provided on the ROM address lines, the contents of the ROM are the address locations for the proper execute cycle.

A microcode instruction contains several fields. For BART, the format in Table 6.30 was chosen.

Next Address Select	Jump Address	Control Signals

Table 6.30

The Next Address Select field provides a criterion for determining how the next address will be determined. Choices include:

- **Next Sequential Location.** This choice corresponds to consecutive locations in the microcode. The next location is obtained by adding one to the current contents of the microprogram address register.
- **Jump Address.** This choice allows microcoded instructions to be executed non-sequentially. The jump address is provided from the ROM by one of the fields in the microprogram.
- **Vector Address.** The vector address is generated from the operation code and address mode of the macroinstruction. This allows each macroinstruction to vector to a unique location in the microcode.
- **Conditional.** When a conditional method is used to determine the next address, a test variable is used to determine if the next address will be the jump address or the sequential address.

The status select signal is generated from a field in the computer instruction and does not require a microcode field.

A possible microcode format for BART would appear as follows.

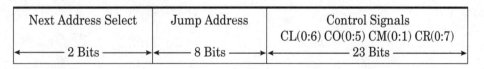

Next Address Select	Jump Address	Control Signals CL(0:6) CO(0:5) CM(0:1) CR(0:7)
◄──── 2 Bits ────►	◄──── 8 Bits ────►	◄──────── 23 Bits ────────►

Table 6.31

Table 6.31 symbolically shows the contents of the microcode fields for a fetch cycle starting at location 0. Using a microprogram, rather than logic, allows the fetch cycle and execute cycle to use the minimum number of clock cycles to perform their microinstructions. This means that an instruction such as ADD immediate will execute faster than an ADD direct since it will require fewer microinstructions.

The symbolic statements shown in Table 6.32 will be included in the next address select microprogram field.

Code	Statement	Comments
00	Inc	The next instruction is in the next sequential location.
01	Jump	The next instruction is at the location specified in the jump address field of the microinstruction.
10	Vector	The address of the next instruction is obtained from the translation table which maps the OpCode/Address mode of the instruction into the starting address for the microcode of the instruction.
11	Decision	The next address is sequential if TEST is not asserted; it is the jump address if TEST is asserted.

Table 6.32

Rather than list the 1s and 0s of the control signal vector, the following tables list the asserted control signals. The jump address field, when not specifically used for a jump address is represented by "—". All microcode addresses are in hexadecimal.

The following tables show microcode for the execute portion of several selected instructions.

Microcode Address	Next Address Select	Jump Address	Control Signals	Microoperations
00	Inc	—	LD_MAR, OUT_PC;	MAR←PC;
01	Inc	—	LD_MDR, LD_PC, LD_MAR, RD_MEM, F_{ALU} = INC;	MDR←MEM(MAR), PC←PC + 1, MAR←PC + 1;
02	Inc	—	LD_IR, OUT_MDR, F_{ALU} = BBUS;	IR←MDR;
03	Vector	—	—	Vector to the start of the execute microcode.

Table 6.33   BART Fetch Cycle

Microcode Address	Next Address Select	Jump Address	Control Signals	Microoperations
04	Inc	—	LD_MDR, LD_PC, OUT_PC, RD_MEM, F_{ALU} = INC;	MDR←MEM(MAR), PC←PC + 1;
05	Inc	—	LD_MAR, OUT_MDR, F_{ALU} = BBUS;	MAR←MDR;
06	Inc	—	LD_MDR, RD_MEM;	MDR←MEM(MAR);
07	Jump	00	LD_AC, LD_SR, OUTA_AC, OUT_MDR, F_{ALU} = ADD;	AC←AC + MDR, SR←STATUS;

Table 6.34   BART ADD Instruction with Direct Addressing

Microcode Address	Next Address Select	Jump Address	Control Signals	Microoperations
08	Inc	—	LD_MDR, LD_PC, OUT_PC, RD_MEM, F_{ALU} = INC;	MDR←MEM(MAR), PC←PC + 1;
09	Jump	00	LD_AC, LD_SR, OUTA_AC, OUT_MDR, F_{ALU} = ADD;	AC←AC + MDR, SR←STATUS;

Table 6.35   BART ADD Instruction with Immediate Addressing

Microcode Address	Next Address Select	Jump Address	Control Signals	Microoperations
0A	Jump	00	LD_AC, LD_SR, OUTA_AC, OUTB_RF, ADDR_RFB F_{ALU} = ADD;	AC←AC + R_i, SR←STATUS;

Table 6.36   BART ADD Instruction with Register Addressing

Microcode Address	Next Address Select	Jump Address	Control Signals	Microoperations
0B	Inc	—	LD_MAR, OUTB_RF, ADDR_RFB, F_{ALU} = $\overline{B}$BUS;	MAR←R;
0C	Inc	—	LD_MDR, RD_MEM;	MDR←MEM(MAR);
0D	Jump	00	LD_AC, LD_SR, OUTA_AC, OUT_MDR, F_{ALU} = ADD	AC←AC + MDR, SR←STATUS;

Table 6.37   BART ADD Instruction with Register Indirect Addressing

Microcode Address	Next Address Select	Jump Address	Microoperations	Comments
24	Decision	26	MDR←MEM(MAR), PC←PC + 1;	Get branch address; jump if test asserted.
25	Jump	00	No operation.	Go to fetch cycle; branch not taken.
26	Jump	00	PC←MDR;	Load branch address in PC; go to fetch cycle.

Table 6.38   BART Branch on Status Instruction

Summary

This chapter presents the basic organizational structure of a programmable digital structure. A key difference from previous digital systems studied is the presence of a series of instructions, or a program. The controller interprets the instructions and prepares a sequence of unique control signals for each instruction.

Each instruction is processed by means of a fetch cycle followed by an execute cycle. The fetch accesses the instruction from memory; the execute performs the algorithm specified by the instruction.

BART is presented as a digital computer with many of the features of more complex machines. An instruction set with multiple addressing modes is specified. Its implementation is presented by developing a datapath and showing the sequence of register transfer operations which perform the fetch and execute operations for many of the instructions. Additional BART features, such as a subroutine call and input/output, are also presented.

A hardwired control unit capable of sequencing the operations necessary to make BART perform is developed. A microprogrammed control unit, which allows greater flexibility in implementing the instructions, is also designed.

For additional information and references on the topics presented in this chapter, access the World Wide Web page provided at: http://www.ece.orst.edu/~herzog/docs.html.

Key Terms

address modes	input/output
Analytical Engine	instruction format
controller	macroinstruction
CPU (central processing unit)	memory mapped input/output
data memory	microprogrammed controller
datapath	nesting
Difference Engine	program memory
execute cycle	programmable system
fetch cycle	stack
finite difference	stack pointer
hardwire controller	vectoring
Harvard architecture	von Neumann architecture
index register	

Exercises

6.1 Compare the von Neumann architecture and the Harvard architecture. What are two advantages and disadvantages of each?

6.2 How can the fetch cycle for BART be shortened? What is the impact of this operation?

6.3 In BART, show how the fetch cycle could be modified if the program counter could be incremented directly instead of requiring the ALU. Could this save any processing time?

6.4   Assume that in a von Neumann architecture, program addresses are provided by a PC and data addresses are provided by the MAR. Instructions are retrieved via the instruction register and data is exchanged via the memory data register. All registers can be connected directly to the address bus and data bus of the memory unit. Show a typical fetch and execute cycle for this situation. Perform the operations in the fewest number of clock cycles.

6.5   Figure 6.14 shows a datapath for a simple 8-bit digital computer. Note there is only a single programming register, the accumulator. The memory is placed directly on the data bus; there is no MDR. All instructions have the same format as BART. In particular an ADD with direct addressing has the 8-bit address in the second byte of the instruction. Show a microoperation diagram (without control signals) for a fetch cycle. Show a microoperation diagram for an execute cycle of an ADD with direct addressing. Use the minimum number of processing blocks in your diagram.

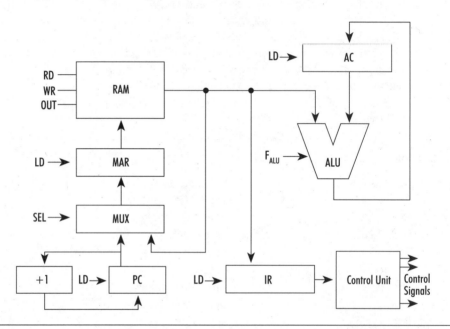

Figure 6.14

6.6   Figure 6.15 shows a datapath for a digital computer. Note there is only a single programming register, the accumulator. The computer uses a single 16-bit bus to interconnect most of the major registers and the ALU. You may choose ALU functions to suit your needs. All instructions are 16 bits long; 4 bits are used for the operation code and 12 bits are used for the addressing.

(a)   Show the register transfer operations performed in each clock cycle for an instruction fetch cycle. Use the fewest number of clock cycles possible. Include comments.

(b)   Show the register transfer operations performed in each clock cycle for an AND operation using direct addressing. Use the fewest number of clock cycles. Include comments.

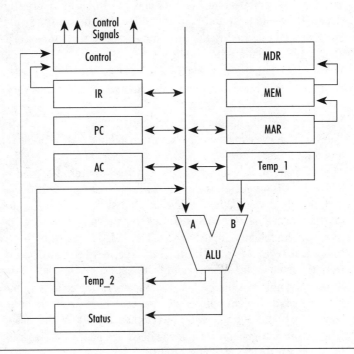

Figure 6.15

6.7   Assuming BART is to use a 16-bit address, what changes would be necessary to the register and data buses? Show a possible fetch and execute cycle for an ADD direct instruction.

6.8   Using the BART instruction set, write the binary code for the following instructions.

   (a) ADD 5A

   (b) ADD #5A

   (c) ADD @R2

   (d) ADD R0

   (e) BOS C,46

   (f) CMP

   (g) STA @R1

6.9   What are the mnemonic representations of the following instructions?

   (a) 5106

   (b) A777

6.10   The following BART program is to begin in location 00_{16} of the program memory. Determine the hexadecimal contents of the memory locations beginning at location 00_{16}. Determine the contents of the accumulator after each instruction is executed.

```
CLA
ADD #01
STA R3
LDA @R3
JMP 1
```

6.11   Write a short program for BART to add two 16-bit numbers stored in multiple register locations—R_0 through R_1 and R_2 through R_3. Put the results in place of the first operand.

6.12   For Example 6.1, modify the program so that if the sum exceeds 255_{10}, the value 00 will be placed in the accumulator.

6.13   Write a BART program to clear locations $C0_{16}$ through FF_{16}.

6.14   Write a BART program to count the number of blank memory locations between 80_{16} and FF_{16}.

6.15   Suggest a possible use for the NOP instruction.

6.16   Show a fetch cycle for a von Neumann machine with a 16-bit word length but 8-bit memory and data links.

6.17   The Institute of Advanced Studies (IAS) computer, designed in 1951, used 40-bit memory width and 40-bit data links, but a 20-bit instruction width. How would this affect the fetch and execute cycles? What restrictions must be imposed on jumps and branches with this machine to assure reasonably simple instruction processing?

6.18   In S_{ft1} of the fetch cycle, both the PC and MAR receive PC + 1. What would be the effect on the execute cycle of an ADD direct instruction if the MAR←PC + 1 operation were omitted from the fetch cycle? What would be the effect on the CMP instruction?

6.19   Show a possible execute cycle to use the operation code 1011 for a NEG, negate, instruction. In this instruction the contents of the accumulator are to be replaced by the two's complement negation of the value. Do not change the characteristics of the ALU.

6.20   BART does not have an Exclusive OR (XOR) operation. How might the operation be implemented with BART microoperations?

6.21   Devise a set of microoperations for BART to implement the SHR, shift right instruction. Repeat for a SHL, shift left instruction.

6.22   Many computers use a skip on status type instruction which causes the next sequential instruction to be skipped if a status condition is satisfied. Formulate a sequence of microoperations to implement SOS for BART. What instruction format should be used? Should the skipped instruction be one byte or two bytes? Why?

6.23   For the BOS (branch on status) instruction, bit I(3) is currently unused. Show how I(3) could be used to indicate direct addressing for I(3) = 0 and indirect addressing for I(3) = 1. Show the microoperations for implementing the BOS with indirect addressing.

6.24   In the BOS instruction, bit I(3) is unused. Suggest a possible use (other than that used in the previous problem) for this bit.

6.25   Suppose that the 74LS382 ALU described in Section 1.3 (Chapter 1) were suggested for use in BART. Would this present any design or performance problems?

6.26   For each of the instructions in the BART instruction set, determine the minimum number of clock cycles required for the execute portion of the instruction. Determine the minimum number of clock cycles required for the BART fetch.

6.27   Assuming that all instructions are equally likely to be executed in a program, how much could performance improve by shortening to their minimum value the fetch time and execution time of the instructions?

6.28 Simple machines such as BART have trouble performing looping operations efficiently. Any counting variable would need to be tested in the accumulator. This would require accumulator contents to be temporarily stored. Design the execute portion of an instruction called increment and skip on zero (ISZ). The instruction is to examine a memory location, increment the value in the location, and skip the next instruction if the value is 0. This should all be done without disturbing the contents of the accumulator.

6.29 Assume that BART had an indirect addressing mode in which the specified address in memory contained the address of an operand. Show an execute cycle for an ADD using indirect addressing.

6.30 Assume that autoincrement addressing is to be implemented in BART. Register R_3 is used to hold the address of the operand. During the execute cycle, the register is to be incremented to point at the next sequential memory location. Show an execute cycle for an ADD instruction with autoincrement. Suggest several possible uses of this addressing mode.

6.31 Why is immediate addressing for STA not useful?

6.32 For the ADD instruction in BART, show a sequence of microoperations to perform the ADD with:

  (a) Indexed addressing

  (b) PC relative addressing

  (c) Stack addressing (The required data is on the stack.)

6.33 For the JMP instruction in BART, show a sequence of microoperations which will perform the ADD with:

  (a) Indexed addressing

  (b) PC relative addressing

  (c) Indirect addressing

  (d) Stack addressing

6.34 In some computers the stack fills upwards in memory; in others it fills downwards. What are the advantages and disadvantages of each method?

6.35 One microprocessor uses an XCHG, exchange, command to swap the contents of the program counter and the accumulator. This feature is used to implement subroutine calls and returns. Describe how this could work. What is a major disadvantage of implementing subroutines in this manner?

6.36 If the instruction decrement stack pointer (DSP) were available, how could the POP instruction be implemented as two consecutive instructions? What changes would be required for POP? How could DSP be implemented?

6.37 Show how a call could be implemented with register addressing.

6.38 Show a possible sequence of microoperations for a CALL instruction using indirect addressing.

6.39 What is the advantage of having a conditional CALL instruction? Using the same status codes used in the BOS instruction, show the microoperations for the instruction CALC, call conditional, which calls the subroutine only if the selected status variable is asserted.

6.40 Write a subroutine for BART in which the contents of R_0 is subtracted from the accumulator. Assume all of the internal registers are available. Show a calling program that uses the subroutine.

6.41 Can subroutines be nested in BART? Is there a limit to the nesting?

6.42 Show a (1/0) control word assignment implementing the add immediate instruction with a microprogram.

6.43 Redesign BART as a 16-bit machine. Use the same instruction set and same instruction set operation codes. What changes in instruction formats would be reasonable? Show fetch/execute cycles for ADD immediate, ADD direct, ADD register, and ADD indirect. What performance improvements are possible?

6.44 Using the microprogrammed controller instruction format for BART as shown in Table 6.33, show the microcode (with comments) for the execute cycles of the following instructions. Use the address indicated as the starting location for the microcode.

  (a) STA direct (30_{16})

  (b) JMP register indirect (40_{16})

  (c) CALL direct (50_{16})

  (d) RET (60_{16})

6.45 Show how nanoprogramming could be used for the microcode generated for the combination of BART fetch, ADD direct, ADD immediate, ADD register, and ADD register indirect. What is the reduction in ROM requirement?

7
Instruction Set Processing

RISC and CISC

RISC and CISC

In ADDs, ORs and reading a disk
Operations must always be brisk
So whatever you're concluding
About modern computing
Nothing is gained without RISC

Prologue

Following the initial pioneering work by von Neumann, Mauchly, and others in the 1940s and early 1950s, computer organization progressed at a leisurely but sustained pace until the 1960s. In 1964, IBM introduced its IBM System/360 family of computers. Ranging in size from a mini-computer to a large mainframe structure, all members of the family used a common instruction set. Within the IBM 360 family, with different sets of microcode, relatively modest hardware structures were capable of running the same instruction set as their larger and much faster brethren. High performance computers used extensive hardware to support the instruction set. Low performance (and low cost) computers implemented the same instruction set with modest hardware and extensive use of microprogrammed control.

With the availability of microprogrammed control, it was possible to design computers with large and complex instruction sets. This allowed the manufacturer extensive price/performance tradeoffs. Machines using microcode to achieve a large and diverse instruction set and multiple addressing modes are known as complex instruction set computers, or **CISC**s.

Researchers at the University of California, Berkeley, and Stanford University began to challenge the basic foundations of CISC architectures in the early 1980s. The proposed alternative was a reduced instruction set computer, or **RISC**. In contrast to CISCs, RISC designs stress very simple and sparse instruction sets. Addressing modes are few and uncomplicated. By simplifying the instruction set, improved processor performance is achieved through pipelining and a hard-wired controller. The Berkeley RISC I was completed in spring 1981. The MIPS (microprocessor without interlocked pipe stages) project at Stanford made several advances in processing instructions for RISC structures.

7.1   Introduction

This chapter considers ways to improve overall performance by modifying the basic machine architecture and instruction set. A reasonable performance measure of a computing device is the time it takes to perform a processing algorithm. There are three components that affect processing time:

1. **The number of instructions required, N.** This is a function of the instruction set of the computer. Computers with powerful and complex instructions may be able to perform a given processing task in relatively few instructions; a machine with a less complex instruction set may require many instructions to perform the same task.

2. **The number of clock cycles per instruction, C.** Uncomplicated instructions generally require fewer clock cycles for their execution than more complex instructions.
3. **The time per clock cycle, T.** The faster the clock can be run, the shorter the overall execution time for an instruction.

The overall processing time, J, is a function of N, C, and T.

$$J = N * C * T$$

To improve performance, one or more of the contributing factors must be reduced. To achieve a short clock cycle, T, fast logic gates and memory components must be used. Fortuitously, logic with fast response, small size, low power requirements, and low cost have been achieved through the innovative work in solid-state electronics, microelectronics, and VLSI design and fabrication techniques. All indications suggest that research and development efforts will provide continuing advances in these areas.

Memory components, especially large memories, have also achieved dramatic improvements in size, performance, and cost. Memories based on magnetic storage principles (such as disks) are considerably slower than memories using electronic media (such as registers). Techniques for integrating memory components into high performance systems will be studied in Chapter 9, "Memory Structure and Management."

The number of instructions required, N, can be reduced by performing more operations in each instruction. Usually instructions that perform more operations also require many clock cycles, C. The tradeoffs between T, N, and C may be complex.

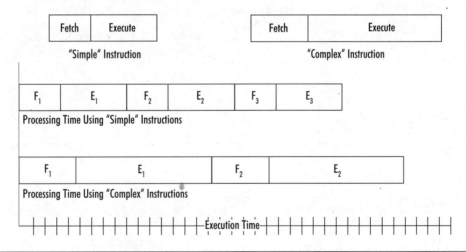

Figure 7.1   Processing Time Using Simple and Complex Instructions

Consider the situation shown in Figure 7.1. A processing task can be completed by two complex instructions or by three simple instructions. The clock rate is assumed to be the same for both. The complex instruction is longer (more bits) than the simple instruction and may require additional memory read operations to acquire the complete instruction; its fetch cycle is longer. The complex instruction also requires more time to perform its processing actions; its execute cycle is longer. The complex instruction spends a larger portion of its total time in the execute cycle; this should make its operation more efficient. In the example shown, overall processing time is less using the simple instructions.

The advantages of complex instructions vs. simple instructions are not completely obvious. The relative advantages of each have caused a division in the approach to computer design between complex instruction set computers (CISC) and reduced instruction set computers (RISC). These issues will be discussed in greater detail later in this chapter. The following sections illustrate some of the modifications that can be made in the structure of computing devices and the resulting influence on performance.

7.2   Overlapped Fetch and Execute Cycles

One approach to shortening the number of cycles required to fetch/execute an instruction is to modify the method of processing instructions. As an example, consider the fetch and execute cycles of BART. Assume that BART is to use only single-byte instructions. This limits addressing modes to register addressing (operand is in a register) and register-indirect addressing (address of operand is in a register). A sequence of ADD instructions with register-indirect addressing are to be processed.

	Operation	Comment
$S_f t_0$	No operation.	No operation.
$S_f t_1$	MAR←PC;	The program counter is moved to the memory address register.
$S_f t_2$	MDR←MEM(MAR), PC←PC + 1;	The instruction from memory is placed in the memory data register. The program counter is incremented and copied to the MAR.
$S_f t_3$	IR←MDR;	The instruction from memory is placed in the instruction register.

Table 7.1

	Operation	Comment
$S_e t_0$	MAR←R_i;	Put the address of the operand in the MAR.
$S_e t_1$	MDR←MEM(MAR);	Get the operand from memory.
$S_e t_2$	AC←AC + MDR, SR←STATUS;	Perform the addition.
$S_e t_3$	No operation.	No operation.

Table 7.2

The fetch and execute cycles are as shown in Tables 7.1 and 7.2. To reduce hardware conflicts, the clock cycle for performing some of the microoperations has been slightly changed from those used in Chapter 6. In the fetch cycle, for example, clock cycle t_0 rather than t_3 has been used for the no_operation. The program counter has been modified to allow it to be incremented without using the ALU. Also, when using single-byte instructions, there is no need to increment the memory address register during the fetch cycle. This microoperation (MAR←PC + 1) has been removed.

If instructions are executed sequentially (no branching), it is possible to begin the fetch operation for an instruction during the execute cycle of the previous instruction. Table 7.3 shows an overlapping of the fetch and execute cycles of several instructions.

	Instruction 0	Instruction 1	Instruction 2
t_0	No operation.		
t_1	MAR←PC;		
t_2	MDR←MEM(MAR), PC←PC + 1;		
t_3	IR←MDR;		
t_4	MAR←R_i	No operation.	
t_5	MDR←MEM(MAR);	MAR←PC;	
t_6	AC←AC + MDR, SR←STATUS;	MDR←MEM(MAR), PC←PC + 1;	
t_7	No operation.	IR←MDR;	
t_8		MAR←R_i;	No operation.
t_9		MDR←MEM(MDR);	MAR←PC;
t_{10}		AC←AC + MDR, SR←STATUS;	MDR←MEM(MAR), PC←PC + 1;
t_{11}		No operation.	IR←MDR;
t_{12}			MAR←R_i;
t_{13}			MDR←MEM(MAR);
t_{14}			AC←AC + MDR, SR←STATUS;
t_{15}			No operation.

Table 7.3

In the table, clock cycles t_0 through t_3 perform the fetch cycle for instruction 0. At t_4 the execute portion of instruction 0 begins. Also at t_4, the fetch cycle for instruction 1 can begin. Note that although there are simultaneous register transfer operations during the execute cycle of instruction 0 and the fetch of instruction 1, there are no conflicts with respect to register usage. In t_6, for example, the contents of the MDR (an operand) is used to perform an addition for the execute cycle of Instruction 0. During the same clock cycle, a new value of the MDR (an instruction) is retrieved from memory for the fetch cycle of Instruction 1. The instruction is not loaded in to MDR until the end of the clock cycle. At this point the previous value, the operand, is no longer required.

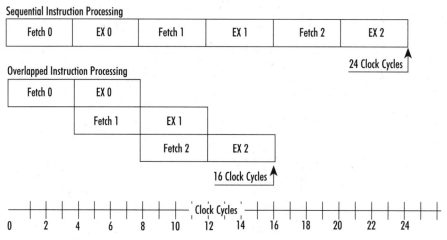

Figure 7.2   Sequential and Pipelined Instruction Processing

Figure 7.2 compares the effective processing time for overlapped and non-overlapped instruction processing of three instructions. The overlapping of the fetch and execute cycles allows the complete processing algorithm to be completed in 16 clock cycles instead of the 24 required for sequential processing. With overlapping, the effective processing time for each instruction after the first is four clock cycles.

While BART performance improves with overlapped fetch and execute cycles, several constraints were also imposed. The timing shown in Table 7.3 is compatible with only two of BART's four addressing modes. When using direct or register-indirect addressing, overlapping is possible only if the number of clock cycles is increased (Exercise 7.3).

Further improvement could perhaps be attained if greater overlap of instruction processing activities could be achieved. In the limit, it might be possible to start processing a new instruction at each clock cycle. The greater the degree of processing overlap, however, the greater the likelihood of conflicting hardware requirements. It is also likely that there will be constraints introduced by the processing algorithm. For example, jump and branch instructions are not suited for overlapped instruction processing.

The overlap process described above with BART instruction processing has many of the features of pipelining. The term pipelining, however, will be reserved for situations in which the hardware and processing algorithm are specially designed to use overlapping processing activities extensively. Pipelined instruction set processing is one of the major features of reduced instruction set computers, RISCs, to be discussed in a later section.

7.3   Memory Structures

High-performance computers require memories capable of providing information at a rate comparable to that which electronic devices in the datapath can accept and process the information. Slow memories will affect both the fetch and execute cycles of the computer. Because of the high cost associated with a large quantity of high speed memory, most modern computers use a **memory hierarchy** able to swap information between large and slow secondary memory structures and relatively small and fast primary memory structures. Algorithms to accomplish this will be presented in Chapter 9.

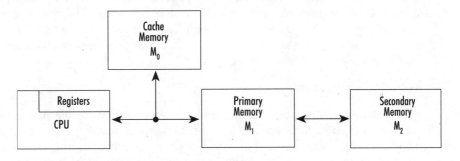

Figure 7.3   A Memory Hierarchy for a Digital Computer

The lowest and fastest level of the hierarchy, as shown in Figure 7.3, is the register set of the processor. While not considered to be a major memory element in early computers, information found in a register eliminates the need for a memory cycle. Using registers extensively in modern computers provides sufficient storage capacity for most processing operations. Registers offer the fastest means of managing information. Information can be accessed from registers with fewer register transfer operations than required for access from conventional RAM storage.

The primary memory is the initial repository for programs and data in computer operations. Primary memory sizes range from a few kilobytes to several megabytes. By using modest-cost solid-state memory technology, data can be accessed in 50 to 100 nanoseconds. For additional expense, primary memory with 10-nanosecond access time is available.

A cache memory is an intermediate memory that functions between the CPU and the primary memory. It is considerably smaller and faster than the primary memory. In its operation, copies of some of the data and instructions from the main memory are kept in the cache for rapid access. A memory access is directed simultaneously at the cache and the main memory. If data for most memory fetches can be found in the cache, the average-memory access time for data and instructions can be significantly reduced.

The secondary memory provides a repository for information too large for storage in the primary memory. The secondary memory, usually implemented with a disk, provides high-volume, relatively low cost-per-bit of storage, and slow access time to the stored information. Higher level memories involving optical disks and archive libraries of magnetic tape provide increased storage but with much longer access times. Methods to manage a multilevel memory hierarchy will be considered in Chapter 9.

7.4   Register Structures

Most early computers were characterized by the use of a small number of general-purpose registers. In many cases, the computer used only a single accumulator for performing all arithmetic and logic operations. The disadvantages of the limited availability of registers is apparent in realistic programming environments. In a looping activity, for example, it is very awkward to both perform calculations and test for conditions that terminate the looping. The programmer must store information from the accumulator in memory to create room to test for loop termination.

Modern computer structures have significantly increased the availability of general-purpose registers within the CPU. These registers can be used to temporarily store computed results or as address registers associated with indexed memory references. The increased availability of registers may have a dramatic impact on overall performance. The availability of registers greatly reduces the need to access the computer main memory. Since memory is significantly slower to access than registers, performance improves.

Register Windows

Modern computer programming and hardware structures place increased importance on modular software. Events such as subroutine calls and interrupts generally require storing the current set of registers in order to provide a clean workspace for the subroutine or interrupt service routine. In the case of a subroutine, parameters often must be passed to the subroutine. In some datapath structures these operations require information to be placed in and retrieved from storage locations and stacks in the computer main memory. They require multiple instructions and involve the relatively slow main memory.

As an alternative, the register environment can be subdivided into **register windows**. The entire range of registers is accessible through a window that selects a smaller group for immediate action. The choice of registers that reside in the window may be changed under software control. Register windows allow the programmer to rapidly change the set of active registers when performing different tasks.

Independent Register Windows

Register banks provide one implementation of the register window concept. As an example, consider the Intel 8051 microprocessor. This device contains a register space of 128 eight-bit registers.

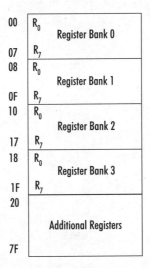

Figure 7.4   Independent Windows in the Intel 8051 Microprocessor

The 8051 architecture has provisions for four register windows, referenced as register bank 0 through register bank 3, as shown in Figure 7.4. These register banks occupy register locations 00_{16}–07_{16}, 08_{16}–$0F_{16}$, 10_{16}–17_{16}, and 18_{16}–$1F_{16}$. The microprocessor can change between the various register banks by executing a single instruction. The instruction set allows explicit reference to registers R_0 through R_7. Once a register bank has been selected, all references to registers R_0 through R_7 refer only to the selected register bank.

The advantage of register banks is clear when examined in the context of a program execution environment. Consider a program with a main portion, two extensive computational subroutines, and an interrupt-service routine. Each can be assigned an individual register bank. When a subroutine is called, the subroutine selects its appropriate register bank. The main program is relieved of the responsibility of storing its registers to create a clean environment for the subroutine. It may use its register bank as a means of passing parameters to the subroutine. At the conclusion of the subroutine, a return to the main program is accompanied by a switch in register banks. This process greatly reduces the need to swap information with the system memory.

An interrupt also requires a set of registers to perform its processing. By assigning an interrupt-service routine to its own dedicated register bank, the interrupt can be serviced without affecting processing in the other register banks. Since no registers need to be stored on the stack, it is possible to react to the interrupt condition faster than would be possible without multiple register banks.

Overlapping Windows

Overlapping windows provide an alternative to the Intel 8051 independent window environment. Researchers at the University of California, Berkeley, used this approach in the design of the RISC I and RISC II processors. Consider, for example, an architecture containing four register windows, each of 32_{10} registers, as shown in Figure 7.5. Each of the register windows has four types of registers.

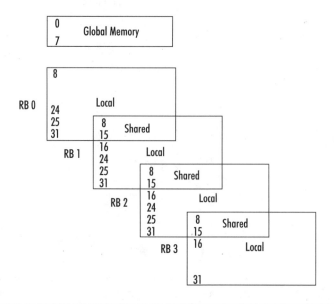

Figure 7.5   Overlapping Register Windows of RISC II

1. **Global Registers.** In Figure 7.5, R_0 through R_7 are global. These registers are common to all register window sets. R_0 of register bank 1, for example, is identical with R_0 of register bank 3. Global registers are useful for storing information which is to be accessed easily by all register windows.

2. **Shared Registers.** These registers are shared between two register windows. The upper shared registers (R_{25} through R_{31}) of register window 2, for example, are identical to the lower shared registers (R_8 through R_{15}) of register window 3. Shared registers are useful for passing parameters between nested subroutines. A subroutine using register window 2 can pass parameters to a subroutine to be run in window 3 by storing the information in its upper shared registers. The nested subroutine can use register window 3 and gain access to the passed parameters via its lower shared registers. Window 3 can then use the lower shared registers to pass results back to the originating program that ran in window 2.

3. **Local Registers.** Access to these registers is limited to code using the register bank. R_{16} through R_{23} in each register bank are local. Local registers can perform computations and preserve results that cannot be easily altered by other parts of the program. In some situations local registers may be considered as private and not accessible by other parts of the program.

An alternative application of register windows is used in the AMD 29000, which will be presented in Chapter 8. The 29000 uses a bank of 64 global registers. An additional 128 registers are addressable relative to a stack pointer. By changing the value of the stack pointer, a wide variety of sizes and relationships can be established between register banks.

7.5   Processor Operations

Modern digital computers have processor functions far superior to the modest operations performed by BART. In addition to integer addition and subtraction, modern processors also provide hardware support for multiplication, division, and floating-point arithmetic.

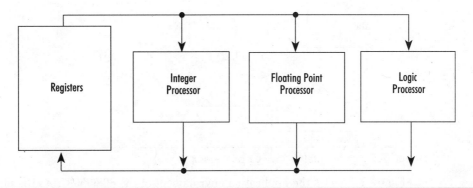

Figure 7.6   *The Use of Multiple Processing Elements in a Digital Computer*

Large scale computers may have separate processors optimized for integer arithmetic, floating-point arithmetic, and logical operations as shown in Figure 7.6. In some instances, the processors include internal registers to hold intermediate results during processing activities. This allows simultaneous scheduling of activities on multiple processors.

Many high-performance microprocessors improve their processing ability through the use of optional **coprocessors**. Coprocessors are independent processing datapaths designed to share processing duties with the main CPU processor. The instruction set of the coprocessor is chosen for a highly specialized set of operations, such as floating-point arithmetic. The coprocessor becomes a hardware replacement for a library of subroutines to perform selected operations.

Coprocessors are designed so that processing actions involving the main processor and coprocessor are coordinated. Each may respond to different instructions from the stream arriving from the program memory. As an example, consider the Intel 8087, a coprocessor designed to complement the Intel 8086 line of microprocessors. Similar and more complex devices are available as support components for most microprocessor families.

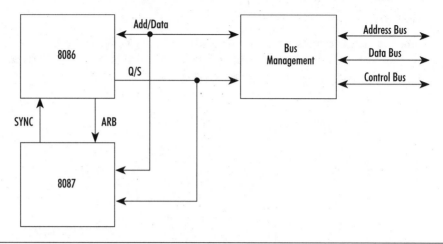

Figure 7.7   An Intel 8087 Coprocessor

Figure 7.7 shows the application environment of the 8086/8087. As the founding member of the Intel 80x86 family, this was one of the first microprocessors to provide optional coprocessors to improve computational performance. The 8087 coprocessor and the 8086 microprocessor both receive the same instruction stream. The 8086 generates all instruction fetches. The instructions are stored in the instruction queue of both devices. Some instructions are intended for the 8086 and others, identified by a first byte with a special escape instruction, are for the 8087. The 8086 enters a wait cycle while the 8087 processes an instruction. The coprocessor informs the 8086 when it has finished by means of a synchronization signal.

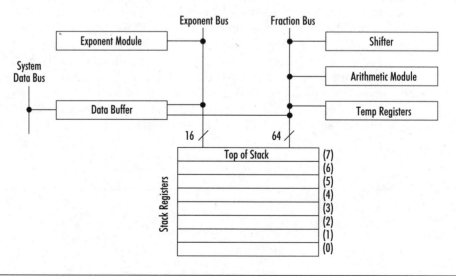

Figure 7.8   Internal Structure of an Intel 8087 Coprocessor

Figure 7.8 shows a portion of the internal configuration of the 8087 numeric coprocessor. Data formats conform to the IEEE floating-point standard. Internal to the coprocessor, all data is stored as an 80-bit temporary real. The 8087 uses an 8-word stack to store operands.

The 8087 has 68 instructions. Most operations access operands from the top two locations on the stack. Results are returned to the top-of-stack location; the length of the stack is reduced by one. Variations on the instructions allow one of the operands to be obtained from memory or from one of the other locations on the stack. No data comes directly from the 8086.

The presence of devices such as the 8087 numeric data processor greatly increases the processing rate for certain operations. A 64-bit floating-point multiplication requires about 2000 microseconds using the 8086 running with an 8 MHz clock. The same operation performed with an 8087 math coprocessor requires only 27 microseconds. A square root requiring 20,000 microseconds on the 8086 requires 36 microseconds with the 8087.

Newer microprocessors incorporate floating-point coprocessors on the same chip as the main processor. This minimizes the need to transport data to external coprocessors and provides high-performance processing.

7.6   Instruction Prefetch

All algorithms for processing instructions involve a fetch cycle to access the instruction from memory. The efficiency of the fetch process can be improved using a first-in-first-out prefetch queue. The prefetch queue is maintained in a fast memory. The CPU uses free bus cycles to access future instructions from the program memory. They are stored sequentially in the prefetch queue. From the prefetch queue they are quickly transferred to the instruction register for use during the execute cycle. Since instruction access is from a fast memory, this technique shortens the effective time duration of the memory read portion of a fetch cycle.

Many modern computers maintain the prefetch queue in high-speed cache memory. This memory is built using the same technology as the processor and can operate with the same effective clock cycle. In some instances, small program loops can be run exclusively from instructions stored in the prefetch queue.

The earliest use of a **instruction prefetch** was included in the IAS (Institute of Advanced Studies at Princeton University) computer in the late 1940s. In this computer, a 40-bit word width was used. A single memory location was used to store one numerical value or two instructions. In this machine, a fetch cycle was followed by two execute cycles. Special programming rules disallowed such operations as a jump to the second instruction in an instruction pair.

The Intel 80x86 family of microprocessors is an example of a device using instruction prefetch. The CPU is effectively divided into two sections. One is responsible for executing instructions; the second, the bus interface unit (BIU), is responsible for fetching instructions in advance of the current value of the program counter. These instructions are stored in an instruction queue until needed. The BIU senses when the system bus is not required for other purposes and attempts to keep the instruction queue full.

Jump and Branch Instructions in Prefetch Systems

Instruction prefetch procedures are complicated by certain types of instructions. Jumps, branches, skips, interrupts, and subroutine calls cause instructions to be executed out of sequence. Instructions in the prefetch queue immediately following one of these instructions may be incorrect. The queue may require flushing to replace unusable instructions with useful instructions at the branch destination.

It is useful, for the purpose of discussing instruction processing, to divide the instruction set into two groups: computational instructions and branch instructions. Computational instructions are executed in a predictable sequential order; address generation requires only incrementing of the program counter. Correct instructions can be fetched in advance of their use by either pipelined processing or instruction prefetch.

Branch instructions may result in program code being executed out of sequence. The correct next address is not known until the execute cycle. There may be two or more possible successor instructions. It is difficult to assure that the proper code has been fetched in advance.

Branch instructions may be further broken down into static branches and dynamic branches. In a **static branch**, the set of possible next addresses can be determined by examining the instruction and knowing the location of the instruction in program memory. For example, the instruction "JMP 0549" clearly indicates that the next instruction will be found in location 0549. Jump instructions and subroutine calls involve a static branch.

The instruction "JZ 0549" (jump on zero accumulator to location 0549) will obtain its next instruction from either the next sequential location or location 0549. Although the exact location cannot be predicted in advance, there are two known possible locations. This is a static branch.

Static branches offer the possibility of performing address generation in advance of instruction execution. In some cases, there may be two possible addresses, but both can be determined in advance. Prefetching can be performed in both directions from the branch. This adds complexity to the instruction-fetch algorithm.

Dynamic branching involves addresses that cannot be determined solely by knowing the instruction or its location. A jump to a location partially specified by a value in an index register is an example. A subroutine return is another example. Instructions with dynamic branches require additional information, such as the contents of a register, before address generation can proceed. They require more complex algorithms to effectively use instruction prefetch.

A static jump or branch instruction, such as JMP, can be effectively serviced by a technique known as the **delayed jump**. Consider, for example, the string of generic instructions shown in Table 7.4. To avoid complexity, only single-word instructions are considered. The program jumps around the data block and then resumes. The programming style, with a data block located in the middle of the program, makes little sense. It does, however, illustrate one of the potential problems encountered in using prefetch queues.

Address	Instruction	Comment
0100	ADD @R0	Computational instruction.
0101	CMP	Computational instruction.
0102	INC	Computational instruction.
0103	JMP 0115	Static branch.
0104	0574	Data storage area.
0105	0321	
...	...	...
0115	AND R3	Computational instruction.
0116	ADD R4	Computational instruction.
0117	NOP	No operation.

Table 7.4

When executing the program in Table 7.4 with a prefetch queue, the location of instructions in the prefetch queue and processor would be as shown in Table 7.5.

Instruction Time	Prefetch Queue Instruction	Processor Instruction
0	ADD @R0	—
1	CMP	ADD @R0
2	INC	CMP
3	JMP 0115	INC
4	0574 <data>	JMP 0115
5	AND R3	<not an instruction>
6	?	AND R3

Table 7.5

In the example program, the prefetch queue cannot determine that the next location following the JMP instruction is not an instruction until after the JMP instruction is interpreted for execution.

There are several possible solutions to this problem. NOP instructions may be inserted following the JMP. Enough NOPs would be added to fill the length of the prefetch queue (in this case, one). This would allow the queue to be cleared in a natural manner and allow the program to resume its effective processing with correct instructions. There would be a slight decrease in performance due to the nonproductive NOP instructions. A second solution would involve flushing the queue whenever a branch instruction is encountered.

A more elegant solution is to use a delayed jump instruction. This is a jump instruction that is not effective until after execution of the next sequential instruction. This delay allows the instruction pipeline to clear itself naturally. In the above example, the delayed jump would be introduced one instruction in advance of the actual jump location. The effect would be to interchange the JMP and INC instructions as shown in Table 7.6.

Address	Instruction	Comment
0100	ADD @R0	Computational instruction.
0101	CMP	Computational instruction.
0102	JMP 0115	Delayed jump instruction.
0103	INC	Computational instruction.
0104	0574	Data storage area.
0105	0321	
...	...	...
0115	AND R3	Computational instruction.
0116	ADD R4	Computational instruction.
0117	NOP	No operation.

Table 7.6

As shown in Table 7.7, using a delayed jump instruction allows the processor to fully use all the instructions in its queue. The amount of delay needed corresponds to the length of the queue. Delayed jumps result in both greater efficiency and less complexity for the control unit. Although the delayed jump may be confusing to the programmer, it can be implemented using high-level languages. As an alternative to a delayed jump, a NOP (no operation) instruction can always be inserted after a jump instruction.

Instruction Time	Prefetch Queue Instruction	Processor Instruction
0	ADD @R0	—
1	CMP	ADD @R0
2	JMP 0115	CMP
3	INC	JMP 0115
4	AND R3	INC
5	ADD R3	AND R3
6	—	ADD R3

Table 7.7

Static branch instructions present a more difficult but similar problem. With a branch instruction the decision to execute the next instruction in sequence or to take the branch address depends on a logical value in the processor. The exact value of the logic variable may not be known until the instruction immediately before the branch is executed. A delayed branch, implemented in a similar way to the delayed jump, can eliminate the need to flush the processing pipeline.

7.7   RISC and CISC Design Philosophy

Instruction sets for digital computers have undergone dramatic changes in the past decade as the technological base for their design has matured. Early machines were programmed using the native instruction set of the computer. In some early computers, programming of problems even involved the manual interconnection of components to achieve the desired data processing structure.

There has been a strong trend in recent years toward the use of high-level languages which insulate the computer instruction set from the programmer. Indeed, one goal of languages such as FORTRAN, PASCAL, BASIC, ADA, and C is to allow the program to be developed independent of the processing hardware.

Complex Instruction Set Computers—CISC

Design techniques, especially the use of microprogrammed controllers, allow the implementation of a wide variety of instructions. Historically this has led to the concept of the complex instruction set computer (CISC). CISC adopts the reasoning that complex instructions with multiple addressing modes offer greater support for the high-level languages programmer's use.

Typical CISC computers offer hundreds of instructions with multiple variations of addressing modes. They may include hardware support for floating-point operations and trigonometric calculations. CISC processors pay a penalty for this complexity in their need for long, multiclock cycle execution times for each instruction.

Reduced Instruction Set Computers—RISC

In contrast to CISC processor design, the reduced instruction set computer (RISC), offers a different approach to achieving high performance. RISC approaches adopt the "lean and mean" philosophy. The instruction set is kept relatively small and uncomplicated. The uncomplicated instruction set allows execution in a small number of clock cycles per instruction. It also allows the use of hardwired controllers, which offer inherent speed advantages and shorter clock periods. Memory operations, which are inherently slower than register operations, are minimized. RISC machines use pipeline principles extensively; they are designed to be very fast.

One characteristic of RISC processors is the availability of a large number of internal registers. These registers replace the main memory of the computer for temporary storage of results. Performance is improved by minimizing the number of inherently slow memory-access operations.

7.8   RISC Instruction Processing

Instruction pipelines effectively reduce the number of clock cycles required to process a computer instruction. The goal of RISC designs is to create a computer structure that can execute at an effective rate of one instruction per clock cycle. This is accomplished by developing a new design philosophy allowing extensive use of instruction pipelines.

RISC Instruction Pipelines

To facilitate pipelining, RISC instruction sets use uncomplicated instruction formats in which all instructions are the same length. This allows one (or more) complete instructions to be accessed in one machine cycle. Each machine cycle contains one or more clock cycles. The processing cycle is divided into several equal-length operations that can be pipelined. The following cycles are typical of those used in RISC pipelines:

- **Instruction fetch (IF).** The instruction is fetched from an instruction cache or a prefetch instruction queue.
- **Instruction decode (ID).** The instruction is decoded and the operands are accessed from register memory.
- **Execute (EX).** The specified operation is performed on the operand or the branch condition is evaluated.
- **Write back (WB).** The result of the arithmetic operation or information transferred from the memory is written back to a register.

If each of the above operations can be performed in one machine cycle, a four-stage instruction processing pipeline, as shown in Figure 7.9, can be used. Data exchanges with system memory cannot conform to the pipelined cycle due to the long access time required. Processing load and store instructions is treated as a special case.

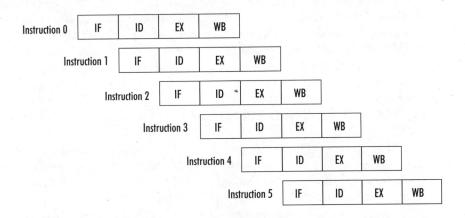

Figure 7.9   An Instruction Pipeline for a RISC Processor

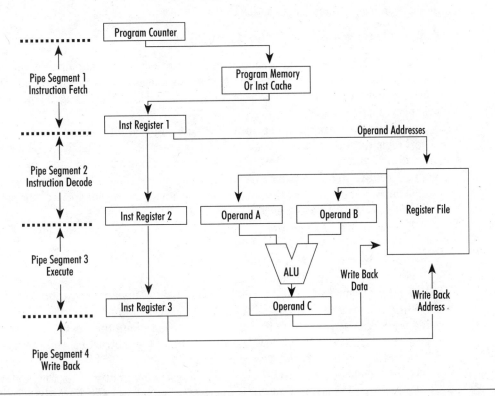

Figure 7.10   Datapath for a Pipelined Processor

Figure 7.10 shows an example datapath structure for a pipelined instruction processor. The instruction fetch pipe segment begins with the address of the instruction in the program counter. The instruction is fetched from either the program memory or an instruction cache. At the conclusion of pipe segment 1 the instruction is in instruction register 1.

In the instruction decode pipe segment, the fields of the instruction are decoded to provide the location of the operands in the register file. At the conclusion of pipe segment 2, the operands are transferred from the register file to the operand registers for the ALU. The instruction is still required for the next segment and is passed on in instruction register 2.

ALU operations are performed in the execute pipe segment. Results of the operations are stored in the Operand C register. The instruction is still required since it specifies the final destination of the processed operand. It is passed on to the next pipe segment in instruction register 3.

The final pipe segment, write back, returns the results of the specified operation to the register file.

The registers in Figure 7.10 pass intermediate results between pipe segments. Three instruction registers are required; one for each of the pipe segments performing a portion of the instruction processing. Each instruction register holds the code for one of the sequential instructions being processed. Unlike structures like BART, four instructions are processed simultaneously.

Control for a RISC machine can be viewed as four separate controllers. Each is responsible for one segment of the pipeline. An overall supervisory function is also needed to synchronize activities between the segments and handle special processing situations such as branching and data dependencies among instructions.

Characteristics of RISC Machines

As with any pipeline operation, the effective rate at which the pipeline can operate is determined by the slowest of the stages. For good performance, each of the identified processing actions must use about the same amount of processing time. This naturally leads to a design philosophy to support pipelines.

- **Register-to-register operations.** By requiring that all instructions use operands residing in registers, access to the main memory is reduced. This effectively reduces memory bandwidth requirements. Memory accesses are much slower than register accesses.
- **Load/Store architecture.** Since memory access is slow, instructions accessing operands in memory require an excessive amount of time and destroy the pipeline efficiency. Therefore, in RISC machines all accesses to memory are of the load or store type.
- **Simple addressing modes.** Complex addressing modes require additional processing time and place additional burden on the control unit.
- **Simple instructions.** All instructions must execute (ideally) in one clock cycle. This eliminates the possibility of using complex and lengthy instructions such as floating-point multiplication. Complex operations can be performed in software or by coprocessors.
- **Simple instruction formats.** All instructions are limited in length to one word. This is necessary for effective use of pipelining. Fields to designate the location of operands are similar in all instructions. This allows easy decoding of the instruction and the access of operands while the decoding operation is in progress.
- **Hardwired control units.** Hardwired control is inherently faster than a microprogrammed control unit. The uncomplicated instruction set allows the hardwired control unit to be designed without significant complexity.
- **Separate data memory and code memory (Harvard architecture).** RISCs are highly pipelined in their operation. Separate memories allow code and data to be accessed simultaneously from the memory.

- **Optimizing compilers.** Since RISC programs are inherently longer than those using complex instruction sets, a good compiler is essential. The compiler must produce tight code and handle delayed jumps and branches. It is not likely that anyone would want to program the machine in assembly language.

Pipeline Stalls

In pipeline structures, the processing period of the pipeline is determined by the *slowest* of the individual segments. For optimum performance, all segments should require the same length of time. This is not always possible since some processing actions are data dependent. The instruction fetch segment is subject to a wide variation in processing times. Instructions that can be retrieved from a prefetch queue or instruction cache can readily be provided to instruction register 1 in a short period of time. If, however, the instruction must be obtained from main memory, the acquisition time may be excessive.

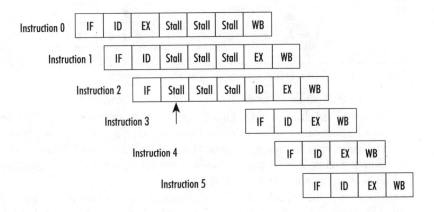

Figure 7.11   Pipeline Stalls Due to Main Memory Fetch

In Figure 7.11, the instruction fetch operation for instruction 2 is extended due to the need to access information from the main memory rather than the cache. To accommodate the extension of the IF pipe segment, all other pipe segments must also be extended. During this extended time no useful operations are performed. This data-dependent event is called a **stall**.

Data Dependency in the Pipeline

In RISC processors it is important to have a smooth flow of information through the processing pipeline. Situations exist in which **data dependency** can cause bubbles in the processing algorithm. Consider the instruction sequence I_0, I_1 as follows:

I_0: $R_0 \leftarrow R_1 + R_2$

I_1: $R_0 \leftarrow R_0 + R_4$

There is data dependency between the two instructions. The write back for I_0 must be performed before the instruction decode (including operand fetch) for I_1. The problem is shown in the top portion of Figure 7.12.

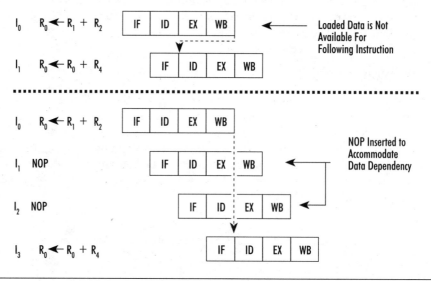

Figure 7.12   Data Dependency in the Load Instruction

The data dependency problem can be treated in several ways.

- **Delayed fetch.** The programmer or compiler can recognize data dependency conditions. Their effect can be eliminated by inserting multiple NOP instructions into the code to allow sufficient time for the required data to be available. This solution is shown in the bottom portion of Figure 7.12. A more efficient solution is to reorder the existing code to fill the delay slots with instructions without data dependency.

- **Hardware interlock.** In this approach, data dependency is detected by hardware. Extra nonprocessing clock cycles are inserted into the processing pipeline to allow sufficient time for data to be available. Additional instruction fetch cycles are delayed until the bubble is passed through the system. This solution is shown in the top of Figure 7.13.

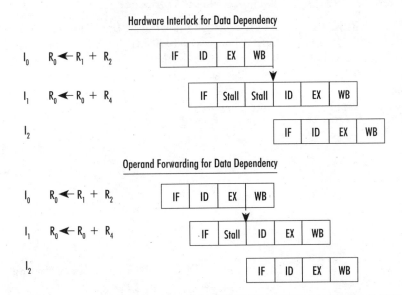

Figure 7.13   Use of Hardware Interlock and Operand Forwarding to Improve Instruction Processing

- **Operand forwarding.** Another possible solution is to modify the hardware of the datapath to allow computational results a short circuit path so they may be used as operands in succeeding instructions. For cxample, in Figure 7.10 the output of the ALU could be routed directly to the operand A register. This would avoid clock cycles required to load the operand C register and then transfer operand C to the register file. If the operand C register were routed directly to the operand A register, bypassing the register file, a single clock cycle would be saved. This solution is shown in the bottom of Figure 7.13.

Branch Instructions

The problems associated with branch instructions in prefetch queues were discussed in the previous section. A similar problem exists in RISC processors. The destination address of a branch is not known until the instruction is processed. While processing the branch instruction, several other instructions (which may not be correct) have entered the pipeline.

The solutions are similar to those used in data dependency problems.

- **Delayed branch.** NOP instructions are inserted following all branch instructions to allow the correct branch target instruction to enter the pipeline. As an alternative, the delay slot following the branch instruction can be filled with useful instructions that have no impact on the branch location.
- **Branch instruction prefetch.** In this technique, two instructions are prefetched: the next sequential instruction following the branch and the target instruction of the branch. One will be correct and can be inserted into the instruction pipeline.

Example 7.1:   The Berkeley RISC I Computer

The RISC I computer grew out of the work of David Patterson. While working at Digital Equipment Corporation (DEC), Patterson studied the problem of making a VLSI version of a computer such as the VAX. Patterson concluded that the complexity of the machine, with its extensive instruction set and multiple addressing modes, would make it impossible to design a workable control unit. Upon returning to Berkeley, Patterson and associates designed and fabricated a machine based on RISC principles.

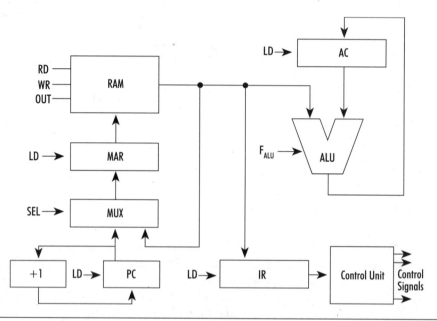

Figure 7.14   Instruction Formats for RISC I Computer

RISC I is a 32-bit microprocessor which was fabricated using NMOS technology. RISC I has only 31 instructions. There is a large bank of general-purpose registers organized in register windows. The RISC I register structure was presented in Section 7.4. Instructions provide three addresses; two for source operands and one for the destination. All locations for operand sources and destinations are within the register space. Figure 7.14 shows the instruction formats for a register mode instruction and a register immediate mode instruction. The instruction format and limited addressing modes allows the use of a hardwired controller.

Table 7.8 includes some representative RISC I instructions.

OpCode	Operation	Comment
ADD	$R_d \leftarrow R_s + S_2$	Integer addition.
SLL	$R_d \leftarrow R_s$ (shifted by S_2)	Shift left.
LDL	$R_d \leftarrow MEM(R_s + S_2)$	Load register from memory.
STL	$MEM(R_s + S_2) \leftarrow R_d$	Store register contents in memory.
JMP	If true, $PC \leftarrow R_s + S_2$ If false, $PC \leftarrow PC + 1$	Conditional jump.

Table 7.8

Within the register file for RISC I, R_0 contains the integer 0. This allows the ADD instruction to be used to perform a move function. For example, $R_5 \leftarrow R_0 + R_3$ moves the contents of R_3 to R_5. The load instruction has modifications that can transfer a single byte or four bytes from main memory to a register.

7.9   CISC vs. RISC Controversy

Debates continue concerning the relative advantages and disadvantages of CISC and RISC approaches to computing structures. While the arguments favoring RISC are strong, CISC computers tend to dominate in most applications. The following are some of the arguments currently being made:

- CISC machines have a huge advantage in software support. It would be expensive and time consuming to convert all programs. There are not sufficient optimizing compilers to allow software written in high-level languages for CISC machines to be recompiled to run on RISCs.
- The main performance advantages gained by RISC machines are in the use of a large register space. CISC machines that use a large register space and get many of the same advantages can also be created.
- RISC machines require an optimizing compiler. CISC machines can also get performance improvement from optimizing compilers.
- RISC machines require a very fast cache memory system to acquire instructions. CISC machines can also benefit from cache memory.

Summary

This chapter is concerned with modifications that can be made to the basic digital computer structure in order to obtain higher performance. A reasonable measure of this performance is *J*, the overall processing time to perform a processing algorithm.

J can be improved by performing more complex activities in each instruction. This is possible because of the great flexibility in the control unit provided by microprogramming. This design approach leads naturally to the concept of CISC, the complex instruction set computer. CISC computers are characterized by large, complex instruction sets with numerous addressing modes.

An alternative approach is to design computers that can execute an uncomplicated instruction set extremely fast. This leads to the concept of RISC, the reduced instruction set computer.

High speed memory is important in both RISC and CISC computation. Extensive multilevel memories including high-speed caches are used for both. Registers form an extension to high-speed memory. Data stored in registers does not require memory-access cycles and is available for high-speed operations. Using register windows improves access to data in registers. Using multiple windows, registers can replace a stack for saving data during interrupt service. Windows can also pass parameters among nested subroutines.

Instruction processing can be improved by using pipelined operations in the fetch and execute cycles and by prefetching instructions. RISC processors use pipeline processing of instructions extensively. The goal of a RISC processor is to complete one instruction each clock cycle. This goal can be approached by using simple instructions that can be extensively pipelined. The use of a high speed clock is possible with a very fast hardwired control unit along with simple instruction formats, simple addressing modes, and extensive register-to-register processing. RISC processors have problems with bubbles in the pipeline due to data dependency and address dependency in the pipeline structure.

For additional information and references on the topics presented in this chapter, access the World Wide Web page provided at: `http://www.ece.orst.edu/~herzog/docs.html`.

Key Terms

CISC	local register
coprocessor	memory hierarchy
data dependency	operand forwarding
delayed branch	register window
delayed fetch	RISC
dynamic branch	pipeline stall
global register	shared register
hardware interlock	static branch
instruction prefetch	

Readings and References

Hennessy, John L., and David A. Patterson. *Computer Architecture: A Quantitative Approach.* San Mateo, CA: Morgan Kaufman Publishers, 1990.

Both Hennessy and Patterson are pioneers in the RISC area. Their text not only provides a technical framework for RISC but also examines the economic issues. The DLX, a pipelined computer, is designed as an example.

Kane, Gerry. *MIPS RISC Architecture.* Englewood Cliffs, NJ: Prentice Hall, 1988.

The MIPS series of RISC processors are among the small group of RISC devices that have successfully made the transition from laboratory to commercial application. This book provides introductory material on RISCs and describes the R2000/R3000 versions of this family.

Patterson, David. "Reduced Instruction Set Computers." *Communications of the ACM*, 7 (1985).

A good tutorial-type article providing insight into the philosophy of RISC architectures.

Exercises

7.1   Show a possible fetch/execute overlap for a BART instruction using register addressing. Use as few processing steps as possible. Allow the program counter to be incremented without using the ALU.

7.2   Show a possible fetch/execute overlap for a BART instruction using direct addressing. What processing improvement is possible? Allow the program counter to be incremented without using the ALU.

7.3   Consider a Harvard architecture with a similar structure to BART. There are two separate memories: one for data and one for code. Show how fetch and execute can be overlapped for a single-word instruction with direct addressing.

7.4   Evaluate the usefulness of each of the following as possible register windows implementations.

  (a) Windows are separated with gaps of several registers between successive windows.

  (b) Windows are fully overlapping with no local registers.

  (c) Windows are fully overlapping with each register a member of four different windows.

7.5   Why can a coprocessor such as the Intel 80x87 perform arithmetic processing actions much faster than the microprocessor ALU?

7.6   Compare the concept of a prefetch instruction queue to a cache memory.

7.7   Intel manufactures both an 8086 microprocessor that uses a 16-bit data bus and an 8088 microprocessor that uses an 8-bit data bus. Both have the same instruction set. Both use instruction prefetch. Explain why the 8086 does not process instructions twice as fast as the 8088.

7.8   Which is the easier way to handle branching problems in instruction processing pipelines, inserting NOP instructions or using delayed jump? Which way gives the best performance?

7.9   Why do RISC processors access memory only with LOAD and STORE instructions?

7.10 What addressing modes are used in RISC processors? Why were they selected?

7.11 How is it possible for RISC processors to execute at the rate of one instruction per clock cycle?

7.12 Branching operations in RISC processors require delayed branching or the insertion of NOP instructions. Give an argument in favor of each.

7.13 In a RISC processor, a five-stage pipeline is used. What modification are needed to the following code to avoid branch problems?

```
ADD   120
JZ    240
ADD   121
JZ    240
ADD   123
JZ    240
```

7.14 In a RISC processor, what advantage is there to using separate data memory and instruction memory?

7.15 Using BART as a model, show how the register-to-register instruction processing, such as ADD R2, could be pipelined.

8
System Software and Representative Computer Architectures

Grace Hopper

A sprightly young woman named Grace
Programmed at a very fast pace
It is said, with a shrug
she found the first bug
and removed it with a handkerchief of lace

Prologue

Grace Hopper, born in the first decade of the twentieth century, was the pioneering personality in the advancement of high-level computer languages. With a Ph.D. in mathematics from Yale, Grace was a professor of mathematics at Vassar College. She enlisted in the United States Naval Reserve in 1943 and was assigned to assist Howard Aiken at Harvard in the development of the Mark I, a relay computer. While working with Mark I, she repaired a malfunctioning arithmetic unit by removing a moth that was entrapped in a relay mechanism. The moth was preserved in her notebook and was the first computer bug.

In a later assignment with J. Presper Eckert and William Mauchly, she became aware of the difficulty potential users experienced with programming computers. All programming at that time was done in the native machine language of the computer. Extensive mathematic and logic skills were required to convert computational problems into coding that the computer could process. There was a realistic concern that an insufficient number of mathematicians existed to provide programming expertise for the rapidly evolving computational machines.

Grace Hopper was a driving force in developing high-level languages. In the early 1950s, a team under her leadership produced a series of **compilers**, including COBOL (Common Business-Oriented Language), which ran on UNIVAC, the first commercial computer. Her success led to other efforts that produced FORTRAN. High-level languages had a tremendous impact on digital computing. Users were no longer required to have extensive background in algorithmic development or a knowledge of the architecture and structure of the computer. The result was that digital computers became accessible for a wide range of business and scientific applications.

Grace Hopper retained a close association with the Navy for almost forty years. She had reached the rank of Rear Admiral when she retired in 1986.

8.1   Introduction

Even the cleverest hardware design is useless unless provisions are made for developing reliable software. This chapter will consider the relationship between the software and the computer. Each level of software, ranging from machine code to compiler code, represents an advancing level of abstraction with respect to the functionality of the computer. Lower levels of programming are closely related to the hardware structure and instruction set of the computer. High level programming attempts to give the programmer a machine-independent interface to the application environment.

Since the goal of this text is to study the design methodology of the digital computer, software that preserves the feel of the actual computer will be emphasized. The use of assembly code represents a reasonable compromise for situations in which the programmer requires complete freedom to control the individual and perhaps unique features of a particular digital computer. This level of control is required, for example, to program the unique hardware environment of an embedded microprocessor. Input/output device drivers for larger computers also require this degree of low-level machine control.

For these and other reasons, assembly code and the use of **assemblers** will be considered in modest detail. This is not to suggest that low-level coding is desirable for all applications. It does, however, give the user a good understanding of the true structure of the machine and complete control of its features.

For a good programmer, knowing how and when to use both high-level and low-level languages is analogous to a good photographer's familiarity with the functions of a modern high-performance camera. For most applications, the "point and shoot" mode produces good quality pictures with a minimum of user decisions. For some situations, the serious photographer will prefer to select the lens opening, exposure time, and other features to specifically tune the camera to produce the desired effect. Both methods are valid. A good photographer will know when to use the automatic technique and when to use the manual technique. A good computer engineer/scientist will know when to use high-level compilers and when to use assembly language.

Programming style, the physical structure and appearance of the program, is an important aspect of good software design. Experience has shown that maintaining and modifying existing programs consumes an appreciable portion of the total software management effort. A consistent and well-documented programming style allows large programs to be subdivided into modules for individual programmers. These modules can then be maintained, modified and reused by other programmers over the lifetime of the code.

For a full appreciation of the programming process, nothing beats actual experience. For this reason the concluding sections of this chapter present instruction sets and an architectural overview of four representative computer families; the Motorola 680x0, the Intel 80x86, the Intel 8051 microcontroller, and the AMD 29000. The instruction set selection and formats are typical of modern computing devices. The structure of the instructions suggests the manner in which the computer designer has related the software and hardware components of the device.

The material in this chapter, accompanied by a real or simulated microprocessor with its software programming aids, can provide a good introductory experience in preparing and executing software on a digital computer.

8.2   Software Hierarchy

Computer programmers must interact with computers at several different levels of software abstraction.

Machine Code

Computer software can be considered at several hierarchical levels. At the lowest level is the machine code. This is the pattern of binary bits that must be present in the program memory of the computer. For many microprocessors, this code is placed in ROM memory chips by special programming devices. Other computers may load this information from a mass storage device such as a disk into a block of RAM memory. The program code from RAM may then be executed by the computer. Regardless of the sophistication of the computer or the computer language used to program it, before execution, program source code must be transformed into machine code.

For example, consider the following program written for the BART computer. The program is to access the data byte from location 25_{16}, decrement the value by five, then store the result in register R_3. In an **assembly language code**, mnemonics are used to express the operation codes, special symbols are used to identify the addressing mode, and addresses are specified explicitly in the program statement.

Memory Location	Assembly Language	Comments	Machine Code
00	LDA 25H	Load accumulator with data from location 25.	01000000
01		[Second byte of instruction.]	00100101
02	ADD #FBH	Decrement the data (add –5).	00000100
03		[Second byte of instruction.]	11111011
04	STA R3	Store in register 3.	00111011

Table 8.1

The assembly language code includes all the information necessary to transform the instruction into the machine code. The starting address of the program, in this case 0, must also be known. By the nature of the instruction and the addressing mode, it can be determined if the instruction requires one or two bytes of memory. The operation code, addressing mode, and address allow the binary machine code to be generated.

Machine language code, while essential for use in the computer, is difficult for a programmer to interpret because it is difficult to follow the algorithm. Programming is very tedious, and mistakes are both likely and difficult to detect.

Assembly Code

The productivity of a programmer can be improved dramatically by providing software support in the form of an assembler. The assembler is a program that accepts an input in a highly regimented programming style—an assembly language program. The assembly language program may be either written by the programmer or generated by a compiler from a high-level language. This relationship is shown in Figure 8.1. The assembler processes the program to produce a machine language program which is directly usable by the computer.

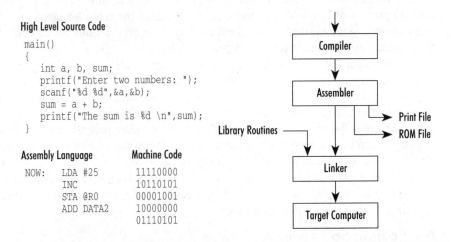

Figure 8.1   Levels of Programming Hierarchy

After assembly, a third processing step is sometimes used to combine the code produced by the assembler with library routines also in assembly language. A **linker** is used to combine the several program modules into a single program and to resolve any address conflicts. The role of the linker will not be considered.

In addition to transforming higher level instructions into machine code, the assembler produces written documentation of the assembly code and programmer comments. It may also produce a file suitable for use by a ROM or EPROM programmer. The resulting memory chips may be used as program memory by the computer. Assemblers also detect some types of programming errors such as improper syntax and improper memory usage.

An assembler is a software program that runs on a computer. Often the assembler runs on a computer different from the target system for the code; this is a cross-assembler. For any target computer, there may be many different assemblers. All produce machine code compatible with the target. Source code written for one assembler may not assemble correctly on a different assembler because of differences in features and syntax.

Compiler Code

Languages such as C, Pascal, and Fortran represent the highest programming level of the software hierarchy. Each uses a special syntax to represent a computer program in a style and structure that provides the programmer with a powerful set of programming features. High-level languages provide a framework for managing complex data elements such as floating-point numbers, complex numbers, and data files. They provide library routines for performing complex mathematical operations such as logarithms and exponentials. They also provide useful programming structures for performing modular programming and managing program flow. A compiler program written for one computer can usually be compiled for a second computer with a minimum of rewriting. This is called **porting** of a program. The goal of the compiler is to make the programming process independent of the host computer.

Converting code written in a compiler language to machine code usually requires two intermediate processing steps. The compiler first translates the compiler language source statements into assembly language statements for the selected target machine. An assembler then completes the process by producing machine-compatible code.

The efficiency of the translation process can vary widely depending on the target processor. Many modern computers have an instruction set especially chosen to allow efficient translation of high-level code. In most cases, an experienced assembly language programmer can produce highly optimized "tight" code that requires fewer (machine language) instructions and runs faster than code programmed with a compiler. An experienced compiler programmer can produce code faster, with fewer errors, and better documented than code programmed with an assembler. In addition, compiler code is easier to port to a new computer than code written with an assembler.

It is common for coding activities to involve a mix of both high-level coding and assembly language programming. The assembly language code is confined to subroutines which must be highly optimized to gain adequate performance.

Application Software

High-level programming languages allow the user to communicate computational ideas independent of the restrictions provided by the computer's hardware structure. Operating systems provide services that allow assemblers, compilers, linkers, and other programs useful for software development and execution to be conveniently used on a host computer. Such software is known as **system software**.

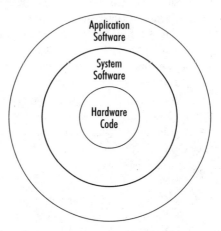

Figure 8.2   Layered Software Support for Computational Hardware

Programs such as word processors and spreadsheets are known as **application software**. They exploit system software to provide services to the user. Figure 8.2 indicates the hierarchy of layered services provided to the hardware core by system software and application software.

8.3   The Assembly Process

Most assemblers have provisions for supporting the following programming structures:

- **Comment fields.** Comment fields allow the programmer to further explain the purpose of a block of code and insight into the programming strategy. The comment field is isolated from the remaining portions of the instruction by a delimiter such as ";".

```
NOP                     ; This instruction performs no operation.
```

- **Use of mnemonics for instruction operation codes.** Mnemonics make code much easier to program and much easier to read. Fewer coding errors are made.

```
CMP                     ; Complement the data value.

INC                     ; This completes the two's complement computation.
```

- **Special symbols to indicate addressing modes.** Addressing modes are indicated by special symbols such as # to indicate immediate addressing or @ to indicate register-indirect addressing.

```
ADD   #27H              ; Add 27₁₆ to the data.

STA   @R0               ; Store the data in the storage array.
```

- **Symbolic addressing.** Rather than refer to operands by their location in memory, assembly language allows the programmer to refer to them by names.

```
ADD   DTA1              ; Add the first data element.

STA   SUM               ; Put the result in memory location SUM.
```

- **Labels.** Symbolic addressing may also refer to locations in the program.

```
      CLA               ; Clear accumulator.

ONE:  INC               ; Increment the accumulator.

      JMP   ONE         ; Loop back to ONE.
```

- **Data conversion.** By convention, unless otherwise indicated, numeric values are assumed to be decimal. They are converted to binary during the assembly process.

```
LDA   #16               ; Place 00010000 (16₁₀) in the accumulator.

LDA   #10H              ; This number is also 16₁₀ but expressed in hexadecimal.

LDA   #00010000B        ; This number, also 16₁₀, is expressed in binary.
```

Code statements in assembly language use four fields of information, as shown in Table 8.2.

Label Field	Operation Field	Operand Field	Comment Field
The label must begin in the first column. Some assemblers require the use of a ":" to terminate the label field.	The field contains the mnemonic description of the operation code.	This field provides a reference to the registers or memory locations where the operands reside. Various addressing modes may be specified. The address may be in symbolic form.	This field allows the programmer to document the instruction in any appropriate manner. It is delineated from the operand field by a ";".

Table 8.2

Assembler Directives

In addition to the statements which get transferred directly to code, most assemblers accept **assembler directives** to provide instructions to the assembler concerning the assembly process. These instructions are not executable by the target computer and have an effect only at the time of assembly. The directives resemble assembly code in form and are called **pseudo-operations**. Following are some typical examples of these assembler directives.

Management of Program Memory

These directives allocate memory in the manner specified by the programmer. It may be necessary to have certain blocks of code at specified locations. For example, some computers have specified locations where interrupt routines must originate or where program execution begins following a RESET or program interrupt. It is also necessary to initialize data such as constants and text strings. Space must be reserved to store computed results.

```
ORG 1000H    ; The following code is assembled starting at address 1000₁₆.
             ; ORG is usually the first statement in the program.

DS  10H      ; Define storage. The following 16 locations are reserved for
             ; data storage purposes, they cannot be used for code storage.

DB  25 31    ; Define byte. Put 19₁₆, 1F₁₆ in the next two memory locations.

DB  'Hello'  ; Put the ASCII codes for the characters H, e, l, l, o in the next
             ; five memory locations.
```

Definition of Symbolic Names

These directives allow the programmer to associate symbolic names with values. The EQU, or equate directive, is a common example.

```
NOW EQU 05H          ; Henceforth every reference to "NOW" will result in value 05₁₆
                     ; being used.

     ADD NOW         ; The value in location 05₁₆ is added to the accumulator.

     LDA #NOW        ; The numeric value 05₁₆ is placed in the accumulator.
```

Using symbolic names and addresses allows the programmer to reference names rather than memory locations. This provides a more friendly programming environment and reduces errors.

Miscellaneous Directives

The directive END is usually required at the physical (not logical) end of the assembly language program. It informs the assembler that there are no more instructions or directives.

Some assemblers allow the definition and use of **macros**. Macros are blocks of code that, once assigned, can be referenced by a shorthand representation. For example, a macro called MULT might multiply the integer contents of R_0 and R_1 and place the (double-length) result back in the same registers. After defining this macro once, the programmer would need to include only the macro name MULT in his code. The assembler would replace MULT with the defined block of code. Libraries of useful macros are available for most computers.

Some assemblers include provisions for controlling the assembly process. Through a process known as **conditional assembly**, it is possible to include in the source code conditional directives that will compile different blocks of code. For example, a certain computer may be supplied in two different configurations. One contains a mathematics co-processor chip; the other lacks the chip. The programmer can specify to the assembler the type of system to be used.

The assembly process requires that the assembly language instructions and assembler directives be translated into machine code that can execute from the program memory of the target device. This consists of several operations:
- Memory must be allocated according to assembler directives such as ORG, DS, and DB.
- Operation codes and address modes must be translated into machine code.
- Actual addresses must be substituted for symbolic addresses.

Manual Assembly

If the programmer understands the underlying architecture and instruction set of a computer, small programs can be manually assembled. Consider the following short program, using the BART instruction set, which adds together two values called HERE and THERE to produce a value NOW. After executing the algorithm, the program halts by entering a tight loop. Recovery to run another program would require a system RESET or other form of intervention.

```
; Example Program
;
          ORG     20H                    ;Start program at location 20H.
START:    LDA     HERE                   ;Get first number.
          ADD     THERE                  ;Add second number.
          STA     NOW                    ;Store results.
FINE:     JMP     FINE                   ;Endless loop for halt.
HERE:     DB      07H                    ;First Number (in Hex).
THERE:    DB      25                     ;Second Number (25D = 19H).
NOW:      DS      01                     ;Reserve 1 byte of storage for result.
          END
```

To change the source language into machine language requires two passes of the source code. During the first pass, the location of all of the symbolic addresses can be determined. In the second, the assembly language instructions can be converted to binary (or hexadecimal) representation.

START, because of the ORG statement, must correspond to location 20H. The first executable instruction (LDA HERE), a two-byte instruction, is placed in locations 20H, 21H. The next three instructions also require two bytes each. FINE, the location of the JMP instruction, must therefore be location 26H. Table 8.3 shows the address associated with each of the labels in the program.

Symbolic Address	Memory Location
START	20H
FINE	26H
HERE	28H
THERE	29H
NOW	2AH

Table 8.3

Using the address table and the source code, each of the instructions can be translated into the machine code to be placed in the program memory. Table 8.4 shows the results of the assembly process. HERE and THERE have contents of 07_{16} and 19_{16}. The reserved space, NOW, at 2A is assumed to contain 00. In reality, the contents are unknown. Memory locations are assigned singly or in pairs to accommodate the length of the assembly language instructions. By including both the source code and the machine code in a program listing, the programmer has documented his work. The listing is also useful for further debugging.

Memory Location	Assembled Code	Source Statement
		;Example Program
		;
		ORG 20H ;Start addr.
20	40 28	START: LDA HERE ;Get first num.
22	00 29	ADD THERE ;Add second.
24	30 2A	STA NOW ;Store result.
26	50 26	FINE: JMP FINE ;Endless loop.
28	07	HERE: DB 07H ;First number.
29	19	THERE: DB 25 ;Second number.
2A	00	NOW: DS 01 ;Reserve space.
		END

Table 8.4

The Assembler

An assembler accepts information from the programmer in the form of a source program. This program is written using a text editor and is usually restricted to standard ASCII characters. Most files produced by high performance word processors are not acceptable since they introduce special control characters related to spacing, type fonts, and general appearance. Word processors usually have a mode that can be selected to produce pure ASCII text.

The source code consists of assembly language instructions for the target computer and directives for the assembler. The assembler reads the source code, follows its directives, and produces a listing and file of the binary contents of code memory locations within the target system. Various program listings are also produced for documentation purposes.

Assemblers process programs using two passes of the source code. In the first, as shown in the flow chart of Figure 8.3, a table is created to associate symbolic names with addresses. The instructions are examined sequentially. An address counter keeps track of the address for each executable instruction in the program. After initializing the address counter to zero, each line of code is examined to determine if it represents an instruction or an assembler directive. If it is an instruction, it is examined to determine if it has a label. Any label is placed in the label table and equated to the current value of the address counter. The address counter is then incremented once or twice to account for the length of the instruction.

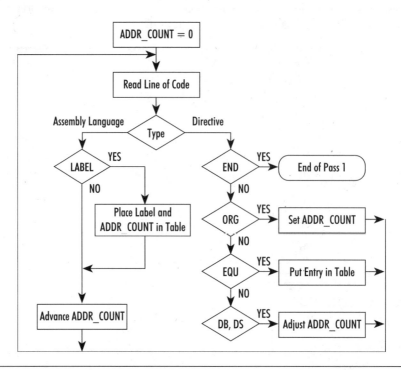

Figure 8.3   Pass One of a Two-Pass Assembler

If the line of code under examination is an assembler directive, an appropriate action is initiated. END terminates pass one. ORG places a new value in the address counter. EQU causes additional labels and values to be inserted into the label table. DB and DS directives allocate space for data and results; their presence requires the address counter to be incremented an appropriate amount.

The processing of pass one continues until all lines of code have been examined. It terminates when the END statement is detected.

During the second pass of the instructions, the values for all labels are known and the code can be fully converted to its machine language compatible form. A possible processing algorithm is shown in the flow chart of Figure 8.4. The original source code is again read sequentially line-by-line. The only assembler directives of interest in pass two are END, which terminates the processing, and DB, which requires the indicated data to be inserted at the specified memory location. Enough information exists from pass one to allow each instruction to be translated into a line of machine code. The value for all symbolic addresses is available in the label table.

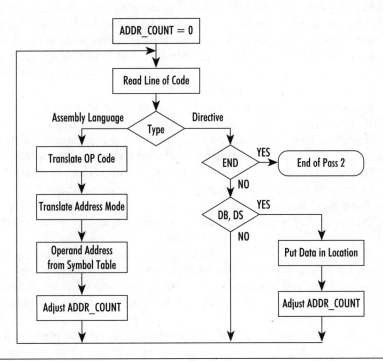

Figure 8.4   Pass Two of a Two-Pass Assembler

8.4   Software Design and Structure

In a software development project the code may be thousands or millions of bytes long. Coding such a system is a complex task. A large project has enormous potential for logic, programming, and algorithmic errors. The debugging and testing of very large programs can be difficult if the software structure is not designed properly.

Large software projects are best attacked in a top-down and structured manner. This technique applies to programs written in either assembly language or in a high-level compiler language. Large programs are subdivided into smaller modules with well-defined requirements. These modules are further divided until a size and complexity manageable by a software engineer or programmer is reached. The goal of modularization is to reduce the program complexity to a level in which it can be understood and reliably written and documented by a single programmer.

Modular structures have many advantages.

- Each module is relatively small and much easier to understand, program, and document than a large program. Each module can be compiled (assembled) individually.
- The logical flow and sequencing between modules is uncomplicated.
- Testing and maintenance is easier on a module than on a large program.
- Precise specification of the requirements for data input and data output from each module provides good documentation.

- Modules may be reprogrammed as necessary without affecting other parts of the program.
- Different modules may be programmed in different languages, as appropriate for the application. Most modules might use a high level language such as C. Highly hardware-dependent modules might achieve higher performance with assembly language.
- Debugged modules can be used as a library for future program development. Library routines can be purchased from outside vendors.
- Partitioning allows a team approach to programming projects.

Modular programming often begins with a sequence of logical steps to the solution of the programming problem. The intent is to describe *what* is to be accomplished rather than *how* it is to be done. A verbal description, called **pseudocode**, specifies the operations to be performed.

The logical sequencing of these high-level modules is also considered. Some rules are used to avoid code with large and uncontrolled sequencing convolutions. Such code is sometimes called "spaghetti code" for obvious reasons. For example, each module should have a well-defined entry and exit point. Entry into the middle of a module is not allowed. Certain structures for looping and conditional operations are also specified.

Each high-level module is, in turn, further divided into smaller modules. The process continues until it is obvious how to implement the requirements in the chosen language. Each module is programmed and tested individually. Creating the complete program involves combining individual modules and testing bigger modules. The process continues until the entire project is finished, debugged, and documented.

As an example to illustrate this approach, consider the programming of a multiplication algorithm for BART. Multiplication is performed by repeated addition of the multiplicand. There are no provisions for accumulator overflow. It is assumed that BART has been modified to include the subroutine CALL and RET instructions. Initially, the multiplier and multiplicand are stored in locations MPLR and MCND. A product is produced and stored in location PROD. An algorithm based on repeated addition is to be used.

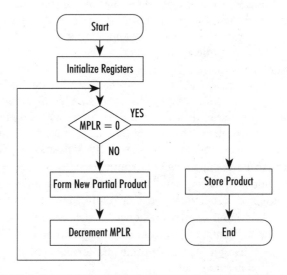

Figure 8.5   *Pseudocode for Multiplication Using BART Instruction Set*

Figure 8.5 is a flow chart of a possible pseudocode algorithm for the multiply operation. A convenient way to implement each module of pseudocode is using a subroutine. Subroutines, by their nature, are well suited for modular activities. They have well-defined entry and exit points. They can be compiled independently of the main program. They require the programmer to make provisions for passing information between the subroutine and the calling program. The following program illustrates the main program structure and the method of calling the various computational modules.

It should be noted in advance that the module size selected for the example is much smaller than would be used in a meaningful software development project. The complexity of the modules is also far lower than would normally be used. The example illustrates the software development technique but is not realistically scaled.

Main Program

For the multiplication example, each block of pseudocode is assigned to a subroutine. The subroutine modules are called INIT, TEST, PAR_PROD, DECRMNT, and STORE. The main program, as shown below, consists of subroutine calls and branching code. The main program is responsible for sequencing program blocks. It maintains the logical flow indicated by the pseudocode flow chart.

```
;Main Program for a Multiplication

;

START:          CALL    INIT            ;Call the initialize module.

LOOP:           CALL    TEST            ;Call the test module.

                BOS     Z, LAST         ;If multiplier is zero, terminate.

                CALL    PAR_PROD        ;Create a new partial product.

                CALL    DECRMNT         ;Decrement the multiplier.

                JMP     LOOP            ;Go back and try again.

                CALL    STORE           ;Store the product.

LAST:           JMP     LAST            ;Halt.
```

Each subroutine may then be further subdivided, if necessary, using additional modular structures involving subroutines. When the original blocks have been suitably decomposed, the code can be written. In this example, one level of decomposition is adequate.

Module INIT

```
INIT:           LDA     MDLR            ;Get multiplier from memory location.

                STA     R0              ;Put multiplier in R0

                CLA                     ;

                STA     R1              ;Clear R1 for partial product.

                RET                     ;
```

Module TEST

```
TEST:           LDA     R0          ;Load accumulator with multiplier to set
                RET                 ;condition flags.
```

Module PAR_PROD

```
PAR_PROD:       LDA     R1          ;Get partial product.
                ADD     MCAND       ;Add the multiplicand.
                STA     R1          ;Store new partial product.
                RET                 ;
```

Module DECRMNT

```
DECRMNT:        LDA     R0          ;Get the multiplier.
                ADD     #FF         ;ADD (-1).
                STA     R0          ;Store the multiplier.
                RET                 ;
```

Module STORE

```
STORE:          LDA     R1          ;Get partial product.
                STA     PROD        ;Store product in PROD.
                RET                 ;
```

There is a performance penalty in the use of highly modular structures. Each level of decomposition usually requires some overhead in the management of data to be passed between modules and in the execution of the subroutine CALL and RET instructions. The advantages gained usually outweigh the slight loss of system performance.

8.5   Program Development and Debugging Tools

In addition to an assembler, other software tools are also useful.

Editor/Word Processor

Program development and documentation requires a large amount of text input. A word processor capable of producing and editing text files is required. To assure compatibility with assemblers, the word processor should be able to produce ASCII text files.

Linker

A linker is a program sometimes used in conjunction with an assembler. If the main program and program modules have been developed separately, there is a strong likelihood that there will be contention for address space among modules. A linker is a software program which accepts input from two or more program modules and combines or links them into a single program. It must resolve all address conflicts. As an alternative to a linker, various program modules could be combined with a word processor before assembly. The assembler would then resolve all addressing of the complete program. A linker is especially useful when using library routines or software that was prepared using multiple source languages. The linker needs access to the assembled code produced by high-order compilers.

Simulator

Many programs are cross-compiled on computers other than the target system. This is especially true for situations involving embedded microprocessors. A **simulator** allows the code to be executed on a machine other than the target. Due to software overhead in the simulation, processing speed is usually much slower than would be achieved on the target processor.

Simulators usually permit the user to modify the contents of critical registers and memory locations. A visual display documents the operation of the computer on an instruction-by-instruction basis. Good simulators have provisions for executing code in a single-step manner to allow the user to closely investigate the processing of an algorithm. Backward execution allows the user to work backwards from an observed error condition to determine events which preceded the error.

Monitor/Debugger

Before a program can be run on the target system, it must be loaded into the program memory of the device. Small microprocessor systems often use a **monitor/debugger program** to allow the designer to load the program into the program memory of the target system, via a serial communication port, and then to control its execution. Due to software overhead, program execution during debugging is much slower than normal processor speed. Monitor commands allow the contents of registers to be examined and modified. Monitors also may allow the program to be executed with break points. This allows the user to specify locations in which the program should be suspended to allow inspection of the contents of critical registers. After debugging and testing, the extraneous hardware and software associated with the monitor are removed and the software is executed from ROM.

8.6   Representative Computer Architectures

The following sections present architectural information and instruction sets for four representative computer structures. The material presented is not intended to be exhaustive nor to emphasize high-performance features of new and complex devices, but rather to illustrate some of the architectural characteristics of real devices that have been highly successful and useful products. It also illustrates the range of instruction sets encountered in modern computers. The encoding of the instruction sets provides insight into the relationship between the hardware and software components of the machine.

- **Motorola 68000.** This 16/32-bit microprocessor was introduced in the late 1970s as a high-performance upgrade to the Motorola 6800 series of 8-bit microprocessors. Its relatively clean and straightforward structure made it appealing for a variety of applications requiring higher performance than existing 8-bit microprocessors. Apple Computer used the 68000 as a processor in its highly successful Macintosh series. The 68000 is the entry-level device into a broad family of high-performance 680x0 devices that preserve the core of 68000 instructions.

- **Intel 8086.** The 8086 16-bit microprocessor was introduced by Intel in the mid 1970s as a high-performance alternative to their 8080 and 8085 series of 8-bit microprocessors. The device achieved instant popularity when a version, the 8088, was selected for IBM's entry into the personal computer market. The 8086 is the entry device of a large family of 80x86 microprocessors. Newer devices, such as the Pentium, still maintain code compatibility with software written for the 8086/8088.

- **Intel 8051.** The Intel 8051 is described by Intel as a microcontroller. The range of applications is intended to be low-level control and monitoring functions rather than high-level number crunching. They are often used for embedded intelligence in consumer products and industrial electronics. The 8051 is a complete computer on a chip; it has provisions for on-chip data memory, program memory, and input/output. There are currently more 8051s in use than any other microprocessor.

- **Advanced Micro Devices (AMD) 29000.** The AMD 29000 is representative of a reduced instruction set computer, RISC. This 32-bit machine uses extensive pipelining of the instruction set processing to achieve a processing rate close to one instruction per clock cycle. The manufacturer claims performance of 20 million instructions per second using a 30 MHz clock. This popular processor has found extensive applications in embedded control applications, especially high-performance laser printers. It supports high-level languages such as C, Fortran, Ada, and Pascal.

The Motorola 68000 and Intel 8086 are entry-level devices into a large and diverse family of computing devices which range from the modest performance of the entry-level devices to increasingly high-performance devices introduced at two- to three-year intervals.

8.7   The Motorola 68000

The Motorola 68000 was first introduced in 1979. It was the first member of a large family of 680x0 microprocessors, which have evolved to incorporate increasingly high-performance hardware features. The 68000 was the first microprocessor to incorporate 32-bit internal registers. It was also the first to provide a non-segmented address space of 2^{24} = 16M bytes. With 32-bit internal address registers, provisions were already in place for successor devices to access an even larger memory range.

Registers

Figure 8.6 shows the programming model of the 68000. Its features include eight general purpose 32-bit data registers (D_0–D_7) and seven general-purpose 32-bit address registers (A_0–A_6). The designers of the 68000 saw the importance of supporting an operating system in addition to a programming environment. This was achieved by allowing two levels of access to machine resources: user and supervisor. Each mode has its own stack pointer, A_7. Upon initialization, the machine is in the supervisor mode. After initializing necessary system resources, the mode is changed to user to accommodate user programs. In dedicated systems this distinction is usually ignored and all programs run in the supervisor mode.

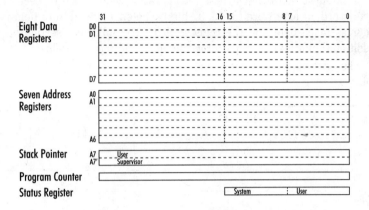

Figure 8.6   Register Set of the Motorola 68000

The eight data registers support data operands of 8-, 16-, or 32-bits. The address registers support 32-bit addresses. In smaller systems, only the least significant 24-bits are available for external addressing. This provides a 16M byte address space.

Memory

Memory is organized as bytes in the absolute address space. Instructions can access data as bytes (8 bits), words (16 bits), or long words (32 bits). A single 32-bit long word can also be accessed as eight 4-bit binary-coded decimal characters.

Instruction Set

The 68000 has 56 basic instructions. Many of them can be modified by selecting an addressing mode. The instruction set includes the following classes of operations:

- Move
- Integer arithmetic (Add, Subtract, Multiply, Divide, Compare)
- Binary-coded decimal operations (Add BCD, Subtract BCD, Negate BCD)
- Logic operations (And, Or, Not, Exclusive Or)
- Shift and rotate (Logical Shift Left/Right, Arithmetic Shift Left/Right, Rotate Left/Right)
- Bit test and manipulation (Bit Test/Clear)
- Jump and branch (Jump, Jump to Subroutine, Branch Conditional/Unconditional)

Table 8.5 shows the instruction format for a single operand instruction, such as the logical complement operation NOT, is as follows:

15 Operation Code 01000110 8	7 Size 00—Byte 01—Word 10—Long 6	5 Effective Address 0	
		Mode	Register

Table 8.5

The most significant byte is the operation code. Bits 7,6 specify if the operation is to affect a byte, word, or long word (8, 16, 32 bits). The effective address field is divided into two fields: the address mode field and the register field. The following section on addressing modes provides additional information concerning the mode and register fields.

For two-operand instructions such as ADD, a binary addition, the instruction format is as follows:

15 12	11 9	8 6	5 0
Op Code 1101	Register	Op-Mode	Effective Address
			Mode	Register

Table 8.6

The register field, bits 11–9, specifies one of the eight data registers. Bits 8–6, the 3-bit Op-Mode field, indicates the final destination of results and the size of the operands. The variations are shown in Table 8.7.

Op-Mode			Operation
Byte	Word	Long Word	
000	001	010	$R_n \leftarrow R_n$ + Contents of Effective Address
100	101	110	Effective Address $\leftarrow R_n$ + Contents of Effective Address

Table 8.7

Bits 5–3 indicate the address mode; bits 2–0 specify the second data register.

Addressing Modes

The available combinations of size, mode, and register fields allow many variations on the instruction set. Table 8.8 shows the addressing modes possible with the mode and register options of the effective address field. In direct addressing, the location of the operand is specified directly by the instruction. The operand may be in a register (register-direct addressing) or it may be in a memory location specified by the instruction (direct addressing). In indirect addressing the operand is always in the data memory. The exact location of the operand is specified indirectly by indicating the register that holds the address (register-indirect addressing) or a location in memory that holds the address of the operand (indirect addressing).

Addressing Mode	Mode	Register	Location of Operand
Data Reg Direct	000	Reg Num	Operand is in the specified data register.
Addr Reg Direct	001	Reg Num	Operand is in the specified address register.
Addr Reg Indir	010	Reg Num	Address of operand is in the specified address register.
Addr Reg Ind, Postinc	011	Reg Num	Address of operand is in specified address register; address register in incremented (1, 2, 4) after data access.
Addr Reg Ind, Preced	100	Reg Num	Address of operand is in specified address register; address register is decremented (1, 2, 4) before data access.
Addr Reg Ind, Disp	101	Reg Num	Address of operand is sum of address in specified address register and (signed) 16-bit displacement in following word.
Addr Reg Ind, Indexed	110	Reg Num	Address of operand is sum of address in specified address register plus (signed) lower byte of the following word plus the index register.
Absolute Short	111	000	Address of operand is in the following 16-bit word.
Absolute Long	111	001	Address of operand is the concatenation of the following two 16-bit words, giving a 32-bit address.
Prog Cntr, Disp	111	010	Address of operand is sum of program counter and following (signed) 16-bit word.
Prog Cntr, Indexed	111	011	Address of operand is sum of program counter plus (signed) lower byte of following word plus index register.
Immediate	111	100	Operand is in lower byte, or word, or two words of following locations (byte, word, or long word specified by size field).

Table 8.8

The wide variety of addressing modes and instructions present a challenging environment for an assembly language programmer. Designers for the 68000 apparently envisioned a machine that would have most of its programs written in a high-level language.

Several variations on the 68000 family continue to provide high-performance computing devices for new personal computers. Each contains all of the features of the previous family member. The 68000 family is used in the Apple Macintosh series, the Commodore Amiga, and in Sun Microsystems work stations.

- **68008.** This is an 8-bit microprocessor that shares the internal structure with the 16/32-bit 68000 but uses 8-bit external components such as memory. This allows fabrication of lower cost devices at slightly decreased performance.
- **68010.** The 68010 has added features that help implement virtual memory. It also has added the ability to operate in a loop mode in which a small set of instructions may be repetitively executed without performing extra fetch cycles.
- **68020.** This is a full 32-bit microprocessor with 32-bit external address bus and data bus. Internally the ALU also operates on full 32-bit operands. An internal cache memory of 256 bytes is capable of storing instructions fetched in advance of their use. This significantly shortens the effective fetch cycle. The microprocessor is also designed to use an external floating-point arithmetic processor.
- **68030.** A 32-bit microprocessor with improved high-performance memory management.
- **68040.** A floating-point data processor has been included on-chip.

8.8   The Intel 80x86

In 1975, Intel introduced the 8086 microprocessor. Although in numeric sequence with its 8-bit 8085 microprocessor, the 8086 provided a greatly expanded instruction set and 16-bit processing. The 8088 shares the same internal structure as the 8086 but uses an external 8-bit data bus. IBM selected the 8088 processor for its initial personal computer, which was introduced in August 1981. A contributing factor was the ability to use existing off-the-shelf memory and support chips developed for 8-bit microprocessors.

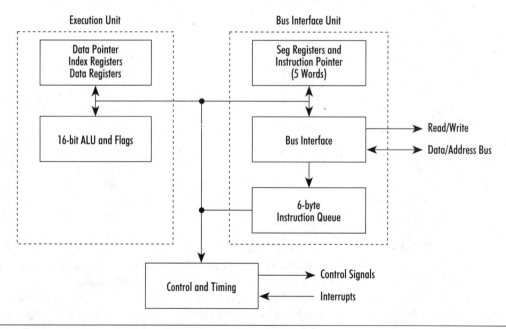

Figure 8.7   Diagram of the Intel 8086 Processor

Figure 8.7 shows the block diagram of the 8086. Note the division of the diagram into two parts. The execution unit, which contains the processing elements of the device is identical in the 8086 and 8088. The bus interface unit is responsible for fetching operands and instructions from the system memory. It is specialized for an 8-bit data bus in the 8088 and a 16-bit data bus in the 8086.

A special feature of the bus interface unit is the 6-byte instruction queue. The bus interface anticipates the need for instructions from locations in advance of the current program counter. It fetches these memory locations during free bus cycles and stores them in the prefetch queue. The execution unit can then obtain instructions from the prefetch queue and avoid the delay associated with a full fetch cycle. During branches and subroutine calls the instruction queue must be flushed to accommodate the new value of the program counter.

Registers

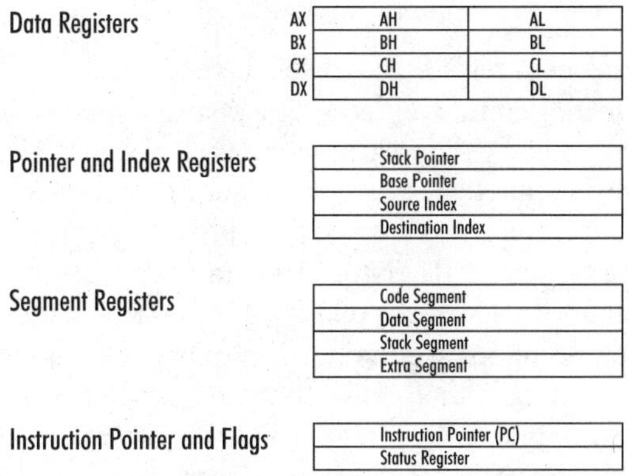

Data Registers			
	AX	AH	AL
	BX	BH	BL
	CX	CH	CL
	DX	DH	DL

Pointer and Index Registers	
	Stack Pointer
	Base Pointer
	Source Index
	Destination Index

Segment Registers	
	Code Segment
	Data Segment
	Stack Segment
	Extra Segment

Instruction Pointer and Flags	
	Instruction Pointer (PC)
	Status Register

Figure 8.8   Register Set of the Intel 8086

Figure 8.8 shows the register set of the 8086. The four 16-bit data registers: AX, BX, CX, and DX can also be accessed as 8-bit registers. Sixteen-bit register AX, for example, is composed of 8-bit registers AH and AL. There are four pointer registers. The stack pointer functions with the memory stack. The base pointer locates the bottom of the stack. Two index registers, useful for moving data blocks from one location to another, are included.

Memory

Within the 8086, all addresses are stored in 16-bit registers; they are capable of accessing a memory range of $2^{16} = 64K$ bytes. External to the 8086, the address bus accommodates a 20-bit address capable of a memory range of $2^{20} = 1M$ byte. The expansion from 16 bits to 20 bits is accomplished by use of the segment registers. The magnitude of each of the segment registers is expanded by appending four zeros to the end of the register. This effectively multiplies the values of the segment registers by 16 before they are added to the appropriate offset address. The segment registers can span the entire 1M byte address space but can only specify locations 0, 16, 32, 48, ... , multiples of 16. The overall memory space of 1M bytes is segmented into regions of 64K bytes as shown in Figure 8.9.

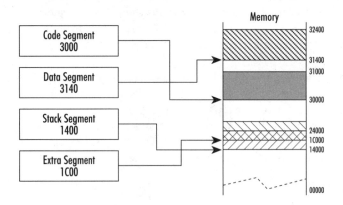

Figure 8.9   Use of Segment Registers in Accessing the 8086 Memory Space

The four segment registers are used as follows:

- **CS—Code Segment Register.** All instructions are located in a segment with a starting address specified by the code segment (CS) register. The effective address of an instruction is provided by adding the 16 bits (with four appended 0s) of the CS register to the 16 bits of the instruction pointer (Intel's name for a program counter).

- **DS—Data Segment Register.** All data, except stack references, are located in memory and are accessed with addresses relative to the DS register (with four appended 0s).

- **SS—Stack Segment Register.** All memory references using the stack, such as PUSH, POP, and subroutine call, are relative to the SS register (with four appended 0s).

- **ES—Extra Segment Register.** The ES register is used for string operations and other operations at the discretion of the programmer.

In Figure 8.9 the extra segment and stack segment partially overlap. Information stored in the overlapping region can be accessed by two different addressing mechanisms.

The segmented memory space of the 8086 is in contrast to the **linear** (nonsegmented) memory space used in the 68000. For systems using large amounts of memory, managing a linear space is less complex. Although it is awkward to manage the memory space in blocks of 64K bytes, segmented memory offers many performance advantages over the smaller memory spaces offered by lower performance microprocessors.

Instruction Set

The 8086 contains an assortment of data move, arithmetic, logic, and program control instructions. There are bit, byte, word, and block operations using 24 different operand addressing modes. The internal hardware can perform a 16-bit multiply and divide as well as less complex arithmetic and logic operations.

The instruction format for the single operand 8086 NOT operation (logical complement of the operand) is as shown in Table 8.9.

7	1	0	7 6	5 3	2 0
Operation Code = 1111011		W	MOD	010	R/M

Table 8.9

The W field indicates the width of the operand to be used.

W	Width
0	Data element is a byte.
1	Data element is a 16-bit word.

Table 8.10

The MOD field indicates the type of displacement which is to be used in calculating the effective address.

MOD	Displacement
00	Memory addressing; no displacement.
01	Memory addressing; next byte is a (signed) 8-bit displacement.
10	Memory addressing; next word is a (signed) 16-bit displacement.
11	The R/M field is used to specify a register.

Table 8.11

Table 8.12 shows the instruction format for the two-operand 8086 ADD operation.

7	2	1	0	7 6	5 REG 3	2 R/M 0
Operation Code = 00000		D	W	MOD	REG	R/M

Table 8.12

The D field indicates whether the reg field is a source of an operand or a destination for results.

D	Comment
0	Operand is from a register.
1	Result goes to a register

Table 8.13

When the MOD field contains 11, the register involved in the instruction is specified by the 3-bit code in the R/M field.

R/M Field Register Code	Byte Operation	Word Operation
000	AL	AX
001	CL	CX
010	DL	DX
011	BL	BX
100	AH	SP
101	CH	BP
110	DH	SI
111	BH	DI

Table 8.14

When the MOD field does not contain 11, the 2-bit MOD field and the 3-bit R/M field collectively specify one of the other three addressing modes: immediate, direct, and register-indirect. The MOD field indicates how a displacement is calculated. The R/M field indicates an assortment of base registers and index registers that are combined with the displacement to specify the actual memory location.

Example 8.2:   Encoding a Register-to-Register ADD Instruction

What is the encoding of the instruction:

```
ADD    AX, BX                    ;AX←AX + BX
```

Solution:

This is a register-to-register operation, therefore MOD = 11. The operands are 16 bits therefore W = 1. When D = 1, register AX is a destination register for the result. The entire instruction then appears as:

Most-significant Byte

Bits 7–2 = 00000   The Op Code for ADD.
Bit 1 = 1   The first operand will be the destination.
Bit 0 = 1   16-bit data.

Least-significant Byte

Bit 7,6 = 11   Register-to-Register operation.
Bit 5–3 = 000   Specify the code for AX.
Bit 2–0 = 011   Specify the code for BX.

The complete instruction is 0000011 11000011 = $03C3_{16}$.

Of the four processors presented in this chapter, the 8086 provides the most addressing modes. Fortunately, most of the complexity of the addressing is obscured by using a good assembler.

Like the 68000 family, the 8086 family has expanded dramatically. Newer versions of the processor have maintained the ability to process code written for the 8086. Some of the variations including the following:

- **80186.** This chip includes the functions of many peripheral components within the chip of the 80186. This includes the clock circuitry, programmable timers, an interrupt controller, and chip-select logic. The goal was to create a single chip requiring only external memory and I/O devices to make a complete computer.

- **80286.** The 286 was created to offer improved performance. Additional instructions are included. Provisions for on-chip memory management necessary for virtual memory implementation are available. The address and data buses are no longer multiplexed; this allows faster acquisition of data and instructions.

- **80386.** The 386 processor provides a full 32-bit data path for instructions and data. It includes an on-chip memory management unit which supports virtual memory. Instructions are executed in a pipelined fashion to improve performance.

- **80486.** The 486 provides an on-chip cache memory to allow very fast access to frequently used data and instructions. It also has an on-chip floating-point arithmetic processor.

- **Pentium.** Introduced in 1993, Pentium preserves code compatibility with its family tree yet offers performance improvements in hardware. Pentium includes two pipelined integer ALUs, an improved pipelined floating-point unit, and improved memory management features.

8.9   The Intel 8051

The Intel 8051 is the core member of a family of microprocessors identified as the MCS-51 family. The 8051 differs from traditional digital computers in several respects. It is an 8-bit microprocessor designed for control applications rather than numeric data processing. Because of its intended uses and extensive input/output facilities, this type of device is called a **microcontroller**. It is a little-known fact that the majority of computer applications use microcontrollers. They are embedded in applications ranging from automobiles to robots to consumer appliances.

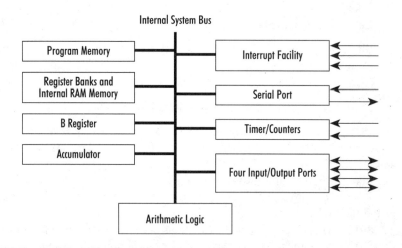

Figure 8.10   Programmer's Model of Intel 8051 Processor

The 8051 is a single-chip computer; all necessary program memory, data memory, and input/output facilities are contained within a single package. Provisions exist for using additional external components to expand the 8051 into a conventional bus-oriented microprocessor. Figure 8.10 shows a programmer's view of the 8051.

Registers

A single 8-bit accumulator is the major processing register of the 8051. It is the source and/or destination register for most arithmetic, logic, and memory-exchange operations. An 8-bit B register supplements the accumulator for operations such as multiply and divide, where a double-length result or operand is encountered.

An internal register file of 128 bytes, the on-chip data memory, provides data storage and the stack. The lower portion of this memory is configured as four register banks, each of eight registers, R_0 through R_7. The instruction set supports extensive operations with these registers.

A major advantage of using the register banks is the ability to change them under program control. While executing one program using the first register bank, for example, the device can change to a second register bank to service an interrupt. The contents of the first bank will remain unchanged and not require storage while the interrupt is being serviced. Upon completing the interrupt, the original register bank can again be selected. The ability to switch register banks greatly reduces the amount of overhead associated with multiple processing tasks.

Memory

The 8051 uses a Harvard architecture; the memory is partitioned into a data memory and a program memory. The data memory consists of the 128 bytes of the internal register file and an optional 64K bytes of RAM external to the chip. The register banks may be accessed directly by referencing a register in the instruction. The remaining internal memory can be accessed using register-indirect addressing. R_0 and R_1 of the active register bank are used as pointers for indirectly accessing the remaining registers in the register file.

Program memory consists of 4K bytes of on-chip ROM; factory programming is required. A companion device, the 8751, provides the on-chip memory as user-programmable EPROM. A third variation, the 8031, requires that all program memory be external to the chip. All versions of the device allow program memory to be expanded up to 64K bytes using external memory components.

Input/Output

The 8051 has four 8-bit input/output ports. These are supported directly by the instruction set. Two of these ports are converted to a system bus if the system is expanded to include external memory or I/O components.

Using pins from one of the 8-bit I/O ports, the 8051 can be configured with an RS 232-compatible serial port capable of simultaneously receiving and transmitting standard 8-bit characters. The serial facility performs all transformations between serial and parallel data. It manages synchronization and the start and stop bits associated with the serial information. A wide variety of data rates can be used.

Two internal timers are available. They can also be configured as counters that respond to external signals. One of the timers is committed to the serial port during data transmission or reception.

Interrupts

Due to the need to provide extensive input/output facilities, the 8051 uses five interrupts. An interrupt causes a program in execution to pause while an interrupt service routine responds to the cause of the interrupt. In the 8051, interrupts can be generated by timer overflows from Timer 0 and Timer 1, two external interrupts, and an interrupt associated with the serial port. Two levels of interrupt priority can be managed. The topic of interrupts is covered in greater detail in Chapter 10, which is concerned with input/output techniques.

Special Features

Since the 8051 is intended for a wide range of control activities, much emphasis has been placed on allowing customization of the device to its application. The data direction of each parallel input/output bit, the baud rate of the serial port, control of the timer/counter, and the configuration of the interrupt facility can all be customized. The programmer can specify the personality of the device through the use of a set of special function registers. These registers are initialized, under program control, to allow the 8051 to provide the desired characteristics.

Instruction Set

The 8051 instruction set includes 51 instructions that can be divided into five functional groupings. Tables 8.15–8.19 show an instruction set summary. The instruction set contains a good assortment of operations using register addressing, direct addressing, and register-direct addressing. An assortment of Boolean, single-bit operations is especially well suited for examining and testing the status and decision variables associated with control and monitoring operations.

Instruction Mnemonic	Description
ADD	Add operands.
SUBB	Subtract operands.
INC	Increment operand.
DEC	Decrement operand.
MUL	Multiply operands (8-bit operands, 16-bit product).
DIV	Divide operands (16-bit dividend, 8-bit quotient. 8-bit remainder).
DA	Decimal adjust.

Table 8.15

Mnemonic	Description
ANL	Logical AND of bits of operands.
ORL	Logical OR of bits of operands.
XRL	Logical Exclusive OR of bits of operands.
CLR	Clear the accumulator.
CPL	Complement the accumulator.
RL	Rotate accumulator to the left.
RR	Rotate accumulator to the right.
SWAP	Swap the nibbles of the accumulator.

Table 8.16

Mnemonic	Description
MOV	Move information from source to destination.
POP	POP contents from the stack to the accumulator.
PUSH	PUSH the accumulator on the stack.
XCH	Exchange the contents of the two locations.

Table 8.17

Mnemonic	Description
CLR bit	Clear specified bit.
SETB	Set specified bit.
CPL bit	Complement specified bit.
JB bit	Jump if specified bit is set.
JNB bit	Jump if specified bit is not set.

Table 8.18

Mnemonic	Description
CALL	Subroutine call.
RET	Subroutine return.
JMP	Unconditional jump.
JZ	Jump if accumulator is 0.
JNZ	Jump if accumulator is not 0.
CJNE	Compare operands; jump if not equal.
DJNZ	Decrement register; jump if 0.
NOP	No operation.

Table 8.19

Addressing Modes

Most of the 8051 instructions are encoded into a single 8-bit instruction. A second and third byte are added to accommodate immediate and direct addressing of operands. To illustrate some of the instruction formats, consider the ADD instruction. One of the operands is assumed to be in the accumulator. The sum is returned to the accumulator. Several addressing mode variations are possible for specifying the location of the second operand. For indirect addressing, code 0110 uses register R_0; code 0111 uses register R_1.

7 4	3 0	7 0
Op Code = 0010	Addressing Mode	Second Byte, used for immediate and direct addressing.

1RRR	Register
0100	Immediate
0101	Direct
011i	Indirect

Table 8.20

Examples

- **Register addressing.** In the register addressing mode, the RRR contains the code specifying one of the eight registers in the current register bank. No second byte is needed.

Binary	Hexadecimal	Assembly Statement	Comment
00101101	2D	ADD A, R5	ADD register 5 to accumulator.

Table 8.21

- **Immediate Addressing.** In this mode, the second byte of the instruction contains the operand for the instruction.

Binary	Hexadecimal	Assembly Statement	Comment
00100100	24	ADD A, #65H	ADD 65H to accumulator.
01100101	65		

Table 8.22

- **Direct Addressing.** The second byte contains the address of the operand. Since the 8051 has a maximum of 256 memory locations for on-chip data memory, 8 bits is sufficient to locate the operand.

Binary	Hexadecimal	Assembly Statement	Comment
00100101	25	ADD A, 7BH	ADD contents of on-chip memory 7B to accumulator.
01111011	7B		

Table 8.23

- **Indirect Addressing.** The 8051 uses register 0 and register 1 as locations for indirect addresses.

Binary	Hexadecimal	Assembly Statement	Comment
00100110	26	ADD A, @R0	ADD the contents of on-chip memory location specified by R_0 to the accumulator.

Table 8.24

The 8051 has had a long and useful life. By using its single-chip configuration, many manufacturers have been able to implement sophisticated control devices with a programmable structure rather than specialized hardware. Modifications to the 8051 have added features, such as more sophisticated communication ports servicing local area networks. The 8051, because of its useful control features and well developed set of software tools, has also been used as an embedded processor for application-specific integrated circuits (ASICs).

8.10   Advanced Micro Devices AMD 29000 (29K)

The AMD 29000 (29K) is the entry-level device in a series of 32-bit RISC processors. Figure 8.11 shows a system diagram of the 29K. A four-stage pipeline is used to process instructions. Although four clock cycles are required to process each instruction, pipelining allows an effective processing rate of one instruction per clock cycle. Pipeline-dependency problems, which were discussed in Chapter 7, are managed by processor hardware. They do not require any action by the programmer or the compiler.

Two high-speed buses allow simultaneous access to the data and instruction memory. To further improve memory access, the 29000 maintains an on-chip branch-target cache to allow program execution of branch looping operations completely from cache memory.

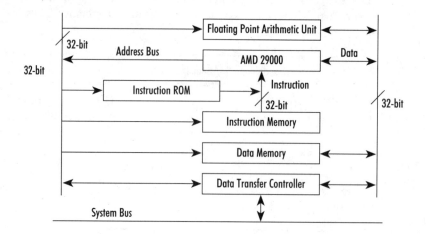

Figure 8.11   Programmer's Model of AMD 29000 RISC Processor

Registers

Absolute Register #	General-purpose Reference
0	Indirect Pointer Access
1	Stack Pointer

2 – 63 not implemented

Global Registers

64	Global Register 64
65	Global Register 65
66	Global Register 66
127	Global Register 127

Local Registers   Stack Pointer = 130

128	Local Register 126
129	Local Register 127
130	Local Register 0
131	Local Register 1
254	Local Register 124
255	Local Register 125

Figure 8.12   Register Structure of AMD 29000 RISC Processor

The 29K uses 192 general-purpose 32-bit registers. These registers can be used to hold addresses or data. Data can be represented in a wide variety of integer and floating point formats. The register organization is shown in Figure 8.12. Registers 0 and 1 are assigned to specific duties as pointers. Registers 2 through 63 are not implemented.

Registers 64 through 127 are global registers. They can be accessed by direct reference to their register number.

Registers 128 through 255 are local registers. All references to the local registers are displaced by the value of a local register stack pointer. By modifying the value of the register stack pointer, it is possible to change the absolute register reference requested by software. This is a form of register windowing and is especially useful for passing data to subroutines and servicing interrupts.

The 29K also contains 23 special-purpose registers. These registers provide control and data for some of the processing actions. For example, special-purpose registers are dedicated to processor status, register bank protection, multiple program counters, and timers.

Memory

With a 32-bit address bus, the 29K supports 4 gigabytes of memory. A sophisticated on-chip memory management unit assists in data and instruction access through techniques such as cache memory and virtual memory. These concepts will be explored in greater detail in Chapter 9.

Instruction Set

Similar formats are used for the 29K's 112 instructions. All except those loading or storing in the memory can be pipelined to execute in one clock cycle. Instructions are 32 bits long and are divided into four fields, as shown in Table 8.25.

31 24	23 16	15 8	7 0
Operation code	Register C (RC)	Register A (RA)	Register B (RB) or immediate data

Table 8.25

The register references in an instruction can be to either a global register or a local register. The right-most field can contain either a register number or data to be used in an immediate instruction. If the last bit of the operation code is 0, it is a register reference; if 1, it contains immediate data. Immediate data fields with positive numbers are filled out with 24 zeros to produce a 32-bit value. Negative numbers are expanded with twenty-four 1s.

While there is no NOT instruction in the 29K, the operation can be performed with a logical NAND with an operation code of 9B. If the value to be complemented is in global register 64, the instruction would appear in hexadecimal as shown in Table 8.26.

HEX Code	Assembly Code	Comment
9B6464FF	NAND RC, RA, FFH	Register C receives the contents of register A NANDed with the 32-bit value FFFFH.

Table 8.26

There are several variations of the ADD instruction. Different operation codes are used to distinguish between ADD with and without carry, signed and unsigned. The instruction to perform an unsigned add, without carry of the contents of global register 64 and 65 with the sum stored in location 66, would appear as in Table 8.27. The operation code is 14.

HEX Code	Assembly Code	Comment
14666564	ADD RC, RB, RA	Register C receives the 32-bit sum of register A and register B.

Table 8.27

Because of the nature of RISC processors, there are no complex addressing modes. Such modes would slow down the processor and require additional processing to compute the effective address.

Summary

A digital computer can be programmed at several different levels. Machine language and assembly language provide maximum control of the features of the machine. In many cases the programming efficiency gained by using higher level compilers compensates for the lower performance of programs written in high-level languages. For most applications, a useful mix of high-level code and assembly-level code provides good performance and high programming efficiency.

Assembly language provides an interesting view into the design of a computer. The manner in which the instruction set is coded provides a great deal of information concerning the internal structure of the machine.

The assembler provides a method of transforming assembly code written by the programmer into machine code that can be executed by the computer. Assembler directives allow the programmer to allocate memory, define names, and control other aspects of program execution. Good assembly code is well documented and modular.

Four representative machines—Motorola 68000, Intel 8086, Intel 8051, and Advanced Micro Devices 29000—provide insight into the register configuration and instruction set selection of a modern digital computer.

For additional information and references on the topics presented in this chapter, access the World Wide Web page provided at: http://www.ece.orst.edu/~herzog/docs.html.

Key Terms

application software
assembler
assembler directives
assembly language code
compiler
conditional assembly
linear memory
linker

macros
microcontroller
monitor/debugger program
porting
pseudo-operations
pseudocode
simulator
system software

Readings and References

Motorola MC68000 16-Bit Microprocessor User's Manual

Slater, Michael. *Microprocessor-Based Design*. Englewood Cliffs, NJ: Prentice Hall, 1989.

Intel Microprocessor Handbook: 1992

AM29000 32-Bit Streamlined Instruction Processor User's Manual

Comer, David J. *Microprocessor-Based System Design*. New York: Holt, Rinehart and Winston, 1986.

This book has material on both the Intel 8086 and Intel 8051. It is especially involved with hardware aspects of the devices.

Exercises

8.1 Why are cross assemblers, which run on computers different from the target for the program, important? What advantages do they have?

8.2 What is the importance of each of the following to an individual who will be using a computer in business applications such as keeping employee records and producing payroll checks? Justify your answer.

(a) Ability to run programs written by others.

(b) Ability to write programs in high-level languages.

(c) Ability to write programs in assembly language.

(d) Ability to write programs in machine language.

(e) Ability to design computers with special instruction sets.

8.3 What is the importance of each of the following to an individual who will be using a computer to gather and analyze data from specialized electronic instrumentation? Justify your answer.

(a) Ability to run programs written by others.

(b) Ability to write programs in high-level languages.

(c) Ability to write programs in assembly language.

(d) Ability to write programs in machine language.

(e) Ability to design computers with special instruction sets.

8.4 What is the importance of each of the following to an individual who will be using a computer in an robot-control application requiring extremely high performance? Justify your answer.

(a) Ability to run programs written by others.

(b) Ability to write programs in high-level languages.

(c) Ability to write programs in assembly language.

(d) Ability to write programs in machine language.

(e) Ability to design computers with special instruction sets.

8.5 The term *computer literacy* is common in today's educational vocabulary. What skills, related to computers, are required by each of the following groups?

(a) A medical physician.

(b) An accountant.

(c) A technician who repairs scientific instrumentation.

(d) A computer scientist managing a corporate computer facility.

(e) A computer engineer using embedded controllers in specialized applications.

8.6   In the two-pass assembler, what action is caused by the ORG instruction?

8.7   Under what conditions would it be possible to implement an assembler with a single pass of the program?

8.8   A disassembler performs the inverse process to an assembler. Starting with machine code, it generates assembly code. How can data and instructions be distinguished by a disassembler? How can labels be determined?

8.9   Which is likely to run faster, an algorithm coded in a high-level language or the same algorithm coded in assembly language? Why?

8.10   Which is likely to require less program memory, an algorithm coded in a high-level language or the same algorithm coded in assembly language? Why?

8.11   Most high-level compilers allow portions of the code, such as subroutines, to be written in assembly language. Why is this an important feature?

8.12   What are some of the trade-offs to be considered when making the choice of a programming language for a specific application?

8.13   How are labels represented in machine code?

8.14   What are some of the trade-offs to be considered when making the choice of a computer for a specific application?

8.15   Would it be possible to write a C compiler for a Turing machine? Extra credit: Write such a compiler.

8.16   Why must assembly language programs be written using ASCII text symbols. Would it be possible to create an assembler that would work with word processor files?

8.17   Compilers and assemblers are computer programs. What language must be used to write them?

8.18   Could an assembler for a microprocessor be written using a high-level language such as C? Explain your answer.

8.19   What is a macro and why are they useful?

8.20   What is the difference between a macro and a subroutine?

8.21   Why do many computers not have a HALT instruction?

8.22   Assemble the following code, which is written for BART. Include a listing of the program labels and their values.

```
              ORG    30H
     START    LDA    ONE
              STA    R0
              CLA
              ADD    @R0
              OR     #5AH
              STA    TWO
     HERE:    JMP    HERE
     ONE:     0CCH
     TWO:     DB     11H
              END
```

8.23 Disassemble the code in Table 8.28, which was written for BART. A hexadecimal listing is provided. It is known that the code starts at location 0.

Memory Location	Contents
00	F8
01	44
02	10
03	38
04	F4
05	39
06	0D
07	38
08	80
09	06

Table 8.28

8.24 What is the effect on program performance and efficiency if modular structures are used instead of a single monolithic program?

8.25 When using program modules, why are subroutine calls rather than jump statements used?

8.26 The 68000 uses separate registers for data and addresses. What are some possible advantages and disadvantages to doing this?

8.27 Most computers allow addressing relative to the program counter. What is the advantage of programming with this addressing mode?

8.28 Assemble the following code for the 68000. Assume that 16-bit addresses are adequate. All ADD operations are to use a 16-bit word. Use the absolute short addressing mode.

```
ORG    1000H
ADD    R0, #50H          ; Assume 16-bit words.
ADD    R4, R5
ADD    R2, @R0
```

8.29 Assemble the following code for the 68000. Assume that 16-bit addresses are adequate. All ADD operations are to use a 16-bit word. Use the program counter displacement addressing mode.

```
ORG    1000H
ADD    R0,#50H           ; Assume 16-bit words.
ADD    R4,R5
ADD    R2,@R0
```

8.30 Assemble the following code for the 68000. Assume that 16-bit addresses are adequate. All ADD operations are to use a 16-bit word. Use the program counter displacement addressing mode.

```
ORG    0FFABH
ADD    R0,#50H           ; Assume 16-bit words.
ADD    R4,R5
ADD    R2,@R0
```

8.31 Assemble the following code for the 8086. Assume that 16-bit addresses are adequate. All ADD operations are to use a 16-bit word. Use the direct addressing mode.

```
ORG    1000H
ADD    R0,#50H           ; Assume 16-bit words.
ADD    R4,R5
ADD    R2,@R0
```

8.32 Assemble the following code for the 8086. Assume that 16-bit addresses are adequate. All ADD operations are to use a 16-bit word. Use the program counter relative addressing mode.

```
ORG    1000H
ADD    R0,#50H           ; Assume 16-bit words.
ADD    R4,R5
ADD    R2,@R0
```

8.33 The 8086 uses a 16-bit data bus. The 8088 uses an 8-bit data bus. If the same clock rate is used by both, the 8086 is not twice as fast as the 8088. Why?

8.34 If the prefetch queue in the 8086 and 8088 were disabled, would the 8086 execute a program twice as fast as an 8088? Justify your answer.

8.35 The 8086 uses a 6-byte prefetch queue. Under what conditions does this work well? Under what conditions does it work poorly?

8.36 Assemble the following code for the 8051. Assume that 16-bit addresses are adequate. All ADD operations are to use an 8-bit word. Use the direct addressing mode.

```
ORG    1000H
ADD    A,#50H            ; Assume 8-bit words.
ADD    A,R5
ADD    A,@R0
```

8.37 Assemble the following code for the 29000. Assume that 16-bit addresses are adequate. All ADD operations are to use an 32-bit word.

```
ORG    1000H
ADD    R0,#50H           ; Assume 32-bit words.
ADD    R4,R5
```

8.38 The 29000 lacks the rich assortment of address modes used on the 68000 and 8086. What is the reason for this?

8.39 The 68000 and 8086 are two-address computers. Instructions are capable of specifying two locations to specify the source or destination of operands/results. The 29000 instruction format has room for specifying three addresses, all of which are registers. Does a three-address computer have any advantages over a two-address computer? What are they?

8.40 BART is a one-address computer. One of the operands is assumed to reside in the accumulator. Does this type of structure have any advantages over a two-address machine? What are they?

8.41 Is a zero-address computer possible? If so, how would operands be located and results stored?

8.42 Compare the local register organization of the AMD 29000 with the register windows described in Chapter 7.

9
Memory Structure and Management

Howard Aiken

Computing began to awaken
At Harvard with Howard Aiken
His components were cruder
Than a modern computer
With relays makin' and breakin'

Prologue

Howard Aiken, a graduate student in physics at Harvard, was one of the first to see the possibility of performing arithmetic and logic operations using electromechanical relays. He authored a paper in 1937 indicating the requirements of such a computing device. It was to be able to handle both positive and negative numbers. It would perform several different calculations. It was to operate fully automatically; no human interaction was needed.

Due to the earlier work of Hollerith and IBM, the technology existed for performing calculations with data stored on punch cards. Such machines were useful for commerce but lacked the performance level necessary for serious scientific calculation.

Aiken headed a Harvard design team funded by IBM. The result was the Harvard-IBM Mark I computer. This was an enormous machine with over 760,000 parts and 500 miles of wire. The Mark I used 72 electro-mechanical ratchet counters for storage. The numerical content advanced with each electrical pulse applied. Numbers were 24 bits with one bit used for a sign. An additional 60 registers were constructed of mechanical switches. They were used for manually entering information. Automatic input to the machine was by punched paper tape. Instructions were composed of three fields: source operand, destination of results, and operation to be performed.

The machine was considered fast by standards of the day. An addition or subtraction required one-third of a second. A multiplication needed five seconds and a division took sixteen seconds. A logarithm to 20 decimal places could be computed in 90 seconds. The Mark I was a success, but its usefulness was quickly challenged by faster and more powerful electronic computers.

9.1   Introduction

Reliable, high-performance digital memory devices are essential to developing high-performance computing systems. A survey of the historical techniques used for data storage illustrates both the ingenuity and frustration caused by a search for the ideal memory implementation. Babbage proposed a computer relying on mechanical storage. Howard Aiken, with the Harvard Mark I, used electromechanical relays to perform both logic and memory functions.

The need for increasingly larger memories in digital computers is well documented. John von Neumann stated in 1946 that 4096 words of memory exceeded a computer's need to solve most problems. He didn't foresee the dramatic technological improvements that would make large memories useful, feasible, and necessary.

Today's computer designer has access to many memory devices. Each has highly individual methods for data storage and data access. Each has different size, performance characteristics, and cost. The primary performance consideration for memory is speed. The access time, t_A, is defined as the elapsed time between a request for memory data and the delivery of the information.

An ideal memory would be large enough to satisfy the needs of the largest databases, fast enough to keep pace with the fastest electronics, small enough to fit the confines of a hand-held computer, and cheap enough to match the low prices associated with integrated circuit devices. Such devices don't exist. Despite impressive accomplishments in memory research and development, the solution for an ideal memory has been elusive. A variety of devices do exist, however, with one or more of the characteristics of the ideal memory. The goal of the designer is to discover techniques that take advantage of the best aspects of each of the memory devices and weave them into a high-performance system at moderate cost.

This chapter examines various ways of implementing memory structures. It then explores techniques for achieving fast access times. It concludes with an examination of strategies involving a mixture of memory structures such as solid-state RAM, very fast cache memory, and large but slow disk memory. Using cache memory and virtual memory techniques, multiple types and sizes of memory can be combined to produce acceptable memory performance characteristics for modern digital computers.

9.2   Memory Technology

Historically, mechanical techniques, electromechanical devices (relays), delay lines, and cathode-ray tubes have all been used as memory components for computing devices. All have major performance problems that make them unsuitable for modern computational devices. This section will provide an overview of the electronic and magnetic phenomena which dominate memory structures for modern computer applications.

Electronic Memory

Advances in integrated circuit fabrication have dominated memory development since the early 1970s. The highly ordered and repetitive structure of electronic memory devices is a good candidate for the fabrication processes of very large scale integration (VLSI).

Static Random Access Memory—SRAM

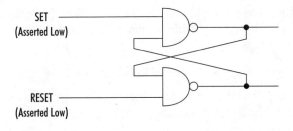

Figure 9.1   A Latch, the Basic Component of an SRAM Memory

Latch structures, as shown in Figure 9.1, are an integral component in static RAM, or SRAM, memories. They offer a stable memory that can be easily read and modified by electronic devices. SRAMs are volatile; when power is removed, the memory contents are lost. They can be fabricated at high densities using integrated-circuit fabrication techniques. SRAM offers high speed with access times in the range of 10 to 25 nanoseconds.

Dynamic Random Access Memory—DRAM

A DRAM differs from an SRAM in its internal method of data storage. SRAM is constructed using stable flip-flop circuitry. DRAM, as shown in Figure 9.2, relies on two levels of charge on a capacitor to represent the two binary states. This method uses fewer circuit elements, takes up less chip area, and is less expensive than SRAM. DRAM also offers moderately high speed with access times in the range of 50 to 150 nanoseconds.

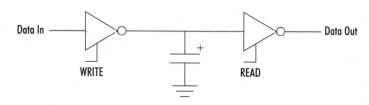

Figure 9.2   Capacitor Storage in a DRAM Memory Cell

DRAM memory is subject to leakage of charge. To assure data integrity, DRAMs must be refreshed at regular intervals. An internal block of DRAM refreshes every time there is a read or write to any element in the block. Failure to adequately refresh may result in data loss. To assure adequate refreshing, a DRAM controller is usually used. This device provides refresh signals to blocks of DRAM at appropriate time intervals. Since normal read/write operations are not possible during refresh, the DRAM controller also provides an appropriate BUSY signal to other system components.

Non-volatile Electronic Memory

SRAM and DRAM memory technologies are volatile; they require constant power to retain their contents. This can be a significant disadvantage in many applications. Battery back-up memories are used to avoid volatility. These devices, sometimes with the battery chip embedded within the same packaging as the memory, can retain their contents for several years in the absence of system power. From the designer's perspective, these devices have identical read/write characteristics as a RAM.

Read-Only Memory—ROM, EPROM, EEPROM, Flash

In contrast to RAM memory, which supports both read and write operations, read-only memory (ROM) is limited to read operations. Originally ROMs were programmed during the manufacturing process. Their contents were fixed. New varieties of ROM, such as erasable programmable read-only memories (EPROMS) can be programmed, erased, and reprogrammed by the user. The programming is performed out-of-circuit in special programming devices. Electrically erasable programmable read-only memory (EEPROM) and flash memories can be reprogrammed under electronic control while in their application environment. In both cases the write cycle is much slower and more complex than the read cycle. Flash memory contents may be changed relatively quickly within the application in which they are used. They are often presented as an electronic replacement for small disk drives.

Charge-coupled Devices (CCD)

Charge-coupled devices are constructed using semiconductor integrated-circuit techniques. An array of microscopic plates in the material functions as a capacitor and is capable of holding charge. Under the influence of electric fields, the pattern of charge can be made to migrate across the material. The large number of plates in the media comprises the storage media.

Data continuously circulates. Data emerging from the CCD array is detected and restructured as a 1 or 0 and then reinserted. CCD memory is volatile; data must be recharged continuously as it emerges from the loop.

Charge-coupled devices, because of the potential for VLSI fabrication, could become a replacement for disks. Since no moving parts are involved, performance and reliability are very good. At present, however, CCDs are still prohibitively expensive compared with disks.

Magnetic Memory

All magnetic memory techniques, including tapes, disks, and cores use an electromagnetic phenomena known as hysteresis. The residual magnetic flux of a material (the extent of its magnetization) is influenced by the applied magnetic field as shown in Figure 9.3. A magnetic field is generated by driving current through the coil of an electromagnet on the write head. Because of hysteresis, a magnetic material subjected to a magnetic field will retain a significant portion of its magnetic flux even after the driving field is removed. The material will retain a positive or a negative residual magnetic flux depending on whether it was last influenced by a positive or negative magnetic field from the write head. This retention is the basis for the memory of the device.

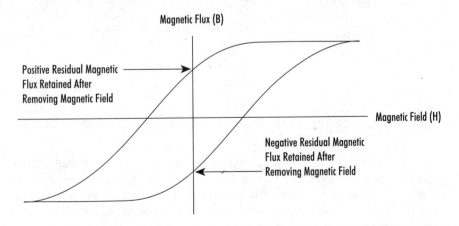

Figure 9.3   *Magnetic Hysteresis in Magnetic Material Used for Computer Memory*

Magnetic Surface Recording

Magnetic surface recording uses a thin magnetic coating on a disk or tape. The individual magnetic domains are magnetized to two different polarities by a write current passed through an electromagnet write head very close to the magnetic surface. The two polarities correspond to the two binary states.

To read the stored information, the magnetic material is moved past a read head consisting of an iron core and a surrounding coil of wire. Faraday's law states that a voltage is induced in a conductor subjected to a changing magnetic flux:

$$E = K \frac{d\phi}{dt}$$

The motion of the magnetic material past the read head induces a voltage (by Faraday's law) whenever there is a change in the polarity of the magnetic flux. A change from a negative flux to a positive flux results in a positive voltage pulse; a change from a positive flux to a negative flux results in a negative voltage. Regions in which there is no change of flux produce no induced voltage.

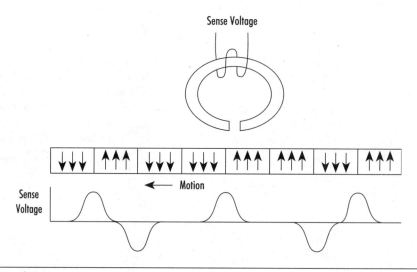

Figure 9.4   Sensing Magnetic Patterns on a Moving Magnetic Storage Media

By sensing and interpreting the induced voltage patterns, as shown in Figure 9.4, the magnetic condition of the surface can be reliably detected. Since long strings of 1s or 0s do not induce voltage, it is necessary to choose a data encoding technique that produces variations in the flux patterns on the surface of the disk even if the logical value of the data remains constant. An alternative method is to use a separate timing track to assure data synchronization.

Magnetic surface recording is nonvolatile. Access time is relatively slow due to the need to mechanically move the magnetic material past a read or write head. There may also be relatively long latency times if the required information is physically far from the read head. Magnetic surface recording provides a very large memory capacity at a very low cost per bit.

Bubble Memory

Bubble memory has become an alternative, but never fully accepted method of achieving non-volatile memory storage. Bubble memory uses an inherently nonvolatile magnetic memory principle involving a string of localized magnetic domains called bubbles. Electric fields applied to a pattern of plates causes the bubbles to move from one plate position to the next. The organization of a bubble memory is similar to that used for charge-coupled devices.

Writing is performed by sending a signal to adjust the alignment of a group of bubbles. A "1" causes a alignment opposed to that of the surrounding material; a "0" leaves the material in the same alignment as the surrounding media. In shift register fashion, the bubbles propagate across the media. Data is read by passing a stream of bubbles past a read head. Information contained in the bubbles is available only as a bubble reaches the edge of the storage material. This can result in a long wait for requested information.

Bubble memories are available on cards with standard interfaces such as Multibus. This relieves the designer from the complex access algorithms necessary to recover data. When used on a standard card, they can perform the same functions as a very slow random access memory. Bubble memories are often selected for applications in high radiation environments which may be unsuitable for electronic memory.

9.3   Memory Structures for Computing Applications

Several memory structures have evolved to serve the needs of computing applications. RAM structures were discussed in detail in previous chapters. A RAM is accessed based on an address of the desired word. For a RAM, t_A, the memory access time, is constant and independent of location of information within the address space.

Magnetic disks, tapes, and bubble memories are examples of **sequential access memories**. In a sequential access memory, the access time depends on the location of the desired information within the storage media. The term **temporal access memory** represents a variety of devices including stacks and queues. In these devices no addressing is used. The location of information depends on when the data was placed in storage.

Disk Memory

A disk is a circular plate of material that has been coated with iron oxide, a magnetic material with a characteristic rust brown color. Floppy disks are made from a flexible plastic material and range in size from two to eight inches. Hard disks use a rigid aluminum plate as the base for the magnetic coating.

Information is stored and written on the disk using a write head able to induce a magnetic flux in the surface material. Reading is accomplished by passing the rotating disk close to a coil of wire in a read head. The changes in flux patterns induce voltages that can be interpreted as bit patterns.

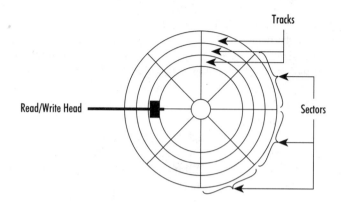

Figure 9.5   A Magnetic Disk

The surface of the disk, as shown in Figure 9.5, is divided into concentric circles called tracks. Each concentric track is divided into sectors. The boundary lines for sectors are established by writing information on the outermost track; this is known as disk formatting. All the data within each sector is read and written as one continuous unit with a fixed number of bytes. Because the tracks are concentric, data written in sectors near the center is compressed more than that written in sectors near the outer edge. This has a slight impact on the reliability of storage and retrieval.

A typical (low-density) floppy disk for a personal computer has 80 tracks on each side of the disk for a total of 160. There are 9 sectors per track. Each sector can store 512 bytes of information. Hard disks, because of their greater dimensional stability, may have 300 to 600 or more tracks per disk. They are capable of writing with a higher density, which allows more sectors (typically 32) per track. Large hard-disk systems may also use multiple read/write heads and multiple platters to improve performance.

Data is accessed from a disk by specifying a surface, sector number, and track number. In small systems using a single read head, the head is first positioned over the proper track. After waiting for the requested sector to arrive, the entire sector of data is read. The data rate depends on the density of the recording and the rotational speed of the disk. Hard disks rotate considerably faster than floppy disks and record at higher data densities.

The **latency time** is defined as the portion of the access time that is dependent on the location of the information. This would include, for example, the time for disk rotation and head movement in a disk system. For a hard disk rotating at 3600 revolutions per minute, the average access time is the time for the disk to rotate one-half revolution, or 8.3 milliseconds. Floppy disks rotate at 300 revolutions per minute, causing much slower access times. In addition to the rotation time, head movement requires an additional 160 milliseconds (worst case) in a floppy disk and up to 40 milliseconds in a hard disk.

Since latency times are often on the order of tens of milliseconds, even in high-quality disk systems, access of a single piece of information is prohibitively slow. Sequential-access memories access information in blocks called **clusters**. Clusters are composed of one to eight sectors. When accessing a cluster, the latency delay is encountered only once; thereafter the succeeding data can be accessed and transferred at a high data rate. In accessing information from a disk, the cluster containing the information is first located. The cluster is transferred to a RAM. The desired information is then accessed from the RAM.

Tape Memory

Information is stored on the magnetic surface of a tape in units that are seven or nine bits wide. Information is read and written as a record. Between records on the tape, there is an inter-record gap (IRG). The IRG provides a delimiter that can be sensed between records. Desired records are located by counting gaps from the beginning of the tape. The gap also provides an area in which the tape can achieve full speed before encountering an area in which data must be read/written. Typical recording density on tape is 1600 bytes per inch with a tape speed of 20 inches per second. This allows a data transfer rate of 32,000 bytes per second.

Magnetic tapes provide a non-volatile, low-cost medium for backup and archive storage of digital information. Because of the sequential and slow access times, they are unsuitable for most primary memory applications.

Stacks and Queues

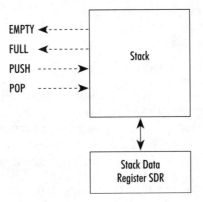

Figure 9.6   Structure of a Stack Memory

Stacks and queues, as shown in Figure 9.6 and Figure 9.7, are examples of temporal access memories. A stack is a last-in-first-out structure. Information is stored with a PUSH operation; the last item stored is retrieved with a POP operation. EMPTY and FULL are stack status signals. Stacks are widely used to implement subroutine and interrupt structures in digital computers. They are key components in calculators and mathematical coprocessors.

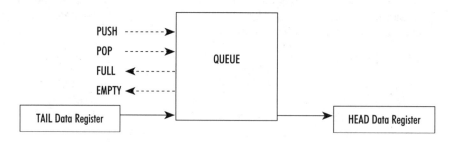

Figure 9.7   A Queue

First-in-first-out operations on data can be performed using a queue. These are especially useful in data buffers, where data arrives in rapid bursts at irregular intervals. It is stored for sequential processing in the order in which it arrives. An example is a printer buffer that stores data from a computer at a high data rate and delivers it to a printer at a lower data rate.

In a stack, all operations are performed at the one end of the structure. In a queue, insertions are made at one end, the tail of the queue, and removals are made at the other end, the head of the queue. Figure 9.7 illustrates the basic structure of a queue. Data enters the queue through the tail of queue data register during a PUSH; it exits through the head of queue data register during a POP. The major obstacle in constructing a queue is the need to develop a strategy to handle data operations at both ends of the structure.

Content-Addressable Memory (CAM)

The search criteria for stored information is called the **key**. In a RAM the data address is the key. Given the address, the corresponding information can be obtained. In many situations the search criteria is not the location, but rather the contents of a portion of the memory cell. The key is a part of the data field rather than the address.

For example, a table in memory stores information about an electrical engineering and computer science building on a university campus. The tabular data contains two fields of information; an office number and the year in which the room was last painted. Both information fields require four decimal digits for representation; data is stored in memory as binary-coded decimal. This requires 32 bits of data. For ease of interpretation, memory addresses and contents are shown in decimal. An abbreviated table of the memory contents for teaching classrooms is shown in Table 9.1.

To determine when room 2500 was painted, the memory must be searched with a key of 2500. The location containing "2500" in the room number field and the corresponding date field is needed. Since it is not known where the desired information is stored (in this case it is in location 03) an exhaustive sequential search of the memory would be required. Although software search techniques are useful, searching a large database may consume an unacceptably long period of time.

Memory Address	Memory Contents	
	Room Number	Year Last Painted
00	0106	1964
01	3210	1992
02	1400	1856
03	2500	1994
04	4404	1939
05	3910	1972
06	1280	1992
07	3325	1990

Table 9.1

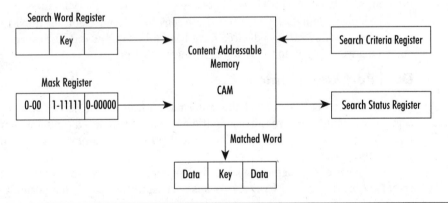

Figure 9.8   A Content-addressable Memory (CAM)

Searches of this type can be greatly simplified and improved by the use of a **content-addressable memory** or CAM. With a CAM, as shown in Figure 9.8, the key is presented via one field of a search word register. A mask register allows selected portions of the search word (the key) to be applied to the CAM. The search criteria register presents information concerning details of the search and may specify how to treat multiple matches. The search status register presents information concerning the search. For example, it indicates if there were multiple matches to the key or no matches to the key. Based on the search criteria, the CAM produces the data word for which there is a match of the specified search bits.

The CAM is required to perform a parallel search on all valid words in its memory to find matches. Unused locations are eliminated from the search. This requires significant logic to perform the bit-by-bit comparison of the key with the proper field of all of the memory contents. To be useful, this must be accomplished in much less time than would be required for a sequential memory search. A high-quality CAM memory can search the entire contents of the memory and retrieve information in one clock cycle.

In the example concerning the painting history of room 2500:

Search word register   0010 0101 0000 0000 xxxx xxxx xxxx xxxx (2500 xxxx)
Mask register   1111 1111 1111 1111 0000 0000 0000 0000 (FFFF 0000)

The search criteria register is loaded with the code indicating a search for an exact match. The search status register indicates a single valid match. The retrieved contents are presented to the matched word register. The retrieved word is examined for the field indicating the last painting date.

Matched word register   0010 0101 0000 0000 0001 1001 1001 0100 (25001994)

If the search criteria had a room number (key) of 1234, the search status register would provide a code indicating there was no match found. If the search were changed to find what rooms were painted in 1992, the search status register would indicate that there is more than one match; the output would provide the first match with a provision for cycling through the remaining matches.

Writing to a CAM does not require an address. The data is stored in the first available unused location. A status indicator is needed to show when the memory is full.

The logic to implement a CAM is quite complex. Because of their importance in memory management applications, these devices have been fabricated using VLSI technology. They are available in modest sizes of up to 1K words at reasonable prices.

Dual Port RAM Memory

The ability to store and retrieve information from two different access ports in a memory is especially useful when two independent devices need to share a common memory. Two computers, for example, may share a table of data. The **dual port memory**, as shown in Figure 9.9, is a way to accomplish this.

Both ports of the memory operate independently. To prevent physical conflicts, a busy status signal is provided to port 1 when port 2 is performing a memory operation. Port 2 gets a busy status signal under similar circumstances with port 1. When the memory is not busy, each may read and write using control signals similar to those required for a single port RAM.

In addition to physical conflicts, logical memory conflicts are also possible. For example, port 2 may be in the process of modifying a data block with new information. If port 1 attempts to read the block during modification, it may read some old values and some new values. To prevent this, two memory locations are preserved to allow the exchange of status between the two ports. This status is in the form of a code word written in the dedicated location. A typical status signal provided by port 2 might indicate that it is in the process of modifying data. Port 1 would read and interpret the status information before attempting to read the data. When port 2 finishes updating the data, it changes the status information and allows access by port 1 to a valid data block.

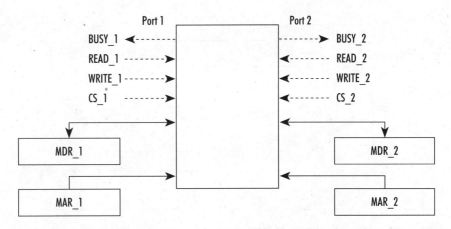

Figure 9.9   A Dual Port Memory

9.4   High-speed Memories

The process of accessing data from a memory, the memory-access cycle, requires three operations as shown in Figure 9.10A. During the first period, the address-presentation time, or t_P, the address is presented to the memory. High-order address bits are decoded to produce a chip-select signal. When both the address bits and chip select have assumed stable values, a read signal and output-enable signal are generated.

The second period, the memory-access time or t_A, is defined as the time lapsed between asserting the read signal and producing a valid data word. It represents the time required by the memory element to decode the internal address and direct the resulting contents to the output bus.

In the third period, the processor data capture time or t_C, the processor transfers the memory information to an internal register. Time period t_C includes the set-up time for the destination register.

In Figure 9.10A, each of the periods in the memory-access cycle is assumed to be of equal length. In Figure 9.10B the three parts of the memory-access cycle are performed in sequence. The total time for a **memory cycle**, t_m, is:

$$t_m = t_P + t_A + t_C$$

There are several possible methods to decrease t_m and provide higher performance memory:

- Make greater use of internal registers instead of memory.
- Acquire a faster memory.
- Improve the effective memory bandwidth by obtaining more information during each memory fetch cycle.
- Use pipelined access of an interleaved memory.
- Use a small, high-speed cache memory for frequently-accessed information.

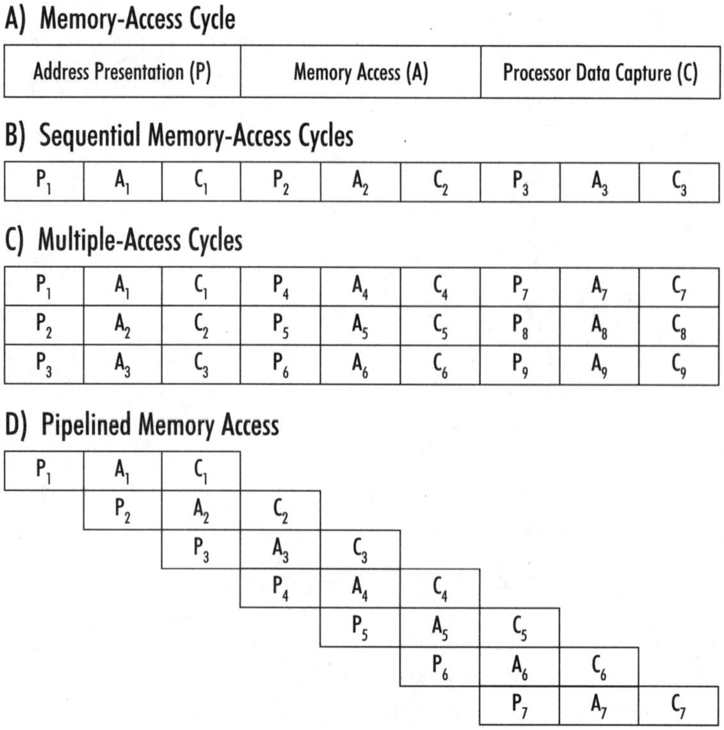

A) Memory-Access Cycle

Address Presentation (P)	Memory Access (A)	Processor Data Capture (C)

B) Sequential Memory-Access Cycles

P_1	A_1	C_1	P_2	A_2	C_2	P_3	A_3	C_3

C) Multiple-Access Cycles

P_1	A_1	C_1	P_4	A_4	C_4	P_7	A_7	C_7
P_2	A_2	C_2	P_5	A_5	C_5	P_8	A_8	C_8
P_3	A_3	C_3	P_6	A_6	C_6	P_9	A_9	C_9

D) Pipelined Memory Access

P_1	A_1	C_1						
	P_2	A_2	C_2					
		P_3	A_3	C_3				
			P_4	A_4	C_4			
				P_5	A_5	C_5		
					P_6	A_6	C_6	
						P_7	A_7	C_7

Figure 9.10   A Subdivided Memory-Access Cycle

Register Storage

The biggest improvement to memory access time is achieved by not using the memory at all. This is accomplished by using alternative storage techniques such as expanded internal registers in the datapath. Improvements are then achieved by programming techniques rather than by improvements in memory technology. RISC processors, as presented in Chapter 7, use increased register storage extensively.

Fast Memory

Up to a point, faster memories with lower values of t_A can be acquired for a premium price. Due to fundamental differences in the circuitry and fabrication methodology, it is difficult and very expensive to obtain devices with memory access times approaching the access times of the fastest registers and logic components.

Since modern computer systems could potentially use enormous memory resources, the cost of memory is a serious consideration. The cost-per-unit of memory increases dramatically as the access time decreases. The quantity of high-speed memory is normally limited by economic considerations.

Interleaved, Multiple-access Memory

The average time for a memory cycle can be decreased by obtaining more than one word in a memory fetch cycle, as shown in Figure 9.10C. Historically, many computers, including the Institute of Advanced Studies Computer at Princeton and the IBM 7094 series, acquired more than one memory word in each memory cycle.

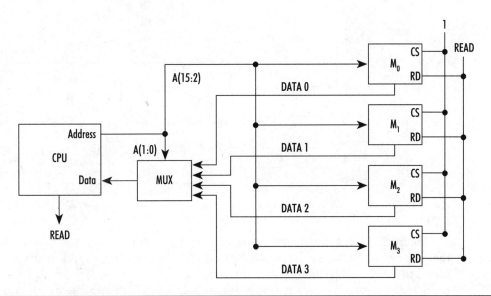

Figure 9.11   A Multiple Word Width Memory

To acquire multiple words in a single cycle, the memory must be wider. This can be done by storing information in an interleaved format with data distributed among multiple memory units. Consecutive memory locations of the physical memory are stored in different memories. For example, Figure 9.11 illustrates an interleaved memory for an address space of 64K, 8-bit words. Groups of four consecutive words are stored in four different 16K × 8 memory elements. Each memory element receives the same read signal. Each also receives the same 14 bits of address $A(15:2)$. The lower two bits, $A(1:0)$, of the address are used by the multiplexer to interconnect the proper memory element with the processor. Access to consecutive address locations requires only a change in the selection bits for the multiplexer and does not require an additional memory cycle.

Interleaved, Pipelined Access Memory

In some situations, the access of multiple words does not produce more useful information. Data, for example, are not always sequentially accessed. Instructions may include jumps and branches requiring memory fetches from nonsequential locations. Sequential accesses of instructions are more likely to be useful than sequential accesses of data.

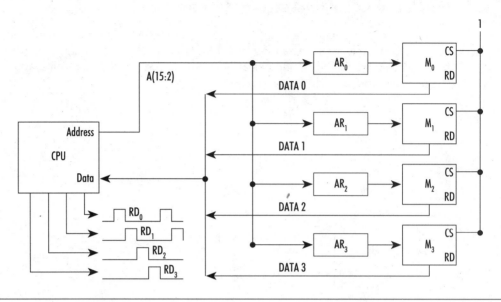

Figure 9.12   A Pipelined Memory-access Structure

Figure 9.10D illustrates data accessed from memory in a pipelined fashion. In Figure 9.12, the memory contents are interleaved among four memory units. Addresses are sequentially moved to the individual address register associated with each memory device. Following reception of an address, each memory unit receives an individual RD control signal to begin data access. Only the memory unit with an asserted RD signal places information on the memory data bus. Other units remain in a high-impedance state. Data is sequentially accessed from each memory unit. By pipelining the address presentation, memory access, and data capture, the effective data-acquisition rate is improved.

There is no requirement that all pipelined memory accesses be from consecutive locations. To operate efficiently, however, consecutive memory accesses should not be addressed to the same physical memory device. To effectively use a pipelined memory, it is necessary to issue memory requests in advance of the time the data is needed. This presents additional complexity to the controller. With the timing shown in Figure 9.10D, the device is capable of producing memory information three times as fast as a nonpipelined approach. If consecutive memory accesses are directed at the same memory unit, the advantages of the pipeline are lost.

The Cache Memory

A cache memory is a small and very fast memory capable of providing information to the CPU of a computer while avoiding the relatively long memory cycle of the main memory. The cache memory holds a subset of the information stored in the main memory. Requests for information are made simultaneously to both the main memory and the cache. If the information is in the cache, a **cache hit**, the effective access time is dramatically reduced. A cache miss requires information to be accessed from the main memory. A key element in using cache memories is implementing a management strategy to assure a high percentage of cache hits with a modest-sized cache memory. Implementing a cache memory will be considered later in this chapter.

9.5 Memory Hierarchy Structures

Digital computers must work within the constraints imposed by three limits associated with their memory. The **logical address space**, A_L, of a computer is the range of memory that can be addressed by a programmer. The size of the logical address space is a function of the instruction set and the registers of the computer. Any address that can be expressed via the machine programming language and for which a memory access cycle can be initiated is within the logical address space.

The **physical address space**, A_P, is the range of memory available to a system via its main memory, M_1. M_1 is the memory connected directly to the computer memory bus. Memory accesses within A_P result in the return of requested information within a period of time determined by t_{m1}, the memory cycle time of M_1. For example, a personal computer using an Intel 486 might have a physical address space of 2^{24} (sixteen megabytes) of storage in solid-state main memory. The physical memory space, A_P, is the only memory space (with the possible exception of a cache memory) that can be directly accessed by the processor.

In addition to the main memory, M_1, a system may contain secondary memory, M_2, in the form of high-capacity disks. Processor access to information from M_2 requires two steps. The required data is first accessed from the disk and transferred to main memory. The processor then accesses the information from main memory. This combined activity requires access time thousands of times longer than that required to obtain information directly from M_1.

The **available address space**, A_A, is the range of memory available to the system via its main memory and any additional active memory components in the system. A personal computer might include a 500-megabyte hard disk and an optical disc to expand the available address space.

The size of the logical, physical, and available address spaces are seldom equal. The above example of a personal computer illustrates a situation with $A_L > A_A > A_P$. Most computers are designed with the expectation that the logical address space is considerably larger than available memory or physical memory. Historically, the growth and use of memory in modern computers has exceeded the predictions of the designers.

Most modern high-performance computers use a large logical address space to accommodate the extremely large programs and databases encountered. The Intel 486 microprocessor, for example, is capable of providing 32-bit addresses. This provides a logical address space of 2^{32} bytes, over four gigabytes (4,000,000,000).

Unfortunately, a main memory of four gigabytes of very fast SRAM is prohibitively expensive. An alternative is to use a modest-sized SRAM along with additional larger, lower cost, lower speed memories, such as DRAM, magnetic and optical disc, and magnetic tape. A method is needed to allow intermixing of a variety of sizes and speeds of memory. This is referred to as a **memory hierarchy,** shown in Figure 9.13.

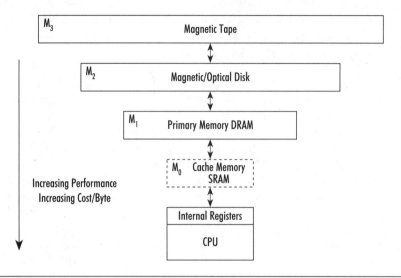

Figure 9.13   A Memory Hierarchy

A memory hierarchy involves using a mixture of memory types and sizes effectively, trying to take advantage of the best aspects of each. Within the hierarchy, M_1 is designated as the physical or primary memory. It is the memory with a direct connection to the registers and data links of the CPU. M_1 is selected in size and speed in such a way that it provides adequate service to the processor at reasonable cost. M_1 is big enough to allow the processor to perform a reasonable set of computations. In an actual computer, M_1 is often a solid-state DRAM memory with a size range of 1 through 16 megabytes. By analogy, M_1 may be considered as your personal collection of technical books that supports your vocation.

M_2 is a larger memory and contains all the contents of M_1 as well as additional material. The contents of M_2 cannot be accessed directly by the CPU. Instead, it is necessary to transfer information from M_2 to M_1 before it can be used. Access to M_2 is slower than access to M_1. Information is usually transferred from M_2 to M_1 in blocks larger than a single word. Economy of size and lower access rates allow the cost per unit of storage to be considerably less in M_2. M_2 might be a magnetic or optical disc memory with a size range from ten to one thousand megabytes. By analogy, M_2 can be compared to the local technical lending library where books can be accessed and brought to your office for reference.

Within your personal library, you probably have short excerpts that you use quite often from various sources. While they are not complete, they clutter your desk and allow very fast access to often-needed information. Within the memory hierarchy an optional **cache memory**, M_0, fills this role. Although the information in a cache may be fragmentary, its high-speed accessibility greatly enhances processor performance.

Within the CPU, the registers may be considered an additional level in the memory hierarchy. An extensive set of registers allows storage and access to data without accessing an external memory. In the library analogy, register memory corresponds to the short-term human memory that can retain some information for instantaneous recall.

Higher order elements in the memory hierarchy are established in a similar manner. M_3 would contain a superset of the information in M_1 and M_2. Its access time would be quite slow, but its cost per byte of storage would be relatively low. In an actual system, it might be a digital tape containing archival information. Information accessed from M_3 would be transferred down the hierarchy for eventual use in M_1. By analogy, M_3 would correspond to a regional lending library. Before coming to your office, the book would be first transferred to the local lending library. From the local lending library it would be transferred to your office for use.

In order to effectively use A_A, the available address space, a methodology is needed to swap information between various levels in the hierarchy. When memory swapping between the main memory and secondary memory is done in a manner transparent to the programmer, this is known as a **virtual memory** system. In such a system the programmer may write programs as if the address space of the combined main memory and secondary memory were available. The virtual memory management system performs information swaps so that actual data access always comes from the physical memory.

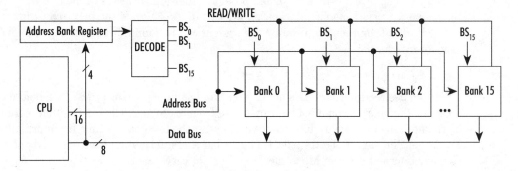

Figure 9.14   Use of Memory Bank Switching for Access to the Full Physical Address Space

Situations also exist in which $A_P > A_L$; the size of the physical memory exceeds the ability of the processor to address it directly. For example, many microprocessors have a logical address space of 64K bytes. This is caused by the relatively small 16-bit address bus. To take advantage of a larger available address space, perhaps a grouping of 16 memory chips each with 64K bytes, a bank switching technique is often used. An example system is shown in Figure 9.14.

External to the CPU, a 4-bit address bank register expands the address from 16 bits to 20 bits. The address bank register is decoded to produce bank-select signals to select one of the 16 banks of memory. The address bank register can be loaded using an I/O instruction. Once loaded, it retains its value until changed. This allows consecutive memory accesses to be made from the same memory bank. For situations in which there is a common data memory and program memory (von Neumann architecture) this presents a potentially serious problem. If the register bank is changed, succeeding instruction fetches will also be directed at the new memory bank. This problem is solved by segmenting the memory space so that part of it is permanently accessed and part of it is bank switched.

9.6   Performance of Multiple-memory Systems

Performance in a multiple-memory system involving units M_0, M_1, ..., M_n depends on several factors including the logical organization of the structure, the location of desired information, and the time to access information from each level of the hierarchy.

S_i = Size of memory unit M_i

Within a memory hierarchy, each level of the hierarchy is larger than the previous.

$$S_0 < S_1 < ... < S_n$$

Each level also includes all the information present in the previous. The highest level of the hierarchy contains all the information accessible by the system. While this is not absolutely required, it simplifies the design of such systems.

Assume that the system in question involves multiple memory units including a cache (M_0), a primary memory (M_1), a disk (M_2), and possibly other higher levels in the hierarchy. Assume that the memory access time for information present in level zero of the hierarchy, the cache memory, is t_{m0}. If all information required by the system can be found in M_0, then t_m, the average access time for information, is:

$$t_m = t_{m0}$$

If the desired information is not present in M_0, but is present in M_1, then before the CPU can access it, the information must be transferred from M_1 to M_0[*]. The information is transferred as a block between one memory level and another. t_{m1}, the total access time for information located in M_1 is then the sum of the transfer time of a block of information from M_1 to M_0 plus the time to transfer information from M_0 to the CPU. In most cases the transfer time from M_0 to the CPU is negligible compared to the block-access and transfer time from M_1 to M_0 and is ignored. For n levels of memory:

$$t_{m0} < t_{m1} < ... < t_{mn}$$

H_i, the **hit ratio** for a given level, i, of the hierarchy is the probability that the desired information can be found in M_i. Since each level of the hierarchy is larger than the previous:

$$H_0 < H_1 < ... < H_n = 1$$

The hit ratio is not a constant for a particular computer but is dependent on the pattern of accesses to program memory and data memory. Estimates for the hit ratio can be obtained by running benchmark programs which have a known mix of various instructions distributed through the address space.

---

[*]   For correctness, it should be noted that data acquired from main memory does not necessarily need to pass through the cache memory prior to access by the CPU. Information acquired from M_1, the main memory, is provided simultaneously to the cache (for future use) and the CPU. Assuming that it does pass through, the cache will simplify the analysis and not have significant impact on the results.

Let $P(M_i)$ be the probability that the required information will be accessed from M_i. Since all information passes through M_0, the probability that the desired information will be accessed from M_0 is:

$$P(M_0) = 1$$

All information not in M_0 must be accessed through M_1.

$$P(M_1) = 1 - H_0$$

The probability of access from M_2, by the same reasoning, is:

$$P(M_2) = 1 - H_1$$

All information resides in M_n. The probability of access from M_n is:

$$P(M_n) = 1 - H_{n-1}$$

The average access time for the system is then:

$$t_m = P(M_0) \, t_{m0} + P(M_1) \, t_{m1} + \ldots + P(M_n) \, t_{mn}$$
$$t_m = t_{m0} + (1 - H_0) \, t_{m1} + \ldots + (1 - H_{n-1}) \, t_{mn}$$

For a three-level memory hierarchy involving a cache, main memory, and disk:

$$t_m = t_{m0} + (1 - H_0) \, t_{m1} + (1 - H_1) \, t_{m2}$$

The **efficiency**, e, of a memory hierarchy is defined as the ratio of the access time of information from M_1, t_{m1}, to the average access time for the entire memory system.

$$e = \frac{t_{m1}}{t_m}$$

$$= \frac{t_{m1}}{t_{m0} + (1 - H_0) \, t_{m1} + (1 - H_1) \, t_{m2}}$$

$$= \frac{1}{\dfrac{t_{m0}}{t_{m1}} + (1 - H_0) + (1 - H_1) \dfrac{t_{m2}}{t_{m1}}}$$

In most three-level hierarchies $t_{m0} < t_{m1} \ll t_{m2}$.

No Cache Memory

If the cache memory is not present, this simplifies to:

$$e = \frac{1}{1 + (1 - H_1) \dfrac{t_{m2}}{t_{m1}}}$$

A high value of H_1 is essential to achieve a high efficiency. For a two-level system, R is defined as the ratio of the access times for M_2 and M_1.

$$R = t_{m2} / t_{m1}$$

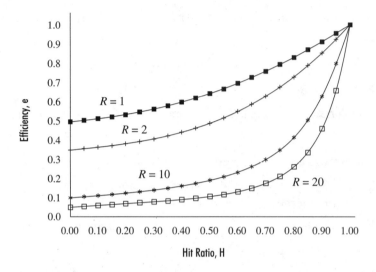

Figure 9.15   Efficiency of a Two-level (No Cache) Memory Hierarchy (R = t_{A2} / t_{A1})

Figure 9.15 is a plot of the efficiency of the system as a function of the hit ratio for several values of R. For high values of R, the hit ratio must be very high to achieve good performance.

With Cache Memory

When cache memory is present, the highest performance is achieved if the required information is found in the cache. Since the memory cycle time for the cache is less than that of M_1, an efficiency greater than 1 is possible.

9.7   Management of Memory Hierarchies

Memory hierarchies can work effectively by exploiting two characteristics of computer programs.
- **Locality of Reference.** When data or instructions are accessed from memory, it is likely that future memory accesses will be from locations close to current locations. This is due to the sequential location of instructions and using sequential blocks of memory to store data tables. By transporting information in large blocks from M_2 to M_1, for example, information necessary for future memory access are likely transferred. This increases the likelihood of hits in future access from M_1.
- **Temporal Locality.** Data and instructions that have been used recently are likely to be used again in a short period of time. This is especially applicable to loops of program instructions and highly repetitive processing on data sets.

A key idea in implementing hierarchical memory is locality of reference. While programs and data necessary for a computing activity may span a wide range of the available address space, there is an uneven pattern of usage. Figure 9.16 shows an example histogram of the frequency of reference of information within the available address space of a computer. Most of the memory references are restricted to a relatively small percentage of the total available address space. By keeping the information most likely to be accessed in M_1 or M_0, the amount of time required for data swaps can be minimized.

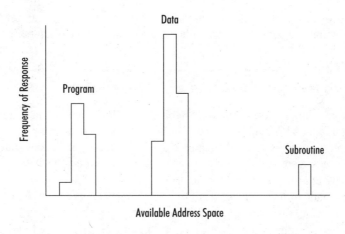

Figure 9.16   Locality of Reference in the Available Address Space

Memory hierarchy structures contain a primary memory, M_1, a cache memory, M_0, a secondary memory, M_2, and perhaps higher order memories, M_i. Such structures can be managed in four ways: by the programmer, by the compiler, by the operating system, or by the system hardware.

Cache Memory/Primary Memory Management

The high-speed operations required to manage the data exchange between a cache memory and primary memory are possible only with specialized hardware. The methodology and algorithms for cache memory management will be considered in more detail in the next section.

Primary Memory/Disk Memory Management

The programmer can manage the data exchange between a main memory and disk by using **overlays**. Overlays are segments of data or instruction code that are moved under program control from a peripheral storage device to M_1, the physical memory. When placed in M_1, it overlays and replaces the former contents of M_1. By this method, the same location in main memory can be used for multiple purposes.

The programmer, however, is often the least prepared to manage an extensive memory structure. Programs involving treatment of hardware features are best handled in machine-specific code. Good programmers are more productive when writing in high-level code. Since each target machine may have a different memory configuration, it is difficult to produce a general-purpose program to work on a wide variety of machines.

It is possible to manage memory overlays with the compiler, but this would be difficult to perform efficiently. Before compiling code, the compiler would need information concerning the exact configuration of the target system. Compiled code could be run only on a specific target system. This would be a very unsatisfactory solution.

System software, specifically the operating system, can manage a memory hierarchy by a technique known as virtual memory. In this technique, the programmer is allowed to program using the entire available address space. This can be as large as the logical address space. The physical memory stores a portion of the total information in the form of memory pages. Each memory page usually consists of 256 to 4K words of storage. Memory pages are always swapped between primary memory and higher order memories as complete units. This topic is discussed in greater detail in Section 9.9.

9.8   Cache Memory Structure and Organization

Even with high-quality DRAM, there is usually a timing mismatch between the CPU and the main memory. Reasonably priced DRAM may have an access time five to ten times greater than that required for a datapath register transfer operation. This mismatch can be bridged by including a cache memory between physical memory and the CPU.

The need for high-speed memory is greatly increased by the dramatic performance improvements made in RISC processors. Any advantages gained in high-speed CPU operation can be lost if the memory presents an access bottleneck.

Cache memory, in contrast to virtual memory, is managed by hardware. This is necessary to attain the high speed necessary to make the memory useful. The goal in using cache memory is to allow memory accesses (from the cache) within one clock cycle. There are no wait states required.

The relationship of cache memory to physical memory is similar to the relationship between physical memory and virtual memory. In both situations, the smaller, faster memory contains a selected subset of the information present in the other. For simplicity of presentation, it will be assumed that the memory system consists of only a cache memory and a physical main memory.

The unit of memory management in a cache system is a **line**. A line consists of several (a power of 2) consecutive words of memory. A line is transferred as a unit between main memory and the cache. A typical line width is one to eight bytes in modern computers such as the Intel 80x86 and the VAX 11/780.

To manage a cache memory, it is necessary to first know if the requested data element is in the cache, and if it is in the cache, determine its location. Most memory management techniques are based on dividing the logical memory address into fields. These fields are then used as part of the management algorithm.

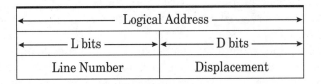

Table 9.2

An individual word of information in the logical memory space can be specified by its line number and its **displacement** within the line. The combination of the two, as shown in Table 9.2, is the address of the word. Information in the cache is stored in tabular form. Consider, for example, a computer using a 16-bit address for a 64K byte main memory. A line size of 8 bytes is used. In this case the address may be broken into a 3-bit displacement and 13-bit line number. In Table 9.3, each line of the cache is identified by its line number. The line number, which is the key used to locate data in a cache memory, is called a **tag**. The division of the tag into blocks of two, three, or four bits is in anticipation of material in the next section.

In Table 9.3, each element in the cache contains the tag consisting of the full 13-bit line number. V indicates a **valid bit** to show that the entry contains a valid line of data. The remainder of the storage row contains the 8 elements of the data line.

Tag	V	Data Line							
0000 0000 00 000	1	05	F3	12	00	46	33	FF	2A
0000 0000 00 010	1	55	32	43	33	22	32	45	33
0000 0000 01 000	1	21	32	57	12	2A	56	63	15
0100 1111 01 101	0	44	2A	84	45	2A	87	27	15
1011 0001 11 010	1	89	5C	22	87	2A	88	81	77
1111 1100 00 000	1	67	CF	11	99	3F	95	11	57
0011 0000 10 110	1	AA	D1	CC	98	34	92	11	83
1111 1111 11 111	0	C3	1A	4F	45	55	4D	00	FF

Table 9.3

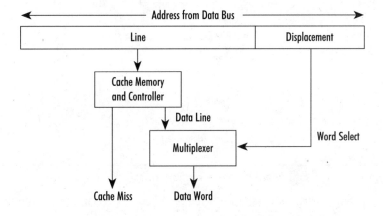

Figure 9.17   A Cache Memory System

Figure 9.17 illustrates the operation of a cache memory containing several lines of code, each identified by its tag. The address presented on the address bus is divided into tag and displacement fields. The tag is presented to the cache memory controller, and if it matches one of the tags for information stored in the cache, there is a hit. The corresponding data line is accessed from the cache. A multiplexer uses the displacement bits to select the requested data word from the line. If there is no match of tags, there is a miss and the data must be retrieved from M_1.

The major problem in implementing the cache is finding a mechanism for quickly determining if there is a hit or miss with the data. This involves comparing the line number of the requested data with all tags stored in the memory. One possible method is to access each of the memory entries sequentially and perform a compare of the line number and tag of the stored data. If it is not known where the data is stored in the cache, this can be a slow process and negate any advantage of having the cache. Three other approaches could possibly lead to viable solutions.

Directly Mapped Cache

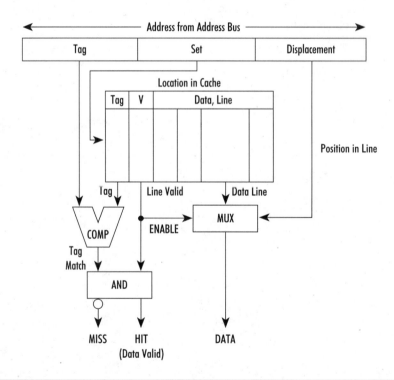

Figure 9.18   Hardware Implementation of a Directly Mapped Cache Memory

For fast cache access, an efficient method of determining if the desired line is in the cache memory is needed. In a directly mapped cache, as shown in Figure 9.18, the line number is divided into a tag field and a set field. The set field determines the physical location to be used for storing the line in the cache memory. In a directly mapped cache, each line of information has only one possible storage location in the cache table. This eliminates the need for a sequential search of all tags.

1. The address on the address bus is divided into a tag, a set, and a displacement.
2. The set is presented as an address to the cache memory, and its contents are accessed. This includes the tag of the stored information, the line of data, and some management bits.
3. The tag from the address field is compared with the tag stored in the set location. If they match, there is a hit; if not, there is a miss and the information must be obtained from M_1.
4. If there is a miss, a new line is obtained from M_1 and placed in the proper location in the cache.

If information from the cache memory of Table 9.3 is placed in the directly mapped cache, the contents appear as Table 9.4. The set number is bits A(5:3) of the original address. This is the right-most 3-bit field of the tag shown in the original table of cache contents. Some of the entries in the previous table have identical set numbers. Since only one line from each set can exist in the cache, only the first entry is included.

Set	Tag	V	Data Line							
000	0000 0000 00	1	05	F3	12	00	46	33	FF	2A
001	—	0	—	—	—	—	—	—	—	—
010	0000 0000 00	1	55	32	43	33	22	32	45	33
011	—	0	—	—	—	—	—	—	—	—
100	—	0	—	—	—	—	—	—	—	—
101	0100 1111 01	0	44	2A	84	45	2A	87	27	15
110	0011 0000 10	1	AA	D1	CC	98	34	92	11	83
111	1111 1111 11	0	C3	1A	4F	45	55	4D	00	FF

Table 9.4

The directly mapped cache table is easy to search since only one entry needs to be examined. The disadvantage is that only one line corresponding to the same set may be stored in the cache. For example, if address 1111 1100 00 000 110 is requested, there will be a miss; the contents of addresses 0000 0000 00 000 000 through 0000 0000 00 000 111 occupy the memory allocated for set 000. If a program requires frequent access to different regions of memory with the same set number, numerous cache misses will result.

Associative Cache

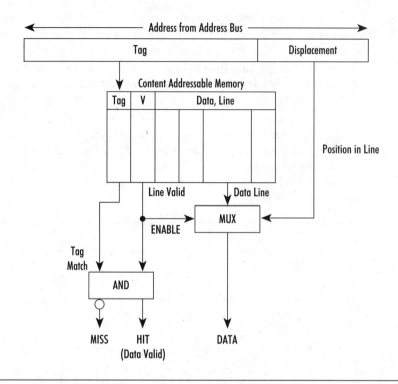

Figure 9.19   An Associate Cache Memory

In an associative cache, the tag consists of the line number, which is all the bits of the address except the displacement. The cache table is stored in a content-addressable memory. Using the tag field as a key, the CAM allows a rapid simultaneous search of all tags in the memory. If a tag match is found, the corresponding line is produced. A multiplexer, using information provided by the line displacement, selects the proper word. Figure 9.19 shows a configuration using a CAM to implement an associative cache.

Set-associative Cache

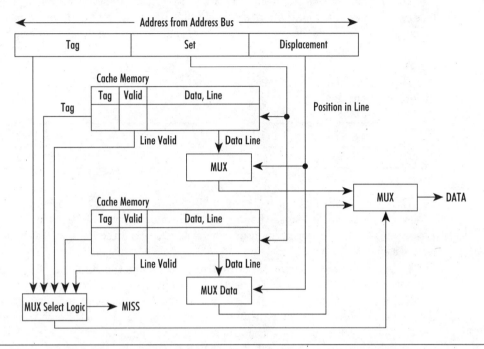

Figure 9.20   A Two-Way Set-associative Cache

Because large and fast CAMs are expensive, set-associative caches offer a good compromise between performance and cost. In an *n*-way associative cache, the lines are divided into sets as in direct mapping. More than one member of each set can be simultaneously present in the cache. The most common implementations involve two-way associative caches. In Figure 9.20, the two-way associative cache has been constructed as two directly mapped caches with additional logic to determine if a tag from either memory matches the tag from the presented address.

Cache Reads

A cache miss requires a read of the required data from M_1. To optimize performance, the data read from M_1 can proceed simultaneously with the cache search. If there is a cache hit the read from M_1 is aborted. If the cache search is unsuccessful, the data is acquired from M_1 and transferred to the CPU. Additional memory cycles access the remaining bytes of the line.

Cache Writes

It is important to maintain coherency between the cache memory and main memory. Every write performed to the cache creates a discrepancy between the contents of the two memories. To maintain cache coherency, writes to the cache must eventually be sent to M_1. One technique is to write through to M_1 anytime there is a write to the cache. With a sufficiently intelligent cache controller, the controller can do the write through to M_1 instead of burdening the CPU.

Example 9.1:   The Intel 82385 Cache Controller

The Intel 82385 is a single-chip controller for a cache memory system of the 80386 microprocessor. It provides fast local storage of up to 32K bytes for frequently accessed data and instructions. All 2^{32} words of the memory space are mapped into the cache. Information found in the cache can be accessed in 30 nanoseconds. This usually allows the processor to run with no wait states.

The 82385 references memory with a 32-bit address. For caching purposes, the address is divided into four fields.

Tag (17 bits)	Set (10 bits)	Displacement (3 bits)	Byte Add (2 bits)

Table 9.5

The cache addressing allows 1024 (2^{10}) lines to be held in the cache. Each line is eight (2^3) words of 32 bits, or 32 bytes. The displacement address specifies a word; the byte address can access one byte of the word. The user can select either a direct-mapped cache or a two-way associative cache.

To the microprocessor, the cache controller looks like the standard 80386 system bus with a very fast memory. During a read cycle, the controller compares the tag with the tag in the on-chip cache directory. If the data is present, it is acquired from the cache memory. If there is a miss, the data is acquired from the main memory. As data from M_1 passes through the controller, it is also placed in the cache, and the cache directory is updated. Following a miss, the controller anticipates further misses by immediately starting a fetch for the next sequential data.

The write cycle uses a **posted write**. Information is written to the cache memory with no wait states. The cache controller passes the information on to the system memory using free bus cycles.

The cache table uses 1024 (2^{10}) storage locations. Each memory location includes:

Tag (17 bits)	Tag Valid (1 bit)	Word Valid (8 bits)	Line (32 bytes)

Table 9.6

Each word within the line contains its own valid bit. This allows the device to function without retrieving a full line of data for the cache.

The 82385 has several special features to improve performance. A flush command can be used to invalidate the complete contents of the cache. Certain areas of memory space can be specified as non-cacheable. This allows devices such as I/O chips to share the memory space. When used in a multiprocessor system, the cache controller can snoop to see if another processor has written to M_1. If this happens, the controller updates the cache table to indicate invalid data.

9.9   Virtual Memory Structure and Organization

A virtual memory system is a multilevel memory system managed by system software with the goal of providing a low average memory cycle time. This service is provided transparently to the programmer. The programmer can assume complete access to a memory structure as large as the largest memory in the hierarchy.

For this discussion, the virtual memory system is assumed to include two levels of memory units—M_1, the physical memory, and M_2, the virtual memory. Information stored in M_2 can be accessed by transferring blocks of information in the form of pages to M_1. While a memory system might have both a cache memory and virtual memory, only the virtual memory will be considered in this section.

Several issues are important in determining performance of such systems.

Page Size

Whenever a miss occurs in a virtual memory system, a block of information must be exchanged between the secondary memory, M_2, and the physical memory, M_1. The block could be as small a single word. This would be inefficient, however, because locality of reference makes it likely that the next piece of required information will be from the next sequential location. Disks generally require information in a disk segment to be the smallest data unit to be transferred. Transferring a disk segment requires only a single latency period followed by a high-speed serial data transfer.

A second possible transfer unit might be a block the size of the physical memory. While this might work well for small programs, it would cause a great deal of memory swapping if the address of program memory and data memory in the logical memory space were significantly different. It might also cause extensive swapping, for example, if a repeated subroutine call was located far from the calling program.

An argument can also be made for using **data segments**, which are blocks of information that conform to natural boundaries of memory usage. A subroutine, for example, might be a useful data segment, since it is quite likely that the subroutine would be used as a complete entity. A data table might also be a useful data segment.

Although data segments correspond closely to patterns of usage, they have several disadvantages. Data segments, although easy to identify by the programmer, are often difficult to identify by system software. They are not of uniform size. This creates extra complexity when swapping data segments between physical memory and virtual memory.

Most implementations of virtual memory use a fixed-size page as the unit of information transfer. The page size is an integer number of disk segments. The number of pages stored in both physical and virtual memory is also a power of two.

Although the page size should be greater than one word and less than the full size of the physical memory, an optimum page size is difficult to determine. In practice, the various parameters such as page size and number of pages are determined by simulating a broad class of representative user programs and choosing values found to perform well.

Page Location in the Main Memory

In a paged virtual memory system, all page sizes are assumed to be 2^d words. If the virtual memory space is 2^v words, there are 2^n pages in the virtual memory space. It is assumed that a two-level memory is used and that M_2 is large enough to span the range of program addressing.

In the virtual memory:

2^v (memory locations) = 2^n (pages) 2^d (words per page)

or

$v = n + d$

The v bits required to specify a virtual memory location can be viewed as specifying a page number (n bits) and a displacement within the page (d bits). Figure 9.21 illustrates the composition of a virtual address.

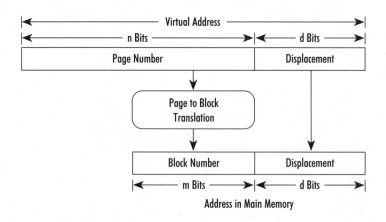

Figure 9.21   Address Translation in a Virtual Memory

The physical memory, M_1, is smaller than the virtual memory space and is assumed to have 2^p memory locations ($v > p$). p bits are needed to specify an address. M_1 is divided into 2^m page size blocks. Each can hold one page from M_2. An address in M_1 consists of m bits to specify the block location and d bits to specify the displacement within the page stored in the block.

In the physical memory:

2^p (memory locations) = 2^m (pages) 2^d (words per page)

or

$p = m + d$

To implement a virtual memory system:

1. A method is needed to allocate the 2^n pages of the virtual memory among the 2^m blocks of M_1.
2. A mechanism is needed to determine if a specified address is located in M_1, where it can be accessed quickly, or M_2, where a page swap is necessary.
3. If the required information is in M_1, its location must be found.
4. When new pages are brought to physical memory, an algorithm is needed to determine their storage location. Displaced pages may need to be copied back to M_2.

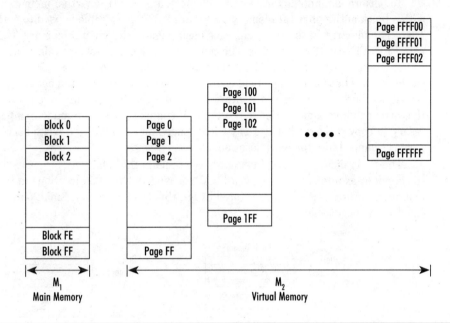

Figure 9.22 Virtual Memory Allocation

Figure 9.22 illustrates an example virtual memory system with a 4-gigabyte virtual memory, $v = 32$, and a 64-kilobyte physical memory, $p = 16$. A page size of 256 bytes was selected, $d = 8$. With these specifications, there are 256 (100_{16}) blocks in M_1, the main memory. Each can accommodate one page with 256 (100_{16}) words. The virtual memory spans an address range of 00000000_{16}–$FFFFFFFF_{16}$. This consists of pages 000000_{16}–$FFFFFF_{16}$.

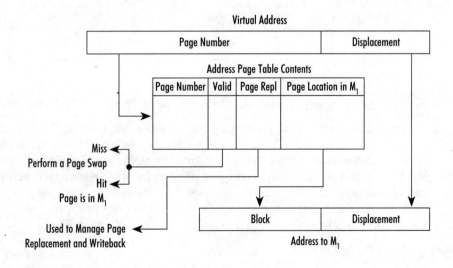

Figure 9.23 Implementing a Virtual Memory System

To perform a mapping for a virtual memory, a page of the virtual memory must be associated with a block of the physical memory. In Figure 9.23 a page table is used to perform the mapping. There is one entry in the page table for each page of the virtual memory. This requires a RAM with $FFFFFF_{16}$ or 16 megawords. The contents of the page table include a valid bit, an entry indicating whether the page is in M_1. If the requested page is in M_1, the page location field indicates the location. The replacement field contains information related to page replacement and will be considered later. When pages are interchanged between M_1 and M_2, it is necessary only to update the corresponding page entries in the table. If changes have been made to a page in M_1, it is necessary to copy the page back to M_2. Clean pages do not require a write-back, even when a new page is swapped into the block location.

The main disadvantage of using a page table is the relatively large-sized RAM required. In the previous example, a 16-megaword RAM is needed. Only 256 locations in the RAM carry information regarding pages that are stored in M_1; this is an extremely small portion of the space allocated to the page table.

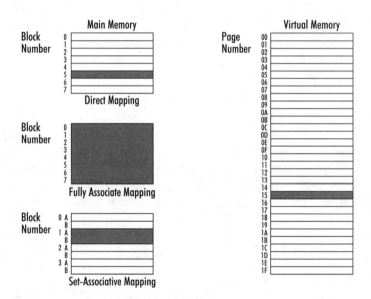

Figure 9.24 Alternative Methods of Performing a Page to Block Mapping

Alternative methods of performing the page-to-block mapping, with greater memory efficiency, are shown in Figure 9.24. The techniques are similar to those used in cache memory management. In **direct mapping**, pages from virtual memory are allowed only in certain positions of M_1. The location is a function of the page number. In **fully associative mapping**, a page from virtual memory can be located in any block of M_1. While this gives good flexibility, it makes it hard to ascertain the exact storage location. In **set-associate mapping**, page locations in the virtual memory may be stored in either sub-block A or B in M_1.

Direct Mapping

Direct mapping provides the simplest method for performing page-to-block mapping, which is accomplished using a page table implemented in RAM. There is one storage location in the table for each of the storage blocks of M_1. To facilitate the search in the table, the location of certain pages from virtual memory are restricted to certain blocks. The pages are assigned to blocks by using bit patterns in page numbers to assign pages to **page sets**. The n-bit page number is divided into a tag with t bits and a set number with s bits.

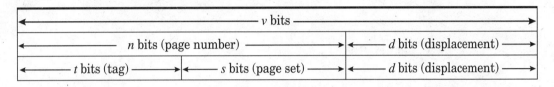

Table 9.7

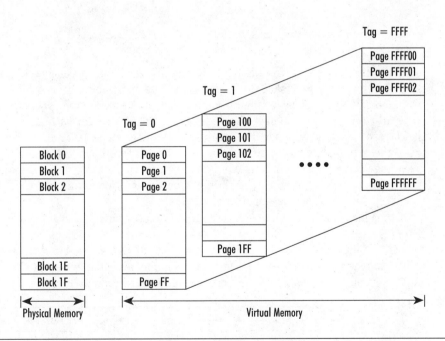

Figure 9.25   *Page-to-set Mapping in a Direct-Mapped Virtual Memory System*

All pages with the same set number are mapped into the same block in physical memory. For the example system, this is shown in Figure 9.25. All page numbers that are multiples of 256 (100_{16}) are mapped into block 00_{16}. Pages 1_{16}, 101_{16}, 201_{16}, etc. are mapped into block 01_{16}. Of the $FFFFFF_{16}$ pages, $FFFF_{16}$ of them are mapped into each of the available blocks in M_1. The user's program is assumed not to require many references to multiple pages in the same set.

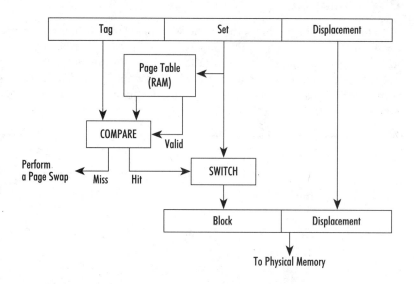

Figure 9.26   A Direct-Mapped Virtual Memory System

With direct mapping, the tag-to-block mapping is simple. A conceptual structure to support the required operations is shown in Figure 9.26. The virtual address is examined to determine its set number. A compact page table is maintained in RAM. For the previous example, only 256 pages are stored in M_1; the page table requires only 256 entries. Each address is associated with the corresponding set number. The location in the page table associated with a set contains the tag of the actual page stored in that location. Since there is only one possible block in which a given page could be located, only one location in the page table needs to be examined. The set number of the address provides the block number for location within M_1. If there is a miss, a page swap is required before accessing the data.

Other fields in the page table include a valid bit, V, which is set to indicate that the information in the table entry is valid, not garbage resulting from a previous activity or a system reset. An additional field is allocated for information related to page replacement and swapping, which will be considered later. Since the page table is implemented in system RAM, its size and organization is strongly influenced by the word size used in the computer.

←—— t bits ——→	←—— 1 bit ——→	←—— r bits ——→
Tag	Valid	Replacement

Table 9.8

Direct mapping is relatively easy to implement. It has the disadvantage that only one page from a given set can be stored simultaneously in the physical memory. If an executing program performs frequent references to two or more pages of the same set, extensive page swapping, called **thrashing**, can occur. Since page swapping is time-intensive, thrashing greatly diminishes performance.

Associative Mapping

In **associative mapping**, restrictions on the mapping of a page from virtual memory to M_1 are removed. Any page from M_2 may reside in any block in M_1. This greatly reduces the possibility of thrashing due to frequent utilization of pages which all must reside in the same block. Since no page sets are involved, as in direct mapping, a page from virtual memory must be identified by its complete page designation. In this situation, the page number becomes the tag, as shown in Table 9.9.

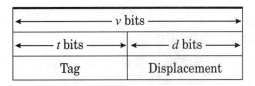

Table 9.9

If associative mapping is used in the example, each of the 256 blocks in physical memory can contain any of the 64K pages of virtual memory. All of the 256 locations (one for each block) of the page table would need to be examined to search for a tag match. If a RAM were used for the page table, this search process would be sequential, very slow, and inefficient.

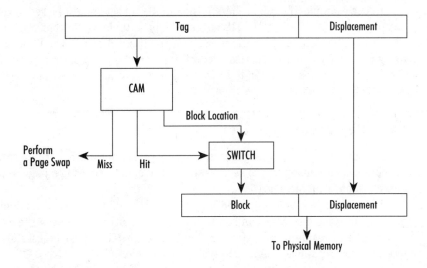

Figure 9.27   Associative Mapping in a Virtual Memory Page Table

An alternative solution is to use a content-addressable memory (CAM) for the page table as shown in Figure 9.27. Such a memory would require a field to store the block number in addition to the tag, valid bit, and replacement information.

← t bits →	← 1 bit →	← s bits →	← r bits →
Tag	Valid	Block	Replacement

Table 9.10

The use of a CAM, although well suited for a virtual memory system using associative mapping, requires a special hardware component.

Set-associative Mapping

Set-associative mapping offers a compromise between direct mapping and associative mapping. The page table is organized by sets to allow rapid searches of the tags. q blocks, rather than the single block used in direct mapping, are allocated for each page set. At most, q sequential accesses of the page table are required to determine if the desired page is in M_1. q is selected to be large enough to allow multiple pages of a set to be simultaneously present in the physical memory. For simplicity, q is restricted to a power of two. If the physical memory is kept at a constant size, the number of pages in a page set must be increased so that the total number of page sets is less than used in direct mapping. In the extreme, if only one set is used, this method becomes associative mapping.

In the example, if $q = 2$ is chosen, the number of sets might be reduced from 256 sets, each of 256 pages, to 128 sets, each of 512 pages. The page table would continue to contain 256 entries—one for each block in the physical memory. Two blocks, A and B, would be allocated for each set. In using the page table, two entries would be searched for the desired page. This process increases the number of pages allocated to a set, but allows multiple pages from the same set to be simultaneously present in the physical memory.

9.10   Page Replacement

When a memory reference is generated for a location not in M_1, a **page fault** occurs. The virtual memory system must perform several sequential actions to bring a page from virtual memory into a block in physical memory.

First, the block to be replaced must be identified. This action is trivial for a direct-mapped system, since the new page can reside only in one location. For one of the associative methods, the new page can reside in any of several locations. Criteria must be established to choose a block location. Several replacement strategies are possible.

Round-robin Replacement

In round-robin replacement, the blocks to be vacated are chosen sequentially. This is equivalent to a first-in, first-out (FIFO) decision. For an associative mapping, all of the blocks in physical memory participate sequentially. For a set-associative mapping, only blocks dedicated to the set involved with the data transfer are candidates for selection. Round-robin replacement is easy to implement since the sequential block location is simple to specify and update. A next replacement address must be maintained for each set. After each memory swap due to a miss, the next replacement address is incremented (modulo n). This technique has the disadvantage that a highly utilized page may be replaced simply because of its position in the rotation criteria.

Random Replacement

This technique is similar to round-robin, but the sequence of blocks selected for replacement is selected randomly rather than sequentially.

Least Recently Used

In the least recently used replacement policy, the page table maintains an age field for each of the pages in physical memory. Any page not used recently is considered a good candidate for replacement. A page might enter physical memory, for example, with an age of 0. For each memory reference in which it is not involved, it ages by 1. When referenced, its age is reduced to 0.

Within the page table, sequential ordering of pages, by age, is maintained. With RAM memory a sequential search is needed to find the page with the greatest age. A CAM, however, might have the ability to search the entire range of relevant memory locations in one memory cycle.

Longest Time Until Next Need

The best candidate for page replacement seemingly would be the page which will not be used for the largest time duration in the future. Unfortunately, this is very difficult to determine. Because of conditional branches in the program, it may not be possible to accurately determine future memory needs. Even without conditional branching, such a technique would require an advance examination of the code to be executed.

Example 9.2:   *Performance of Page-replacement Algorithms*

Assume that a virtual memory system uses fully associative mapping. There are 16 blocks in the virtual memory and 4 blocks in the main memory. The sequence of block requests is as follows:

{1 5 7 1 B 3 7 1 B 1 4 B 7 1 5 4 1 1 5 B}

Required:

For round-robin and least recently used replacement algorithms, find the contents of M_1 after each block access. Determine the hit ratio.

Analysis:

A table showing the contents of the page table will be created for each of the situations.

Solution 1: Round-robin Replacement Algorithm

In the round-robin replacement, a register will be used to keep track of the replacement item pointer.

Time	Page Request	Memory Location Block Contents				Round-Robin Pointer	Result
		0	1	2	3		
0	1	—	—	—	—	0	miss
1	5	1	—	—	—	1	miss
2	7	1	5	—	—	2	miss
3	1	1	5	7	—	3	hit
4	B	1	5	7	—	3	miss
5	3	1	5	7	B	0	miss
6	7	3	5	7	B	1	hit
7	1	3	5	7	B	1	miss
8	B	3	1	7	B	1	hit
9	1	3	1	7	B	1	hit
10	4	3	1	7	B	2	miss
11	B	3	1	4	B	3	hit
12	7	3	1	4	B	3	miss
13	1	3	1	4	7	0	hit
14	5	3	1	4	7	0	miss
15	4	5	1	4	7	1	hit
16	1	5	1	4	7	1	hit
17	1	5	1	4	7	1	hit
18	5	5	1	4	7	1	hit
19	B	5	1	4	7	1	miss

Table 9.11

With the round-robin replacement, there were 10 hits in 20 accesses. $H = 0.50$.

Solution 2: Least Recently Used Algorithm

In this algorithm, the table is required to keep track of the contents of a block location as well as the age of the block. Block contents start with an age of 0. Each time there is a miss, a new block is brought in with an age of 0 and (if necessary) the oldest block is removed. Each time a block currently in M_1 is accessed, its age is reduced to 0. For each memory cycle in which a block is not referenced, its age advances by 1.

Time	Page Request	Memory Location Block Contents				Result
		0	1	2	3	
0	1	—	—	—	—	miss
1	5	1/0	—	—	—	miss
2	7	1/1	5/0	—	—	miss
3	1	1/2	5/1	7/0	—	hit
4	B	1/0	5/2	7/1	—	miss
5	3	1/1	5/3	7/2	B/0	miss
6	7	1/2	3/0	7/3	B/1	hit
7	1	1/3	3/1	7/0	B/2	hit
8	B	1/0	3/2	7/1	B/3	hit
9	1	1/1	3/3	7/2	B/0	hit
10	4	1/0	3/4	7/3	B/1	miss
11	B	1/1	4/0	7/4	B/2	hit
12	7	1/2	4/1	7/5	B/0	hit
13	1	1/3	4/2	7/0	B/1	hit
14	5	1/0	4/3	7/1	B/2	miss
15	4	1/1	5/0	7/2	B/3	miss
16	1	1/2	5/1	7/3	4/0	hit
17	1	1/0	5/2	7/4	4/1	hit
18	5	1/0	5/3	7/5	4/2	hit
19	B	1/1	5/0	7/6	4/3	hit

Table 9.12

With the least recently used algorithm there were 12 hits in 20 memory accesses, $H = 0.60$. Although the above example demonstrates the replacement algorithms, the sequence of memory requests should not be considered as a realistic test of the algorithms.

Memory Coherency

It is important to maintain coherency between information in physical memory and virtual memory. Coherency is lost when there is a write operation to physical memory not written through to virtual memory. Because information must be transferred in blocks, a write-through requires that an entire page from physical memory be written to virtual memory, even if only one word was changed. This is very inefficient.

A more efficient method of maintaining coherency is to perform write-throughs for information in a block only when a new page is to be transferred to the block. This is a **memory swap**. The page in the block to be replaced is returned to virtual memory. The new page is transferred from virtual memory to the vacated block. A write-through is not required if the information in the physical memory has not been changed. A **change bit** is maintained in the replacement information field to indicate any write operations to the page. A valid bit might also be maintained in the virtual memory to indicate a particular page has been changed in M_1 but not yet updated in the secondary memory.

Summary

This chapter is concerned with system memory. For modern computers, memories are developed using either electronic flip-flops or magnetic-surface materials. Memories can be classified as non-volatile or volatile, depending on their behavior when power is removed. Memories can be static or dynamic, depending on their need for periodic refreshing. Memories can be random access, sequential access, or temporal access, depending on their addressing characteristics. Content-addressable memories allow access based on the contents of a portion of the data entry.

Memory speed can be increased by either using faster devices or increasing the data bandwidth of the memory system. Memory bandwidth can be increased by multiple-word access or by pipelining the access process. Both methods introduce problems associated with branching and nonsequential memory access.

Memory hierarchies allow the designer to use a mix of memory technologies to attempt to gain the advantages of each while minimizing the disadvantages. A complete hierarchy might include a small and very fast electronic cache memory, a moderately large solid-state RAM memory, a large sequential-access magnetic disk, and a very large bulk-storage magnetic tape. The performance of a memory hierarchy can be calculated if the hit ratios for various elements in the hierarchy are known. High ratios lead to better overall performance.

Cache memory is a very fast memory that stores a subset of data stored in the main memory. It is managed by system hardware. If data is in the cache, it can be acquired at access speeds similar to the processing rate of the datapath.

Virtual memory is an attempt to create a mix of memories that appear to the programmer as a single large fast memory. Good performance is based on techniques that use locality-of-access principles. Management of the system is performed by the operating system, which attempts to swap required pages of memory from secondary storage to main memory. Good page-replacement algorithms can improve overall system performance.

For additional information and references on the topics presented in this chapter, access the World Wide Web page provided at: http://www.ece.orst.edu/~herzog/docs.html.

Key Terms

associative mapped cache	logical address space
associative mapped virtual memory	memory coherency
available address space	memory cycle
cache hit	memory hierarchy
cache memory	memory swap
change bit	overlays
content-addressable memory (CAM)	page fault
clusters	page replacement
data segments	page sets
direct mapped cache	physical address space
direct mapped virtual memory	sequential access memories
disk memory	set-associative mapping
displacement	tag
dual port memory	tape memory
hit ratio	temporal access memories
key	thrashing
latency time	valid bit
line	virtual memory

Readings and References

Rosenberg, Jerry M. *The Computer Prophets*. New York: The Macmillan Company, 1969.
This very readable and concise book follows the intertwined developments of a wide variety of clever inventors whose activities led to the development of the modern digital computer.

Exercises

9.1 A certain disk memory spins at 1200 RPM. It has 80 tracks with 9 segments per track and 512 bytes per segment.
 (a) What is the disk capacity?
 (b) What is the maximum data transfer rate?
 (c) What is the average rotational latency?

9.2 A hard disk memory has 512 tracks and rotates at 3600 RPM. There are 32 sectors per track; each sector has 1024 bytes. Read head movement between tracks requires 20 milliseconds. How long would it take to read all of the data on the disk?

9.3 A complete file often requires many consecutive sectors on a disk surface. After reading the first sector, data must be stored and decisions made concerning the reading of the next sector. What happens if the storage and decision time is greater than the time to advance to the next sector? What can be done to help this problem?

9.4   In several computers in the 1950s, a rotating magnetic drum was used as a program memory and data memory. Individual words were addressable from the drum. What is the major performance limitation of this system?

9.5   Magnetic tape recording density is 1600 bytes per inch. Records consist of 64K bytes and are separated by an inter-record gap of 0.25 inches. How much data can be stored on a 2400 foot reel of tape?

9.6   What is the difference in compression between a record written on track 0 at a location of 5 inches from the hub and a track written at 4 inches from the hub?

9.7   A file for a hard disk that rotates at 3600 RPM is allocated as shown in Table 9.13. Assume that average time to move a head is 40 milliseconds and average time for the first access of a sector on a track is 8 milliseconds. There are 17 sectors per track. What is the time required to access the complete file?

Record	Track/Sector
1	1/2
2	1/3
3	2/5
4	2/15
5	4/15
6	4/16
7	4/17
8	5/1
9	23/1
10	23/2

Table 9.13

9.8   In the previous problem, if a reallocation of storage could be made, what is the minimum access time that could be achieved?

9.9   Many credit cards have a magnetic stripe on the back with the customer's name and identification number. From what you know about magnetic storage, how do you think the information is read from the card?

9.10   If magnetic disks are nonvolatile, is it possible to erase them?

9.11   A relay is an electromagnetic device. When energized, a set of switch contacts that are normally open (NO) are closed. A second set of switch contacts that are normally closed (NC) are opened. Determine what happens to the circuitry shown when it responds to the pattern of SET and CLEAR inputs shown in Figure 9.28. Describe the memory characteristics of the device.

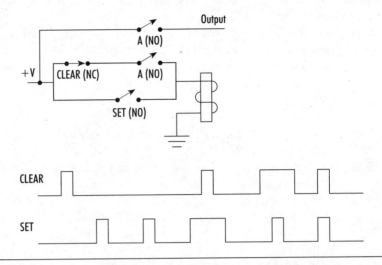

Figure 9.28

9.12 Many floppy disks have a small hole punched in the disk. What is the probable purpose of this hole?

9.13 In a stack memory, what is the purpose of the EMPTY and FULL status signals? Is it possible for the device to perform without them?

9.14 An alternate approach to Figure 9.6 in constructing a queue is to establish a fixed location for the head of the queue. In this case a decision must be made concerning the storage location for new data during a PUSH. This arrangement is called a fall-through queue, for obvious reasons. Show a possible datapath for implementing a fall-through queue.

9.15 Sometimes it is necessary to gain access to both the head and the tail of the queue to store and retrieve data. Such a device is called a double-ended queue, or deque (pronounced deck). Four operations are possible; PUSH_HEAD, PUSH_TAIL, POP_HEAD, and POP_TAIL. Two status signals, FULL and EMPTY, are needed. Show a possible datapath for implementing a deque.

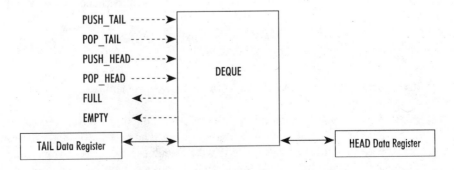

Figure 9.29   A Double-Ended Queue (DEQUE)

9.16 Could an 8K × 8 content-addressable memory be used as a data memory for a small computer such as BART? Explain your reasoning.

9.17 Show how a queue and deque can be designed using a dual port RAM.

9.18 Show a datapath and register-transfer statements for implementing a queue with a RAM.

9.19 Show a datapath and register-transfer statements for implementing a deque with a RAM.

9.20 Computer A and computer B share an 8K × 8 dual port data memory. Show a flow chart for a process in which computer A transports a 1K block of data to computer B. Pay special attention to the status signals.

9.21 Could a dual port memory serve simultaneously as a program memory for two computers? Explain your reasoning.

9.22 Is it possible to have a computer in which the available address space, A_A, is larger than the logical address space, A_L? Describe the characteristics of such a computer.

9.23 Dynamic memory can be refreshed by reading or writing. Consider a 64K dynamic memory system. Each read/write refreshes all 256 bytes having the same first 8 bits of address. During each instruction cycle, there is one clock cycle not requiring use of the system buses. Would it be possible to use this spare clock cycle to refresh the DRAM? Assume that memory must be refreshed every 20 milliseconds. The system clock is 6 MHz. All instructions use six clock cycles.

9.24 Consider a situation in which the memory access time, t_A, is significantly greater than both the address presentation time, t_P, and the processor data capture time, t_C. This is often the case with large random access memories where the access time can be a factor of ten greater than t_P and t_C. How does this affect pipeline memory access performance?

9.25 There is a problem in bank switching. When switching banks for data, successive instruction fetches are also made to the new bank. Suggest several possible solutions for this problem.

9.26 In Figure 9.11 a multiplexer is used to select the proper byte from the 4-byte databus. Instead of using a multiplexer, could an individual READ be generated to gate the proper memory element onto the data bus? What effect would this have on the effective access time of the memory?

9.27 A computer has a two-level memory system consisting of a main memory (M_1) and a disk memory (M_2). The access time for $M_1 = t_{m1} = 0.1 \times 10^{-6}$ seconds. The access time for $M_2 = t_{m2} = 20 \times 10^{-3}$ seconds. What must the hit ratio be to achieve an average access time of
(a) 1×10^{-3} seconds?
(b) 1×10^{-6} seconds?
(c) 0.15×10^{-6} seconds?

9.28 Under the same conditions as the previous problem, what must the hit ratio be to achieve an efficiency of 90%?

9.29 A memory system for a computer consists of a single physical memory with 100M bytes of DRAM with 100 nanosecond access time. The cost is $1000. It is proposed to replace the DRAM with a 1M-byte SRAM with 50-nanosecond access time and a cost of $100 along with a 100M bytes disk, access time 15 milliseconds, cost $200. What must the hit ratio be for this to be a good economic decision without losing performance?

9.30 In a virtual memory system, the main memory is of fixed size. Indicate two methods which might be used to improve the hit ratio.

9.31 In what computational environment will additional virtual memory improve system performance? In what environment will its addition decrease system performance? Why?

9.32 In what computational environment will the addition of cache memory improve system performance? In what environment will its addition decrease performance? Why?

9.33 Compare the concept of a page and a segment in a virtual memory system.

9.34 What is the major advantage and disadvantage of using a directly mapped cache memory rather than a fully associative cache memory?

9.35 In some cache memory systems, the unit of data swapped between the cache and main memory is one word. Why is the unit always much larger in swaps between M_1 and M_2?

9.36 A system with M_1 and M_2 has $t_{m1} = 150$ nanoseconds, $t_{m2} = 20$ milliseconds. The resulting overall access time is 250 nanoseconds. It is proposed to add a cache memory with $t_{m0} = 50$ nanoseconds to try to achieve an overall access time of 75 nanoseconds. What is the efficiency of this system before and after the introduction of the cache? Is this performance improvement reasonable? Explain your conclusion.

9.37 What is the advantage of using a random page-replacement policy in a virtual memory system?

9.38 What would be the effect of having memory mapped I/O in the range of the cache memory?

9.39 In the Intel 82385 cache controller, how is coherency maintained between the cache memory and the main memory?

9.40 In a virtual memory system, assume all page tables are stored in memory with a 50-nanosecond access time. Block access from M_2, a disk memory, is 30 milliseconds (worst case). How much time is required for memory access in a virtual memory system in the following cases?

(a) Direct mapping is used and there is a hit.

(b) Direct mapping is used and there is a miss.

(c) Associative mapping is used and there is a hit.

(d) Associative mapping is used and there is a miss.

9.41 In the example of Section 9.10 regarding replacement algorithms in virtual memory systems, it was stated that the sequence of memory requests was not reasonable. In what way is it not reasonable? Would a more reasonable sequence provide a higher hit ratio or a lower hit ratio? Explain your conclusion.

9.42 Repeat the virtual memory example in Section 9.10 using a replacement algorithm which uses a longest time until next read as the basis for preemptive storage allocation.

10
System Input/Output and Interfacing

Henri Jacquard

A Frenchman named Henri Jacquard
Thought his weavers were working too hard
He cured the stagnation
With new automation
By inventing the data punch card

Herman Hollerith

An American by the name of Hollerith
Had an excessive shirt collar width
Census analysis was hard
So he used a punched card
Exactly the size of a dollar width

Prologue

Herman Hollerith, the son of German immigrants, was born in the United States in 1860. As a young engineer working for the United States Bureau of the Census, he was responsible for developing techniques to decrease the excessive manual labor and time required to tabulate the census data.

The census, a counting of the entire population, is mandated by the United States Constitution to be performed every ten years. The 1880 census took almost 7.5 years to tabulate. The 1890 census, with a population approaching sixty million people, would have exceeded a decade to process without technological intervention.

Hollerith, in researching the problem, was exposed to the work of Henri Jacquard and his punch-card controlled weaving apparatus. He adapted the idea of the punch card to enumerate attributes of the population. Holes in the card were sensed using a wire sensor to complete an electric circuit when a hole was present. Composite totals were accumulated by electromechanical ratchet counter devices.

To speed up the process, Hollerith made his punched card 3.25 inches by 6.625 inches, exactly the size of an 1860 dollar bill. This allowed him to use the same apparatus to move his punched cards as the Treasury Department used to mechanically move currency. This size remains the standard for the punch card, although the size of the currency has decreased slightly.

Hollerith's project was successful. The 1890 census required only two years to tabulate. Hollerith went into business for himself, establishing the Hollerith Tabulating Company. In 1911, his company merged with several other companies to form the Computing-Tabulating-Recording Company. In 1924, this company was renamed International Business Machines, IBM, with Thomas J. Watson as the president.

10.1   Introduction

Up to this point, we have been concerned with systems with the following constraints:

- All components are located in a compact area spanning no more than a meter.
- All data sources and storage components share a single clock.
- All data links have high bandwidth; data placed on a data bus in one portion of the system are available without distortion and with negligible delay time in another portion of the system.
- All data sources provide valid data within a known period of time.
- All data destinations accept data within a known period of time.
- All interconnection links are parallel and are capable of transmitting all data bits of a word simultaneously.
- There is no contention for use of the data buses.
- All systems have a single controller.

These characteristics are encountered in the majority of small computing structures. In this chapter, the effects of relaxing some of the constraints will provide the following special cases.

- **Unsynchronized systems.** Two or more systems need to share information and resources. Each has its own independent and unsynchronized clock, controller, and datapath.
- **Spatially separated systems.** Systems physically separated by more than a few meters should not share a clock due to clock skew problems. Serial interconnection structures are less costly than parallel data structures. Radio, telephone, and optical communication links are inherently serial.
- **Systems containing components with unpredictable response times.** Devices such as analog-to-digital converters and memory units may have response times that are data dependent or data-location dependent. For example, the time to acquire data from a disk depends on the relative position of the data to the disk read head.
- **Distributed systems.** Distributed systems are characterized by a spatial distribution of components and a distribution in the overall processing algorithm. Since there may be considerable autonomy in the individual devices, there can be conflicting demands for data buses and other shared resources.
- **Event driven systems.** Many systems must respond to external events, called **interrupts**, which are caused by uncontrolled real-time actions. Such events occur at unpredictable times, are not synchronized with the system clock, and may be of uncertain duration.
- **Digital computer input/output.** Digital computer I/O structures must perform a wide variety of data exchanges and responses to interrupt signals and direct memory access (DMA) requests.

This chapter is concerned with data transmission and synchronization of systems and digital computers that have one or more of the above characteristics. Such systems will require the development of more sophisticated techniques for exchanging information. This will include using handshakes for data synchronization and bus arbitration for sharing communication links.

10.2   The Information Exchange Model

Figure 10.1 shows a structure for two systems, or modules, required to exchange information. The modules can be digital information-processing systems or digital computers. Each has its own system controller, datapath, and system clock. The clocks for the two systems are neither synchronized nor of the same frequency. There are two aspects to the flow of information. There must be an interconnection link for the data, and there must be a means of synchronizing and controlling the exchange of data.

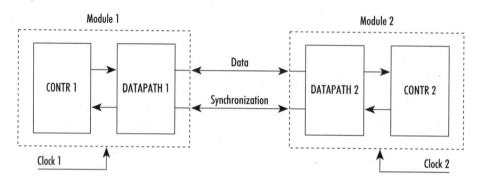

Figure 10.1   Information Exchange between Two System Modules

The **talker** is the source of the information; the **listener** is the destination. The device that issues control signals to initiate and control the data transfer is the **master**; the device that responds and provides status signals is the **slave**. Systems can have only one master, but may have several slave units. In a successful data transfer, a word from a register in the talker is reliably transferred to a register in the listener.

The interchange of information between two devices to coordinate the data exchange is known as **handshaking**. Handshakes are interlocked. Each participant senses a transition in a control signal from the other before asserting its own control signal. Assertion levels are maintained until a change in the handshake partner is detected. Since the duration of each handshake signal is not specified, fast devices are able to exchange information with slow devices. In some situations, a maximum duration, or **time-out** of a handshake assures that a malfunction doesn't cause lockup in operations.

Talker-initiated Data Transfers

Data transfers involve the cooperative actions between a talker module and a listener module. In the most general case, a data transfer involves four sequential steps. In many situations examined in later sections one or more of the actions may be omitted. The data-transfer unit is a word. Figure 10.2 shows the sequence of operations in a talker driven data transfer. The talker-module is the master; the listener is a slave.

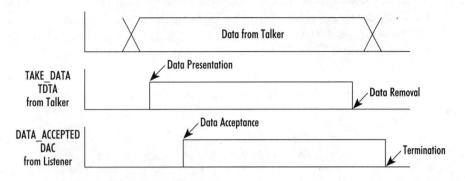

Figure 10.2 Timing Diagram for a Talker-driven Handshake

- **Data presentation.** The talker provides a unit of data. As a master, it asserts a TAKE_DATA (TDTA) control signal to indicate that a data transfer from talker to listener is to take place.
- **Data acceptance.** By noting the TDTA control signal, the listener determines that valid data is being presented. It accepts the data and stores it in a register or memory location. It acknowledges reception of the data with DATA_ACCEPTED (DAC).
- **Data removal.** When the talker detects the DAC status signal it removes the TDTA control signal. This acts as a confirmation to the listener that the acceptance of the data has been noted. The data is removed from the data link.
- **Termination.** The listener removes any signals that have been issued, indicating data acceptance. All conditions are as they were before the presentation of data.

Listener-initiated Data Transfers

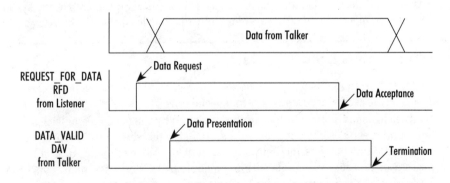

Figure 10.3 Timing Diagram for a Listener-driven Handshake

Figure 10.3 shows the sequence of operations in a listener-driven data transfer. The listener is the master; the talker is the slave.

- **Data Request.** The listener requests data from the talker by asserting REQUEST_FOR_DATA (RFD).

- **Data presentation.** The talker provides a word of data and asserts DATA_VALID (DAV).
- **Data acceptance.** The listener accepts the data, stores it, and acknowledges reception by removing the RFD data request.
- **Termination.** The talker removes the DAV signal and the data. All conditions are as they were before the data transfer.

Synchronizing Handshake Signals

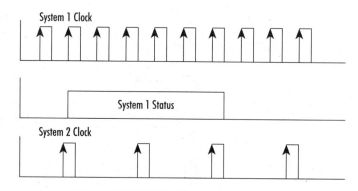

Figure 10.4   Synchronizing Status Signals between Asynchronous Systems

An essential characteristic of a handshake is the ability of the talker and listener to reliably determine the status/control signals of the other. This requires that the signals provided by system 1 be reliably captured in a system 2 register using the system 2 clock, and vice versa. This is known as **synchronizing** the status/control signals. A necessary and sufficient condition to accomplish this is shown in Figure 10.4. A signal generated by system 1 must have a duration of at least two clock cycles of the system 2 clock. This assures that the signal will be present for at least one full clock cycle and thus meet any data setup requirements of the system 2 register. By a similar argument, signals from system 2 must last at least two clock cycles of the system 1 clock.

10.3   Synchronous Data Transfers

The model presented in the previous section applies to the most general case of information interchange. This section examines several special cases.

Synchronous data modules are characterized by the use of a single clock in the talker and listener and a set of control signals and status signals which are synchronized to the clock. Clock skew and data-transmission delays are assumed to be negligible. It is not necessary to synchronize status and control signals within a module since they are already synchronized to a common clock. Most data exchanges within a single system are synchronous.

Fully Synchronized Data Transfer

In **fully synchronous data transfers**, both the talker and listener have all control, status, and data synchronized to the same clock. Most data transfers are completed within a single clock cycle. For slow devices, more than one clock cycle may be allocated. In all cases, the time duration required for a data transfer is known. Most data transfers between registers and memory elements within a single module are fully synchronous.

Talker-driven Synchronous Data Transfer

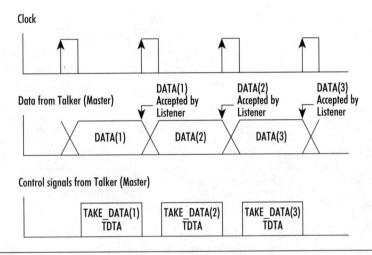

Figure 10.5   A Talker-driven Synchronous Data Transfer

The four-step data transfer process can be considerably simplified when performing synchronous data transfers. Figure 10.5 shows the timing for a talker-driven data transfer between two synchronous modules on a dedicated data link. The talker provides data and the TAKE_DATA (TDTA) control signal. The listener responds by accepting the data at the end of the current clock cycle. In the fully synchronous data transfer, response time of the slave is fully predictable. The talker is aware that the listener will accept data in a known period of time, usually one clock cycle. There is no need for an explicit data accepted (DAC) status signal, since there is no uncertainty to be overcome.

Listener-driven Synchronous Data Transfer

Figure 10.6 shows a listener-driven synchronous data transfer. The listener requests data by issuing a REQUEST_FOR_DATA (RFD) control signal. The talker places data on the data link. There is no need to provide a DATA_VALID signal since the listener knows that the data will be provided in the current clock cycle. The listener accepts the data at the end of the clock cycle. A status signal indicating acceptance is not necessary. The talker removes the data at the end of the clock cycle without informing the listener.

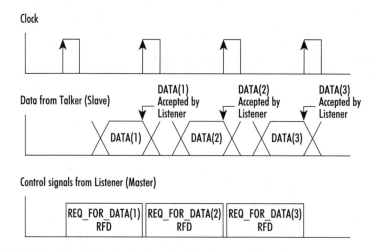

Clock

Data from Talker (Slave)

DATA(1) Accepted by Listener

DATA(2) Accepted by Listener

DATA(3) Accepted by Listener

DATA(1)   DATA(2)   DATA(3)

Control signals from Listener (Master)

REQ_FOR_DATA(1) RFD

REQ_FOR_DATA(2) RFD

REQ_FOR_DATA(3) RFD

Figure 10.6   A Listener-driven Synchronous Data Transfer

Semi-synchronous Data Transfer

A **semi-synchronous data transfer** is also performed between system modules driven by a single clock. All data transfers are clocked and take place on a clock transition. In contrast to fully synchronous data transfer, the exact clock cycle for one or more of the data transfer operations is uncertain. In the talker-driven data transfer, the clock cycle in which the listener will accept the data is uncertain. The listener must identify the clock cycle by providing a DATA_ACCEPTED (DAC) status signal.

Similarly, in the listener-driven data transfer, the clock cycle used by the talker to return requested data is uncertain. The talker must identify this cycle by asserting a DATA_VALID.

Talker-driven Semi-synchronous Data Transfer

Figure 10.7 illustrates a talker-driven semi-synchronous data transfer. The talker provides the data and the TAKE_DATA (TDTA) control signal maintains them, if necessary, for several clock cycles. The listener may be involved in another processing action and unable to sense or respond immediately to the TDTA. When the listener has captured the data, it informs the talker by asserting a DATA_ACCEPTED (DAC). The talker then de-asserts the TDTA. Data is removed and the operations terminate in the next clock cycle.

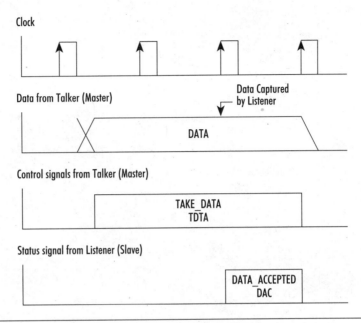

Figure 10.7   A Talker-driven Semi-synchronous Data Transfer

Listener-driven Semi-synchronous Data Transfer

The companion listener-driven semi-synchronous data transfer is shown in Figure 10.8. The listener requests data with REQUEST_FOR_DATA (RFD) but must wait an uncertain number of clock cycles until the data is provided by the talker. This delay may be due to other operations performed by the talker. It might also be caused by a time delay necessary to retrieve the requested data, perhaps from a disk, an analog to digital converter, or a RAM memory device with a long access time. The RFD control signal is maintained until data is provided and a DATA_VALID status signal is received. The listener captures the data on the next clock pulse and the data transfer ends.

Both forms of semi-synchronous data transfers have potential problems. After receiving a data-transfer request from the master, the slave may fail to respond. The master may stall in the processing algorithm while awaiting a response from the slave. If the slave is not present or out of service, a response may never come. A possible solution is to use a time-out procedure. The data-transfer algorithm is designed such that the master never waits for longer than a fixed length of time for a response by the slave. The processing algorithm can consider the proper alternative action to take if a time-out occurs.

The lack of response by a slave unit might also indicate an error condition. For example, if a slave-listener unit receives erroneous data (perhaps a parity violation) it can reject the data by failing to acknowledge its arrival, forcing the master to retransmit the data. Transmission errors are corrected. If there are three consecutive no-response situations, for example, the slave is assumed to be nonfunctioning and the processing algorithm is terminated.

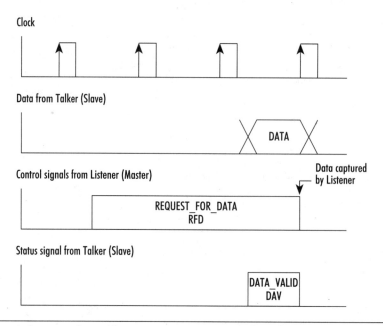

Figure 10.8   A Listener-driven Semi-synchronous Data Transfer

The Wait State

The semi-synchronous listener-driven data transfer described in the previous section is sometimes used in a digital computer to request data or instructions from the system memory. In the presentation of the previous section, the talker issues status signal DATA_VALID when it provides data. As an alternative, the talker can issue a WAIT status signal when it doesn't have data but is actively working on acquiring it, as shown in Figure 10.9. If a WAIT status is not received by the listener in the first clock cycle, it assumes that data will be provided in that clock cycle; no delay is necessary. If a WAIT is received, the listener waits until the first clock cycle after the WAIT is removed and then captures the data.

A WAIT state is sometimes used with memory components. A manufacturer may, for example, use slower and less expensive memory devices by adding circuitry which generates one or more WAIT states every time there is a memory access. This is far preferable to slowing down the system clock to accommodate slow memory devices within a single clock cycle.

In the use of dynamic memory chips, memory refresh operations are required at relatively infrequent time periods. A WAIT status can be issued during refresh periods when the memory cannot respond to data requests.

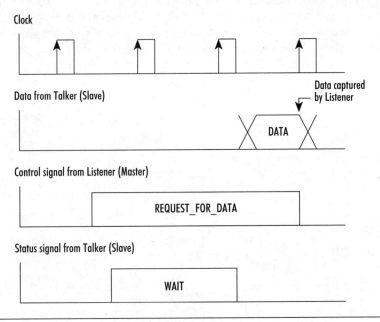

Figure 10.9   *Semi-Synchronous Data Exchange Using a WAIT State*

10.4   Asynchronous Data Transfers With Handshakes

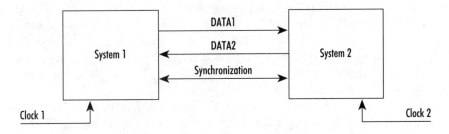

Figure 10.10 *Data Exchange between Asynchronous Systems*

Figure 10.10 shows two autonomous systems wishing to interchange information. Each has its own controller, processing architecture and clock. The clocks are not synchronized and may have widely different clock rates. Both systems may participate in extensive activities independent of the other. At times, however, they want to exchange information. An essential characteristic of a successful data exchange is that each be aware of the status of the other.

Talker-initiated Two-line Handshake

Figure 10.11 illustrates the hardware structure of a system using a talker-initiated two-line handshake. The timing of the handshake transitions was shown in Figure 10.2. The talker (system 1) provides the data and then asserts the TAKE_DATA or TDTA1 control signal. The listener (system 2) monitors the talker's TDTA1 signal. This signal is captured into the TDTA1* register. The "*" indicates the signal is now synchronous within system 2 since it was captured with the system 2 clock. The listener, upon sensing the TDTA1* transition, processes the valid data from the synchronized DATA1* register. When processing or storage is complete, system 2 indicates acceptance by asserting the DATA_ACCEPTED, or DAC2, control output.

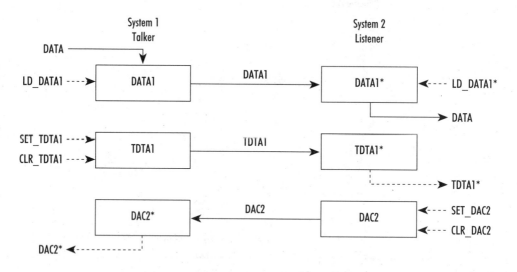

Figure 10.11 Architecture to Support a Source-driven Handshake

The talker recognizes that the data has been accepted by monitoring its DAC2* register, the synchronized DAC2 signal. When DAC2* is detected, the talker first removes the TDTA1 control signal, and a short time later removes the data. The talker is free to engage in other activities.

When the talker turns off TDTA1 it acknowledges receipt of the listener's DAC2 signal. The listener responds by turning off its DAC2 signal. This completes the handshake and the listener may now engage in other activities.

A potential problem with this data exchange is that the talker does not know if the listener is physically capable of responding to the data and the TDTA1 signal. There could be a considerable, perhaps intolerable, delay between the TDTA1 assertion and the DAC2 response. If the response never comes, the talker may be prevented from performing other useful activities. Again, a time-out can be imposed to prevent this.

Figure 10.12 shows a microoperation diagram to support a talker-driven handshake. Outputs from system 1 are status inputs for system 2 and vice versa. The interlocking nature of the hand-shake is evident from the multiple testing blocks. The listener, for example, remains in processing block W until TDTA1 is provided by the talker. Each system remains in a looped test condition until a transition by the other is detected.

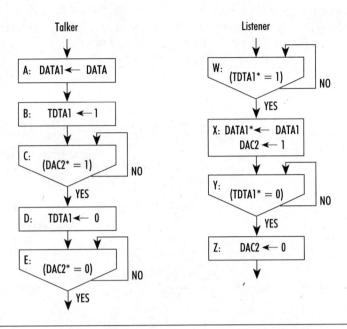

Figure 10.12 A Microoperation Diagram for a Talker-driven Handshake

Listener-initiated Two-line Handshake

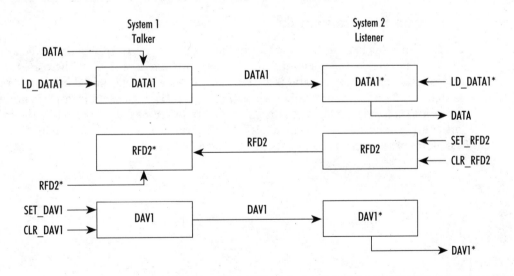

Figure 10.13 Architecture to Support a Listener-driven Handshake

Figure 10.13 shows hardware to support the listener-driven handshake. The listener makes the initial action by asserting a REQUEST_FOR_DATA (RFD2). In response, the talker provides data and then asserts its DATA_VALID (DAV1). When the listener has sensed the DAV1, it captures and processes the data and then turns off its RFD2. The talker completes the handshake by turning off the DAV1 in response to the removal of RFD2.

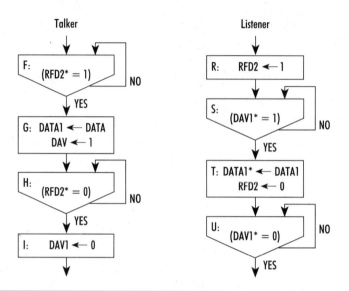

Figure 10.14 Microoperation Diagram for a Listener-initiated Handshake

Figure 10.14 is the microoperation diagram for a listener-initiated handshake.

The Three-line Handshake

The three-line handshake, as shown in Figures 10.15 and 10.16, can be viewed as a combination of the features of the talker-driven and listener-driven handshakes. There are six sequential actions involving control and status signals. The listener provides initial information concerning its ability or desire to receive data with a RFD2 signal (1). If the data transfer is initiated by the source, RFD2 is interpreted as the physical ability to accept data provided by the talker. If the data transfer is initiated by the talker, RFD2 is interpreted as a demand for data from the source. The source provides data and asserts its DAV1 signal (2).

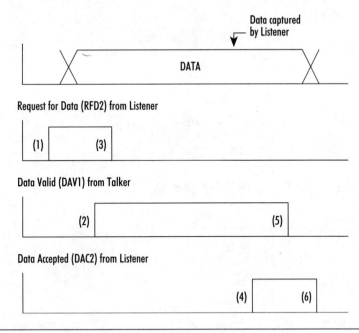

Figure 10.15 Timing Diagram for a Three-line Handshake

The listener acknowledges availability of the data by turning off its RFD2 (3). In this condition, with DAV1 asserted and neither RFD2 nor DAC2 asserted, the listener acknowledges the availability of the data but has not yet completed capturing and processing the data. Hence, the talker is required to maintain the data and the DAV signal.

When the listener completes the capture of the data, it asserts its DAC2 signal (4). The talker responds by turning off its DAV1 and removing the data (5). The listener responds by turning off its DAC2 (6).

With the three-line handshake, both the talker and listener have a more extensive knowledge of the physical and logical status of the other. The data transfer can be initiated by either the talker or the listener. This handshake is used in the IEEE 488 instrumentation bus.

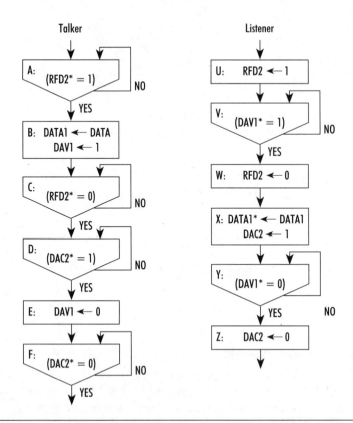

Figure 10.16 Task Diagram for a Three-line Handshake

10.5   Asynchronous Data Transfers to Multiple Listeners

In some situations a single talker must communicate asynchronously with multiple listeners. In one possible configuration, all devices share a single data bus. Additional shared buses are used for synchronization and handshake signals.

Single Talker to Single Listener

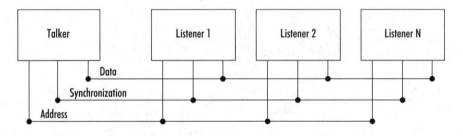

Figure 10.17 A Talker with Multiple Addressed Listeners

In Figure 10.17 a single talker shares a data bus, address bus, and synchronization bus with multiple listeners. Data transfers originate from the talker and are directed to a single listener. An address bus selects the listener. Each listener is assigned a unique address and is responsible for decoding its address from the bus. Only the selected listener participates in the data-transfer handshake.

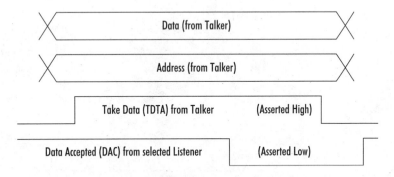

Figure 10.18 Timing Diagram for an Addressed Data Transfer

Figure 10.18 is a timing diagram for the data-transfer handshake. The talker provides the data and address and asserts the take data (TDTA) handshake line. All listeners receive the TDTA signal but only the selected unit responds by accepting the data and asserting the data accepted (DAC).

The individual DACs are connected to a DAC bus through open collector bus driver circuitry. With open collector drivers, multiple signals can be on the bus simultaneously. Because of the characteristics of the circuit electronics, the "low" signal dominates other signals on the bus. Any driver can drive the bus to a low condition, even if all other devices are attempting to drive it to its high condition. This is an OR for signals asserted low. Only when all devices are in their high condition is the bus in its high condition. This is an AND function for signals asserted high.

When used with a single listener, the low bus condition is associated with the assertion of the DAC. Therefore, the talker receives a DAC signal whenever any of the listeners provides one; they are ORed on the DAC bus. Only the addressed listener responds by providing a DAC, asserted low, at the proper time in the handshake. This procedure allows devices to be added and removed from the bus without affecting the bus logic of the talker or listener.

Single Talker to Multiple Listeners

The interconnection structure shown in Figure 10.17 may also be used to simultaneously broadcast information from a talker to multiple asynchronous listeners. A broadcast type data transfer requires that all of the listeners participate in the handshake.

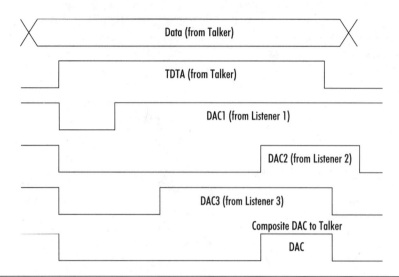

Figure 10.19 Timing Diagram for Broadcast Data Communication

The data is provided by the talker and the TDTA signal is asserted. Each of the listeners responds immediately to the TDTA by de-asserting its DAC signal, indicating that it has not yet captured the data. Each unit then captures the data and asserts an individual DAC signal. The talker must maintain DATA and TDTA until all devices have responded affirmatively with DAC. This requires a logical AND of the DAC assertions which is accomplished by asserting the DAC signal high to perform the bus AND operation. This is shown in the timing diagram of Figure 10.19. The DAC bus remains in its non-asserted, low condition until all DAC lines from the individual listeners are asserted high. This informs the talker when the data transfer to all listeners has been completed.

10.6   Multiple Talker Channel Access Protocols

The previous section presented systems with multiple listeners and a single talker. A system involving multiple talkers and multiple listeners sharing a single data bus presents a more complex problem. Electrical interactions prohibit more than one talker from successfully using the bus simultaneously. A discipline must be imposed to assure that access to the data channel, address bus, and synchronization bus is shared in a fair and reasonable manner. A set of rules to accomplish this is known as a **channel access protocol**. These rules may be implemented either by the talker units themselves, known as a **distributed protocol**, or by a separate **bus master**, known as a **centralized protocol**.

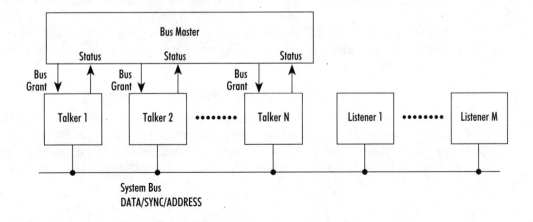

Figure 10.20 A System with Multiple Talkers and a Bus Master

Figure 10.20 shows implementation of a centralized protocol. The bus master is responsible for issuing BUS GRANT control signals to the individual talkers to indicate they have permission to use the shared buses. In allocating the BUS GRANT, the bus master uses status information from the talkers. When BUS GRANT is asserted, the single selected talker may use the system buses to send data to one or more listeners. Nonselected talkers must maintain high-impedance tri-state connections to the buses to prevent interference with activities of the selected talker. When the BUS GRANT signal is removed, the talker must terminate its bus activity within a fixed period of time. The BUS GRANT to the next talker is then asserted.

Time-share Access Protocol

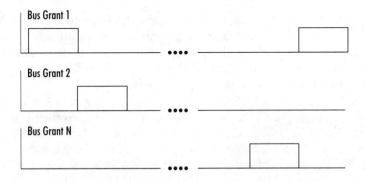

Figure 10.21 A Time-share Access Protocol

Time-division multiplexing, as shown in Figure 10.21, is based on a centralized access protocol. The available time is apportioned among the participating talkers in constant increments. The status of the talkers is not considered. The duration of individual BUS GRANTs is long enough for a talker to perform significant activity but not so long that appreciable time delay is introduced between successive grants. These conditions are obviously dependent on the application environment.

Possible variations of this access protocol include provisions for granting longer periods of bus-access time to selected talkers. Some talkers may also be given multiple time slots within one full access period.

Time-share Access Protocol with Common BUS BUSY

With a time-share system, some units may be granted access to the communication channel for fixed time intervals even when they have no information to send or insufficient information to occupy the full allotted time slot. This obviously is an inefficient use of a resource. In Figure 10.22, a single shared BUS BUSY line has been added between the talkers and the bus master.

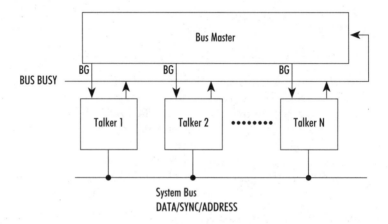

Figure 10.22 A Round-robin bus Allocation with a Common BUS BUSY Status Signal

Talkers are granted the bus in a round-robin sequential manner. When granted the bus, the individual unit asserts the BUS BUSY only for the duration of its needs. If BUS BUSY is not asserted within a predetermined period, this is interpreted by the bus master as an indication that the bus is not required. This allows the bus master to grant use of the channel in a more efficient manner than is possible with time sharing.

In Figure 10.23, a BUS GRANT is issued to talker 1. Talker 1 asserts its BUS BUSY signal, by driving it low, and proceeds to use the bus. A low assertion level is used to perform an OR action on the BUS BUSY bus. When finished, talker 1 releases BUS BUSY. The bus controller then issues a BUS GRANT to talker 2. This talker does not respond, indicating that it does not want the bus or is not functional. The BUS GRANT 2 is promptly removed and a BUS GRANT is issued to talker 3. Talker 3 asserts its BUS BUSY and retains the bus until concluding its data transfer activities.

This method of bus allocation allows each talker to keep the bus as long as necessary to finish its communication activity. Provisions may be needed to prevent a talker from hogging the bus to the exclusion of other devices.

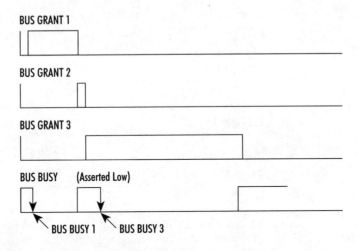

Figure 10.23 Timing for a Round-robin Allocation Protocol with BUS BUSY

BUS REQUEST Access Protocol

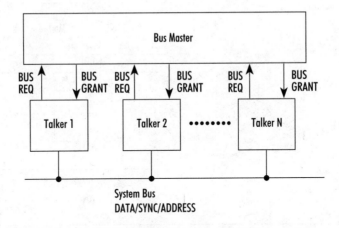

Figure 10.24 Individual Bus Requests in a Multi-talker Environment

When using time-share access techniques, time on the bus resource is lost by attempting to grant bus control to units not needing the bus or perhaps not even functional. Greater efficiency is possible using BUS REQUEST information, as shown in Figure 10.24. The bus master allocates BUS GRANTs only to those units that have requested them. This allocation can be based on time share or other rules, which may recognize different priorities of the requests.

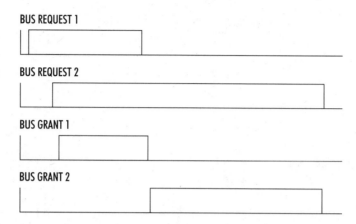

Figure 10.25 Timing for Round-robin Protocol with Individual Bus Requests

In Figure 10.25, talker 1 and talker 2 request to use the bus. Talker 1 is granted the bus and maintains its BUS REQUEST until its data-transfer operation is finished. Talker 2 is granted the bus when talker 1 has released its BUS REQUEST.

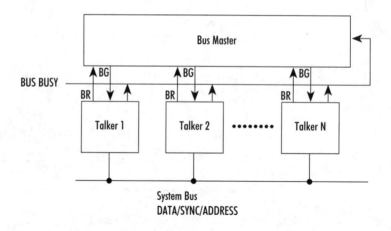

Figure 10.26 A Multi-talker System with BUS REQUEST and BUS BUSY

Combining the individual BUS REQUESTs with a shared BUS BUSY as shown in Figure 10.26, could be more advantageous. By monitoring BUS BUSY, a unit need not request to use the bus unless it determines that the bus is available. This allows additional flexibility for the talker units.

10.7   Input/Output Structures for Digital Computers

Many of the data-exchange mechanisms studied in the previous sections are applicable to the input and output functions of a digital computer. To be useful, a computer must be interfaced to the printers, instrumentation, communication ports, and memory devices that comprise its working environment. Such interfacing is based, to a limited extent, on the instruction set of the machine. The major feature affecting the way in which a computer interacts with its environment is the input/output structure.

The computer I/O facility has four major goals:

- **Synchronization.** The computer and the external environment use different clocks. The I/O port must provide a way to synchronize the external signals to the clock of the computer. This is essential to assure that external data are not altered while being accessed by CPU components.

- **Flow control.** When there is a flow of data to or from the computer across the I/O boundary, it is important that input information be read once, and only once, before the external data source changes it. Output information must be maintained until the external device has had the opportunity to read it. These flow-control operations are accomplished with handshake signals or flags exchanged between the processor and I/O devices.

- **Rapid response to time-critical events.** Some events associated with control activities, data acquisition, or safety functions require the immediate attention of the processor, regardless of its current processing activity. This is accomplished by the use of an **interrupt facility** for the processor.

- **Off-loading of repetitive and time-consuming I/O activities.** Some I/O activities, such as refreshing a video screen, are very I/O intensive. If managed completely by the processor, very little additional processing activity can be accomplished. By using **direct memory access (DMA)** and cooperating **input/output processors** a significant amount of these specialized I/O operations can be performed in parallel with normal processing activities.

The I/O structure of a digital computer can be implemented in several ways depending on the organizational complexity and application requirements.

Dedicated I/O Registers

Many early computers and several modern microcontrollers use one or more internal registers dedicated for I/O operations. Such an organization is shown in Figure 10.27. These registers reside on the internal bus structure of the computer, where they can be serviced with special I/O instructions. Through these registers, data can be transferred across the I/O boundary. While offering an immediately functional I/O interface, this type of organization lacks the flexibility required for most high-performance operations.

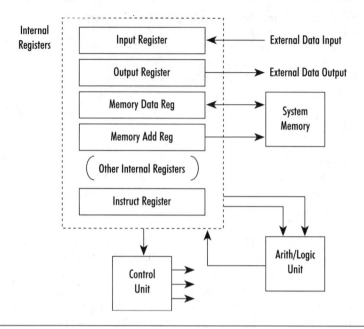

Figure 10.27 A Digital Computer with Dedicated Input/Output Registers

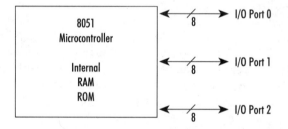

Figure 10.28 The Intel 8051 with Dedicated Input/Output Registers

The Intel 8051, presented in Chapter 8, is typical of the single-chip microcontrollers used extensively in embedded control applications. As shown in Figure 10.28, I/O is provided by three dedicated, 8-bit I/O registers which are directly connected to external pins on the device. Under program control, information can be transferred to or from the registers.

System Expansion Buses

Most modern microprocessors and computers use system expansion buses rather than dedicated I/O registers to easily add additional memory and other resources. These buses consist of three components: An **address bus** supplies addresses to memory and I/O components, a **data**

bus provides a data link to and from the computer, and a **control bus** contains control signals from the system controller necessary to transfer data.

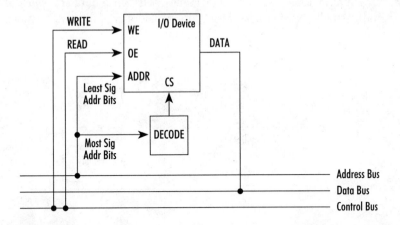

Figure 10.29 Connection of a Device onto a Computer System Bus

Expansion buses generally do not contain the system clock. Instead, they contain a variety of control signals such as READ and WRITE, which are synchronized with the system clock. Figure 10.29 shows a typical interface structure for a device on a digital computer system bus. It could be either a memory or an I/O component. The device is selected for participation in a system bus operation using an address provided on the address bus. The higher order bits are decoded to select the device; the lower order bits are used to select a location within the device. Without the assertion of a chip select (CS), the device is inoperative and provides a high impedance to the system buses.

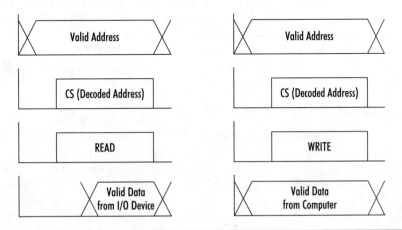

Figure 10.30 System Bus Read and Write Cycles

Figure 10.30 shows a read and write cycle for a system expansion bus component. In contrast to register-transfer operations within the datapath, the READ and WRITE control signals include timing information; a system clock is not used. Since data-transfer operations are synchronized with the single system clock, these data transfers are fully synchronous.

Memory Mapped I/O

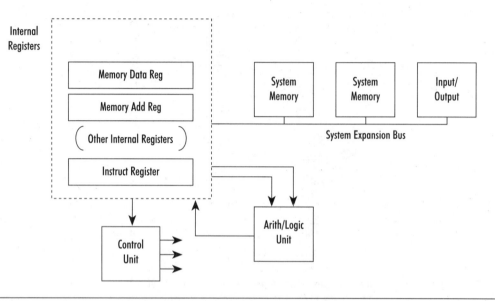

Figure 10.31 I/O Expansion Using the System Expansion Bus

Figure 10.31 shows the system memory bus implementing an I/O structure. Known as **memory-mapped I/O**, I/O registers are mapped into addresses in the computer memory space. The same instructions are used to transfer information to and from memory, and to and from I/O devices. By sharing an address bus, memory-mapped I/O decreases the number of external interconnection pins to be provided by the processor. This is an important advantage when packaging and chip fabrication are considered.

Dedicated I/O Address Space

In systems supporting greater complexity, separate address lines and data buses may be used for memory and I/O operations. Using separate resources for memory and I/O purposes has several advantages. Simultaneous operations among memory and I/O ports are possible and such operations as memory caching and virtual memory are also less complicated.

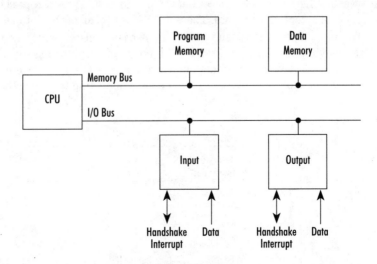

Figure 10.32 A Computer Structure with Separate Memory and I/O Buses

In Figure 10.32 the CPU uses both a memory-expansion bus and an I/O expansion bus. The memory-expansion bus contains address information from the memory address register and a data bus connected to the memory data register. Control signals associated with memory operations, such as READ and WRITE, are also included.

The I/O expansion bus contains a similar assortment of data bus components. An address bus contains address information to allow selection of an I/O device; a data bus provides a connection to an internal I/O data register.

10.8   Programmed I/O

There are several techniques that the computer can use to manage I/O operations. In **programmed I/O**, the I/O system is managed completely by the CPU through the execution of its instruction set. The following discussion assumes that registers associated with I/O operations have been mapped into the memory space of the computer.

Deaf-and-Dumb I/O

The notable characteristic of deaf-and-dumb I/O is the lack of any handshaking or flags to aid in data transfer across the I/O interface. The interface neither accepts status information from the external device (deaf) nor provides any control information to the interface (dumb). Figure 10.33 shows a possible implementation of such a port.

The output data port consists of a data register which accepts and latches information from the processor data bus and provides an output to the external environment. A decoder detects the address for each of the registers. Transferring information to the data register is similar to transferring to a memory element. The output operation is controlled by a WRITE control signal.

An output operation is accomplished by writing data to the I/O address of the output port using an instruction such as store accumulator. Since no status information is interpreted before the transfer, there is no assurance that the previous data was accepted or that the external device is capable of accepting new data. There is no flow control of the data transfer.

The input data port consists of an input data register, which accepts information from the external environment and presents it to an internal computer register via a memory-read instruction such as LDA (load accumulator). An input operation is accomplished by a read from the address associated with the input port. There is no testing to assure that the data is stable, valid, or current.

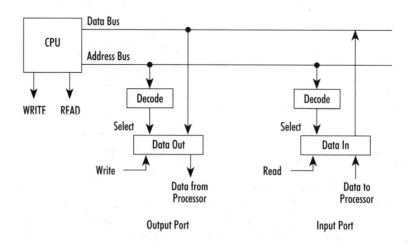

Figure 10.33 Implementation of Deaf-and-Dumb I/O Registers

Handshake I/O

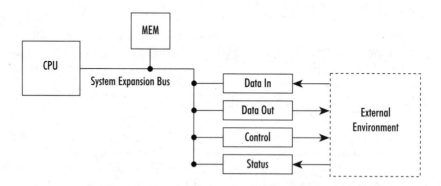

Figure 10.34 Use of Status and Control Registers to Provide Handshake Signals

In handshake I/O, shown in Figure 10.34, additional registers on the expansion bus provide synchronization and flow control for I/O data transfers. A control register is added to the port to provide transfer-control information from the processor to the external environment. Typical uses of this register include handshake signals and signals to act on control points of external devices.

It should be noted that there is no physical difference between the data out register and the control register. Both receive their information from the processor. The difference is in the intended use of the information.

A status register is placed in the port to provide a channel for status information from the external environment. This might include handshake signals or status signals associated with external devices. Loading the status register is under the processor's control to assure synchronization. The content of the register may be sensed by reading the status register and interpreting its contents by means of a program. The status register and data in register are similar in implementation. Both provide a data link from the external environment to the processor.

Data Output

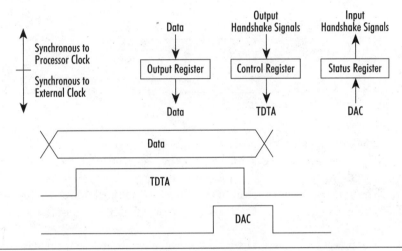

Figure 10.35 Handshake I/O Using a Status Register and Control Register

Figure 10.35 shows an output data transfer handshake initiated by the talker (CPU). One bit of the control register is assigned as the take data (TDTA) bit. One bit of the status register is assigned as the data accepted (DAC) signal from the external device. A data transfer consists of the following actions, each of which is performed by instructions from the processor:

1. The next piece of data is moved to the data out register.
2. The TDTA signal is asserted by writing to the control register.
3. The external device senses TDTA, captures the data, and asserts a DAC signal. The status register captures the DAC. Since the status register is controlled by the processor clock, the signal is synchronized with the processor.
4. The processor repeatedly reads the status register and examines the DAC bit. This process is called **polling**. When the processor determines that the DAC bit is asserted, it terminates the polling activity.
5. The processor, knowing that the transfer has been successfully completed, removes the TDTA signal.
6. The external device, upon sensing the change in the TDTA signal, removes the DAC signal.

Data Input

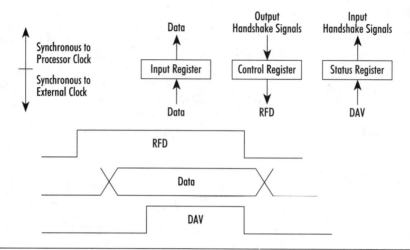

Figure 10.36 Expansion Bus Input Handshake

Figure 10.36 shows the companion data input operation using the control register to provide handshaking signals with the data source.

1. The processor requests information by asserting the request for data (RFD) handshake signal.
2. The data source provides data and then asserts the data valid (DAV) status line.
3. The processor detects the DAV by polling. When it is asserted, the processor captures the data.
4. The processor removes the RFD signal. This indicates that the data has been captured.
5. The data source removes the DAV and the data.

Handshake I/O offers a great deal of flexibility in dealing with a wide variety of I/O devices. It can easily adapt to more complex handshakes. It assures that information flow between asynchronous devices will proceed with a well-defined flow control. One major problem with handshake I/O, when used in a programmed data transfer, is the relatively slow process of managing the data transfers. Each change of a control signal and each acquisition and testing of a status signal requires execution of one or more processor instructions. Polling time may be extensive, and the process may be unacceptably inefficient in some situations.

I/O Flags

I/O flags are used to speed up some of the operations associated with programmed data transfers. Flags are information bits that are cooperatively managed by both the source and destination to transfer data. Figure 10.37 illustrates the role of an output flag in operations for an output data transfer.

1. The computer writes new data to the output data register. The write operation automatically sets the output flag flip-flop.

2. The external device senses when the output flag has been set and accepts the new information. It then clears the output flag flip-flop to acknowledge receiving the data.

3. The computer polls the output flag and waits until it is cleared before writing new information to the output register.

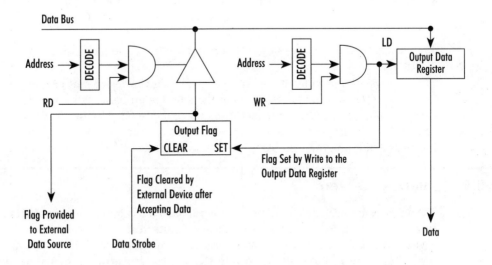

Figure 10.37 Use of a Flag to Synchronize Output Operations

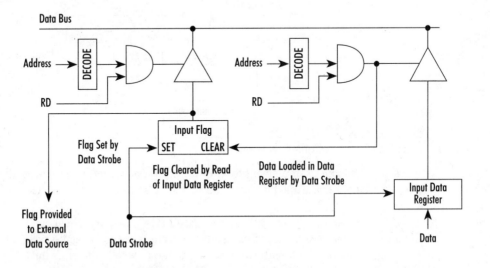

Figure 10.38 Use of a Flag to Synchronize Input Operations

A data input operation would proceed in a similar fashion, as shown in Figure 10.38.

1. The external device provides data. A data strobe is asserted to clock the data into the input data register. The same strobe sets the input flag flip-flop.
2. The computer polls the input flag by reading the contents of the address of the flag in the I/O address space.
3. When the computer detects that the input flag is set, the data is read and transferred to an internal register for processing. The read operation of the input data register automatically clears the flag.
4. The external device, sensing that the flag is cleared, prepares to send additional data to the computer.

While programmed I/O minimizes complexity for the designer, it provides relatively poor performance. While waiting for assertion of the flag, the processor is hopelessly locked in a nonproductive wait loop. If the programmer attempts to do useful activities while waiting for the flag to be asserted, the program becomes considerably more complex. While not useful for high-performance systems, programmed I/O is often suitable for dedicated microprocessor systems.

10.9   Interrupt Driven I/O

The primary performance limitation of programmed I/O is overcome using **interrupt-driven I/O**. Instead of polling to test for an event, the computer can perform useful activities until an I/O device requesting service asserts an interrupt. A processor that can support interrupts has additional hardware facilities designed into the processor control unit. A programmable flag, the **interrupt enable**, allows the processor the option of responding to or ignoring interrupts.

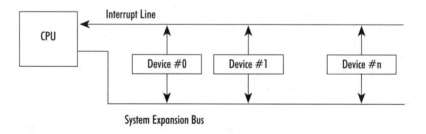

Figure 10.39 A Computer with an Interrupt Facility

Figure 10.39 shows implementation of an interrupt structure in which all devices are connected on a single interrupt bus. An interrupt is asserted by an I/O device through a tri-state driver.

Figure 10.40 shows the processor's response to the interrupt. No action is taken on a pending interrupt until the processor has finished the current fetch and execute cycle. First the interrupt-enable flag is checked. If the interrupt facility is not enabled, a fetch is initiated. If the facility is enabled, the interrupt bus is tested to see if an interrupt is pending. If no interrupts have been asserted, a fetch is initiated and the interrupt cycle is bypassed.

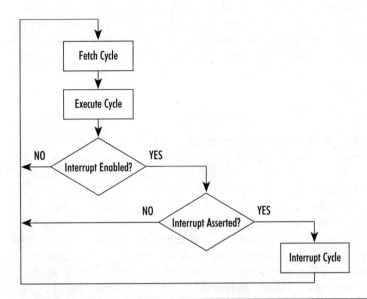

Figure 10.40 The Interrupt Cycle

If there is an enabled interrupt, the interrupt cycle is entered. There are many variations in response to an interrupt. The following operations are typical of those performed under hardware control in response to an interrupt:

1. The interrupt is temporarily disabled. This action is performed by hardware and prevents further interrupts from being serviced until the current interrupt service routine is finished. Most computers allow an interrupt of higher **priority** to take precedence over a lower priority interrupt.

2. The program counter is placed on the stack as part of the hardware response to the interrupt. Other registers, such as the accumulator, may also be placed on the stack. Most computers do this under program control; a few use hardware control.

3. The address of the interrupt service routine is placed in the program counter. The exact method used to accomplish this varies and will be discussed in later sections. The service routine is software written by the programmer specifically to service an interrupting device.

4. The source of the interrupt is identified. This operation is trivial if only one interrupting device is present. It may be quite complex if many possible devices are present. In some computers hardware performs device identification; in others software performs it.

5. Software, the interrupt-service routine written especially for this purpose, services the interrupt.

6. After completing the interrupt-service routine, registers are restored from the stack.

7. The interrupt is enabled. This operation has a delayed response to allow a return to the original program before another interrupt can be sensed.

8. The program counter is restored from the stack and the interrupted program is resumed with a fetch cycle.

An interrupt has many of the same characteristics as a subroutine call. Both cause a change in the normal program flow of the machine. But a subroutine is initiated by software at a known location in the processing algorithm and is under direct control of the programmer, whereas an interrupt is initiated by hardware at an unpredictable location in the processing algorithm. An interrupt has much greater potential to cause unanticipated alterations to the contents of registers and memory.

The Intel 8051 Microprocessor

The 8051 has several interrupts, each associated with one of the features of the device. For example, interrupts can be generated by a counter overflow, a character received on the serial data port, or an external interrupt request; only the last situation will be discussed.

The 8051 has one input pin dedicated as an external interrupt-request line. The processor, under program control, can enable or disable the response to the request. It also can specify if the interrupt facility is to be sensitive to signal transitions or signal levels on the interrupt line.

When an enabled interrupt is received, the current program counter is pushed on the stack. A value of 0003_{16} is placed in the program counter under hardware control. The first instruction executed after the interrupt is from location 0003_{16}, which is the start of the service routine for this interrupt. After completing the service routine, the RETI (return from interrupt) restores the program counter and allows a fetch of the next program instruction.

Following a reset or power-on, the 8051 begins executing code at location 0000_{16}. The three bytes available before location 0003_{16} allow insertion of a jump instruction to bypass the area allocated for interrupt-service routines. Placing the interrupt-service area at the beginning of the program memory space allows maximum unobstructed room for the main program.

The Motorola 6800

The Motorola 6800 microprocessor, an 8-bit predecessor of the 68000, offers a variety of interrupts including a standard external interrupt. In addition, the device includes a non-maskable interrupt that software cannot disable. The programmer can even insert a software-interrupt instruction in the code.

After asserting an (enabled) external interrupt, the address of the next instruction is obtained from a fixed location in an address table located at the top of the program memory space. A two-byte address is transferred from the program memory and placed in the program counter. The other interrupts determine their starting address in a similar manner.

After detecting an enabled interrupt, the 6800 automatically stores most of its internal registers, including two accumulators, index register, and status register on the stack. This is done under hardware control. A return from interrupt automatically restores all of the registers.

10.10   Multiple Interrupting Devices

When a system has several possible interrupting devices, the source of the interrupt must be identified in order to enter the appropriate service routine. Several general methods are used.

Polling

Polling is used when all interrupts are bused on a single shared interrupt input line on the processor. When an interrupt is received, the processor cannot directly identify the source. The processor begins executing code at a predetermined location. It is the code's responsibility to identify the source of the interrupt and to vector to the appropriate servicing software.

Software identifies the source of the interrupt by polling the status registers or flags of each of the devices. The interrupt-request bit is stored in the status register in much the same manner as a flag or a handshake signal. The order of polling may be adjusted by knowing the priority of the interrupt or perhaps the likelihood of a given interrupt.

Polling is relatively slow if many devices are present, but it is still vastly superior to programmed I/O since polling is performed only when it has been determined that at least one I/O device is requesting service. In programmed I/O, polling proceeds without knowing whether a device has requested service.

Multiple Interrupt Lines

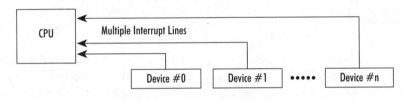

Figure 10.41 Identifying Interrupt Sources Using Multiple Interrupt Lines

Some computers, for example the DEC VAX series, use multiple interrupt lines, as shown in Figure 10.41. If one device is connected to each line, the source of the interrupt can be easily determined. Each interrupt line causes a vectoring to a different fixed location in program memory. If more than one device is connected to an interrupt line, polling is required to identify the source of an interrupt.

Interrupt Priorities

Sometimes several interrupting devices, each with a different priority, must be managed. An interrupt generated by a power-failure sensor, for example, would be more important to service than an interrupt associated with a printer flag. Most computers can service nested interrupts. High-priority devices can interrupt the service routines of low-priority devices. Devices with the same priority cannot interrupt each other.

Interrupt priorities are most conveniently managed with multiple interrupt lines. All devices with the same priority use the same interrupt-request bus. The processor can then easily determine if an active service routine should be interrupted. Within a group of devices sharing the same interrupt line, priority can be established by the order of polling.

Programmable Interrupt Controller

Because of the inherent complexity of managing an interrupt system, a special hardware-management device, the programmable interrupt controller (PIC), has become very popular in microprocessor systems. The PIC is implemented as a single-chip device. It is designed to interface easily into the memory space of a microprocessor system. A typical device, the Intel 8153, will be briefly described.

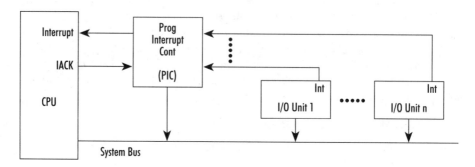

Figure 10.42 Use of a Programmable Interrupt Controller to Manage Interrupts

The 8153 has provisions for accepting eight interrupt-request lines, as shown in Figure 10.42. Internally, the PIC has several registers that the host processor programs to specify operating modes and responses. The processor can specify one of eight priority levels for each interrupt, individually enable and disable each interrupt, store an identifying code for each of the devices, and set a specified interrupt-priority threshold.

When the 8153 receives an interrupt from an I/O device, it determines if the interrupt is enabled and if it exceeds the current value of the interrupt-priority threshold. Only then is the signal propagated to the processor. The processor, upon receiving the interrupt, generates an interrupt-acknowledge signal in response. The PIC responds by providing the processor with the identifying code for the interrupting device. A PIC greatly reduces the hardware and software complexity associated with implementing an interrupt system.

10.11   Direct Memory Access

Programmed I/O and interrupt-driven I/O adequately handle many I/O tasks. They are not well suited for high-speed transfers of large blocks of sequential data. Transferring a block of data from memory to an I/O device, for example, might require a loop of three to four instructions: one to transfer the data, one to increment a data pointer, one to test if the operation is complete, and one to jump to the beginning of the loop operation. Each of these instructions requires multiple register-transfer and memory-access operations to perform the respective fetch/execute cycles.

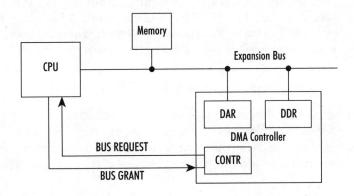

Figure 10.43 A Direct Memory Access Input/Output Structure

An alternative I/O method suitable for high-speed transfers of a block of data is direct memory access (DMA). The data to be transferred is first placed in the system data memory. It is accessible via the system expansion bus. In DMA, an external device, the DMA controller, as shown in Figure 10.43, requests control of the computer system bus by asserting a BUS REQUEST signal. The computer finishes executing the current instruction, then responds by suspending its processing operations, asserting a BUS GRANT, and relinquishing control of the external system bus; all lines from the computer are tri-stated.

The DMA controller, by providing memory addresses and memory control signals, gains access to the contents of the data in the computer memory and performs a read or write operation to its own DMA data register (DDR). From the DDR the data may be used for a specialized I/O. The DMA controller may also contain a counter to count and test the number of data-transfer operations it performs. After completing the data transfer, the BUS REQUEST is released, and the computer releases the BUS GRANT and resumes its operation.

A DMA operation to transfer a data word requires only a few microinstructions, as shown below.

```
t0:          DDR←MEM(DAR),        /* Get data using DMA registers       */
             DAR←DAR + 1;         /* Increment DMA address register     */
t1:          Ri←DDR;              /* Local storage of data              */
```

The entire sequence of microoperations associated with the transfer of one data word can be performed in one clock cycle if pipeline techniques are used. Additional cycles might be required to test the total count of data items and to store the acquired information. If the CPU were used to control the same transfer, using three instructions, as many as 12–24 clock cycles would be required to transfer one data word.

There are two types of DMA transfers. In a **block DMA transfer**, the DMA controller requests a block of time from the CPU to perform a data-transfer operation. With complete control of the system buses, the DMA controller completely processes the I/O action without interference from the CPU or other devices.

In contrast to block DMA, **cycle-stealing DMA** requires the DMA controller to time share the system buses with the CPU. The CPU informs the DMA controller when a computer cycle is not using the system buses. The DMA controller uses the spare cycles to perform DMA operations. DMA operations proceed without affecting the normal processing of a computer program.

10.12   Computer System Expansion and Interfacing

Most digital computers require a degree of customization to make them suitable for their application environment. They are customized by interfacing a wide variety of peripheral devices to the computer. Interface techniques range from device-dependent interconnections to standardized interfaces, which are applicable to a wide range of computers.

Custom Interfacing

Custom interfacing, as the name suggests, involves designing a specialized interface for the specific application. Custom interfacing is most often used when a microprocessor and peripheral devices are physically designed on the same circuit board. Selected electronic components are attached directly to the buses and control signals of the selected computer. Interface signals for the IBM PC bus are presented later in this section.

Custom interfacing requires a great deal of careful design to assure that interfaced components meet current limits, power-supply limitations, timing requirements and thermal constraints. Custom engineering design effort for a complex computer-controlled system can be substantial. Custom interfaced hardware generally requires some customized software. This may assure high performance but may require additional effort in software preparation, documentation, and testing.

Expansion Bus Interfacing

Some microprocessors, such as the Intel 80x86 family, have been successful components in personal computer systems. IBM, when designing the popular IBM PC, created an expansion bus suitable for use with Intel microprocessors. Many of the design issues were incorporated into the bus, which was routed to a series of card connectors inside the chassis. Most of the signals (address lines, data lines, and control signals) available at the card connector are the same as those available in custom interfacing. Where necessary, buffers and additional logic components are used to group signals and pin assignments that are standardized for the PC.

There are many advantages to using expansion bus interfacing. Manufacturers are able to design custom cards meeting the specifications of the interface for many peripheral activities. The user can choose the appropriate set of cards for the application. Because the interface is standardized, some software support may be available for the interface through the operating system provided by the computer manufacturer.

Standardized Interfaces

Several standard interfaces exist for digital computers. These interfaces create a standardized application interface environment that is independent of the digital computer. They allow manufacturers to design equipment compatible with the standard interface rather than equipment that must be compatible with a specific computer expansion bus. The usual form of a standard interface is a card that plugs into the expansion bus of a specific computer and creates a standardized interface for the user.

Standardization is extremely important to manufacturers of communication equipment, printers, measurement devices, and laboratory instrumentation that must work with computers. It greatly reduces the amount of engineering design required to create usable products.

The Centronics Printer Port

One of the most often used expansion ports for a digital computer is the interface connection for a standard printer. For historical reasons, the Centronics interface is the printer interface most commonly used. It is named after the printer manufacturer that pioneered its use.

Printers present a special interface problem due to the wide variety of data rates that must be accommodated. Older daisy-wheel type printers use a relatively slow rate of eight characters per second; dot-matrix printers print several hundred characters per second. Modern laser printers can print at rates exceeding 10 pages per minute.

The Centronics interface is designed to be relatively independent of the printing mechanism. Although there is no true standard, most Centronics interfaces use a 36-pin connector with a distinctive shape. The interface signals are as shown in Table 10.1.

Pin	Signal	Source	Comment
1	DATA_STROBE	Computer	Strobe data to print buffer.
2–9	DATA(1:8)	Computer	Data lines.
10	ACKNOWLEDGE	Printer	ACK data transfer.
11	BUSY	Printer	Status: Printer cannot take data.
12	PAPER_OUT	Printer	Status: Out of paper.
13	SELECT	Printer	Status: Printer online.
14	SIGNAL_GROUND	—	—
17	CHASSIS_GROUND	—	—
19–30	Paired with 1–11	Computer	Used for noise immunity.
31	INITIALIZE	Computer	Initialize printer.
32	FAULT	Printer	Status: Printer error.

Table 10.1

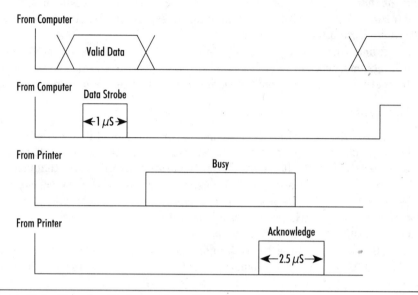

From Computer

Valid Data

From Computer

Data Strobe

←1 μS→

From Printer

Busy

From Printer

Acknowledge

←2.5 μS→

Figure 10.44 Data Transfer Timing with a Centronics Interface

Figure 10.44 shows a timing diagram for a data-transfer handshake. After the computer presents the data, a DATA_STROBE signal lasting one microsecond is provided. The printer immediately responds with a BUSY status signal. BUSY indicates that the printer is no longer able to accept information. After the character has been printed or stored in a printer buffer memory, ACKNOWLEDGE is asserted and BUSY is removed. This allows the computer to send another character.

Most computers are capable of sending data much faster than it can be printed. An internal printer buffer memory can store a large number of characters awaiting printing. When the buffer is full, BUSY is asserted to stop the flow of data until the printer has decreased the contents of the buffer. Additional status signals such as PAPER_OUT, SELECT, and FAULT may be used by the computer to monitor the printing process.

Serial Port, RS 232

Serial data exchange is a widely used and useful concept for many situations. Serial techniques require considerably fewer wires and hence offer savings in equipment and space. Data transfers are much slower than parallel transfer methods due to serial transmission, extra complexity, and overhead associated with managing a serial channel.

Synchronous or asynchronous serial techniques can be used. Synchronous techniques include a clock signal along with the serial data. In some situations the clock signal may be encoded with the data. Long-distance data transmission and local area networks use synchronous techniques to achieve high performance. Synchronous communication will be considered in greater detail in Chapter 12, "Digital Communication and Networking."

Asynchronous data exchange is historically related to techniques developed for the printing telegraph. The serial data port is now the most frequently used interface in digital computers, test and measurement devices, and communication equipment. While asynchronous serial communication lacks the performance of parallel or synchronous systems, its wide usage makes it an important I/O component.

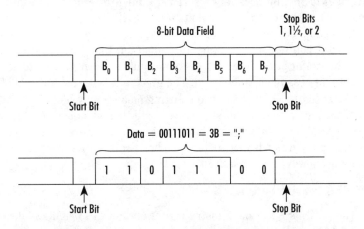

Figure 10.45 Data Formats for Asynchronous Serial Communication

The data unit used in asynchronous serial communication is eight bits long, sufficient to encode an ASCII character. The data format is shown in Figure 10.45. In the data field, the least significant bit is transmitted first, resulting in an apparent reversal of bits between the serial representation and the parallel representation of the data. The 8-bit data field is preceded by a start bit (asserted low), which is used for synchronization purposes. Following the data field is a stop bit (asserted high). The stop bit is either 1, 1½, or 2 bit periods in minimum width. The stop bit may be extended for any length of time to correspond to periods when the data-exchange mechanism is idle.

An essential feature of asynchronous communication is that both the source and the destination agree on the bit rate for sending information. Data rates of 300 baud (bits per second), 1200 baud, 9600 baud, and 19,200 baud are standard.

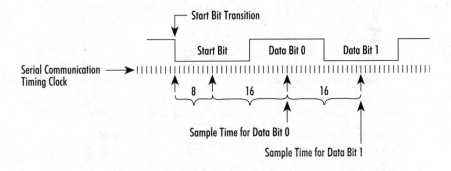

Figure 10.46 Reception of a Serial Signal by a UART

An LSI (large-scale integrated circuit) device, the universal asynchronous receiver/transmitter, or UART, accepts an 8-bit parallel word and converts it to the proper serial representation, including start and stop bits. The receiver portion of the UART receives serial information and converts it to an 8-bit parallel word. The parallel-to-serial and serial-to-parallel conversion is performed by hardware logic components in the UART. Figure 10.46 shows the reception process. The UART is supplied with a timing clock that operates at a rate sixteen times higher than the bit rate used. Upon sensing the start bit transition, the UART first times out eight clock periods to get to the center of the start bit. It then counts sixteen periods to get to the center of data bit 0. The data bit is sampled and stored. The process is repeated until all of the data bits are acquired.

The encoding clock of the transmitting module may be at a slightly different frequency than that of the receiving module. The receiver will then sample the data either slightly earlier or later than the true center of the data bit. Small frequency deviations are easily tolerated. Since resynchronization occurs with each start bit, baud-rate errors are not accumulative. Large baud-rate variations, however, will result in errors.

IEEE 488 Instrumentation Bus

The IEEE 488 standard instrumentation bus was originally developed as a means to interconnect high-performance electronic instrumentation into a single system. Devices participating in IEEE 488 data exchanges are known as talkers, listeners, and controllers. It is possible for devices to be both talkers and listeners. It is also possible for the role of controller to rotate among several devices. This discussion will concentrate on the data transfer among a single computer, as the controller, and several talkers and listeners. The talkers and listeners will be data-acquisition and data-storage devices.

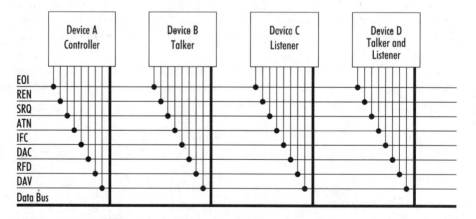

Figure 10.47 Control, Status, Data, and Handshake Signals in IEEE 488

Data transfers in IEEE 488 use an 8-bit data bus. Eight additional signals are used for control and handshake purposes. Data transfers are 8 bits parallel with a single talker and many listeners. The general organization is shown in Figure 10.47.

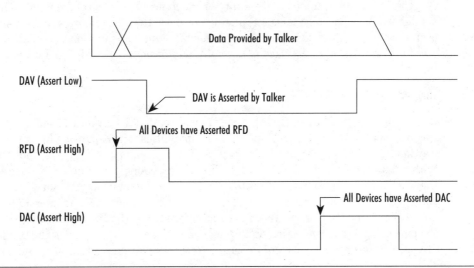

DAV (Assert Low) — DAV is Asserted by Talker

RFD (Assert High) — All Devices have Asserted RFD

DAC (Assert High) — All Devices have Asserted DAC

Figure 10.48 Data-transfer Handshake for IEEE 488

The data-transfer handshake for IEEE 488 is similar to the three-line handshake shown in Figure 10.15. The major variation is that several listeners are able to simultaneously receive data. Three signals are used—REQUEST_FOR_DATA (RFD), DATA_VALID (DAV), and DATA_ACCEPTED (DAC). The handshake signals are driven by open collector electronics; this allows one device to pull the logic to the low condition, regardless of the logic status of the other devices. All must be in a high condition before the corresponding handshake line goes high. Figure 10.48 shows a data-transfer handshake. The RFD line on the data bus is not asserted until all the individual devices are in their asserted (high) condition. A similar situation exists for the DAC signal.

In addition to the three handshake signals, Table 10.2 shows five bus-management signals.

Signal	Comment
ATN—Attention	This signal indicates that the data bus is to be used for a command. The controller uses commands to assign active or inactive status to listeners and talkers.
IFC—Interface Clear	This signal initializes the system to a known condition.
SRQ—Service Request	This signal is similar in action to an interrupt.
REN—Remote Enable	The controller uses this to remotely configure operating conditions and modes on the talkers and listeners.
EOI—End or Identify	This is used by the controller for polling purposes. It also is used to identify the last byte of a multibyte data transfer.

Table 10.2

The IEEE 488 is unique in its ability to accommodate to a mix of fast and slow components. It is intended for modest systems comprising not more than 15 talkers and 15 listeners distributed over a region less than 25 meters. It can achieve a data-transmission rate of about 1 megabyte per second. The IEEE 488 interface is normally purchased as a card that interfaces directly to the internal bus of the computer.

IBM PC Bus

When IBM made its entry into the personal computer market in 1982, it provided an expansion bus as a way to enhance the machine's capabilities. The expansion bus gave manufacturers a way to interface add-on products, such as additional memory, serial ports, and input/output ports. With 8 data lines and 20 address lines, the PC bus strongly resembles the microprocessor bus of the Intel 8088. It has extra features to support expansion memory, DMA data transfers, and interrupts. The PC Bus has since been expanded; the PC/AT Bus accommodates 16-bit data and 24-bit addressing. Table 10.3 shows the signals associated with the PC Bus. The letters associated with the pins designate a connection on the component side or circuit side of the printed circuit. Table 10.4 shows the expansion connector used to provide additional addressing and data width.

Pin	Signal	Comment
A1	I/O CH CK	I/O channel check.
A2–A9	D(7:0)	8-bit data bus.
A10	I/O CH RDY	I/O channel ready.
A11	AEN	Address enable.
A12–A31	A(19:0)	20-bit address bus.
B1	GND	Ground.
B2	RESET DRV	System reset.
B3	+5 volts	Power.
B4	IRQ2	Interrupt request.
B5	−5 volts	Power.
B6	DRQ2	DMA request.
B7	−12 volts	Power.
B8	CARD SLCTD	Card selected.
B9	+12 volts	Power.
B10	GND	Ground.
B11	MEMW	Memory write.
B12	MEMR	Memory read.

Pin	Signal	Comment
B13	IOW	I/O write.
B14	IOR	I/O read.
B15	DACK3	DMA acknowledge.
B16	DRQ3	DMA request.
B17	DACK1	DMA acknowledge.
B18	DRQ1	DMA request.
B19	DACK0	DMA acknowledge.
B20	CLOCK	Processor clock.
B21–B25	IRQ(7:3)	Interrupt requests.
B26	DACK2	DMA acknowledge.
B27	T/C	DMA terminal count.
B28	ALE	Address latch enable.
B29	+5 volts	Power.
B30	OSC	Oscillator.
B31	GND	Ground.

Table 10.3

Pin	Signal	Comment
C1	SBHE	Bus high enable.
C2–C8	LA(23:17)	Unlatched address.
C9	MEMR	Memory read.
C10	MEMW	Memory write.
C11–C18	SD(8:15)	Data bus high byte.
D1	MEM CS16	16-bit memory chip select.
D2	I/O CS16	16-bit I/O chip select.
D3–D7	IRQ(10:14)	Interrupt requests.
D8	DACK0	DMA acknowledge.
D9	DRQ0	DMA request.
D10	DACK5	DMA acknowledge.
D11	DRQ5	DMA request.
D12	DACK6	DMA acknowledge.
D13	DRQ6	DMA request.
D14	DACK7	DMA acknowledge.
D15	DRQ7	DMA request.
D16	+5 volts	Power.
D17	MASTER	DMA bus control.
D18	GND	Ground.

Table 10.4

Although not originally intended as a system bus, the IBM PC bus quickly became one. Most modern PCs offer several 8-bit and 16-bit expansion slots to accommodate many peripheral devices such as game ports, CD-ROM interfaces, sound cards, and instrumentation interfaces.

Multibus

Multibus is typical of a class of backplane expansion systems that emphasize highly modular card-level devices. Although Multibus I was originally developed for Intel 8-bit microprocessors, newer versions such as Multibus II will function with 32-bit data units.

With Multibus systems, the processor is purchased on a standard-sized card. Data lines, control lines, interrupts, and power lines occupy specified pin locations. The original card with the processor includes only very standard processor components. Multibus expansion cards are purchased from a variety of vendors to customize the system to an application. Cards are available for data acquisition, communication, memory, serial ports, terminals, displays, and a variety of other uses. Cabinets, complete with power supplies, are available off-the-shelf to support these custom developments.

The goal of Multibus is to allow the user to assemble a unique computing environment using modular components. Special software-development packages are also available to allow customization of operating systems and operational software. Systems are easily upgraded by replacing the processor board or any other board in the system.

There are a wide variety of competitors to Multibus, including VMEbus, STD Bus, and Futurebus. Each is moderately tuned to a specific family of microprocessors. These buses differ in the number of data bits and the number of address bits they can accommodate. Data transfers may be synchronous or asynchronous. Data transfers may use signals closely related to Intel or Motorola microprocessors. Most also have a method of serving multiple bus masters. Table 10.5 shows the major characteristics of several system buses.

	IBM PC/AT Bus	Multibus	Multibus II	STD Bus	Futurebus	VME Bus
Data Bus	8 for PC 16 for AT	16	32	8	32	16/32
Address Bus	20 for PC 24 for AT	24	32	16	32	23/31
Data Transfer	Semi-sync	Async	Semi-sync	Semi-sync	Async	Async
Multiple Masters	N	Y	Y	Y	Y	Y
IEEE Standard	N	IEEE 796	IEEE 1296	IEEE 961	IEEE 896	IEEE 1014
Interrupts	6	8	1	1	1	7
Connector Pins	62 for PC 62 + 36 for AT	86 + 60	96 + 96	56	96	96 + 96

Table 10.5

Summary

Data exchange within a digital system is very important. Exchanges become more complex when components are not synchronized to a single system clock. This happens when devices are developed as separate modules or when the distance between units exceeds a few meters.

Data transfers can be synchronous, semi-synchronous, or asynchronous depending on the relationship of the modules. Reliable transfers between asynchronous talkers and listeners require interlocked handshakes. Many handshakes are possible depending on the device that initiates the data transfer and the number of listeners involved. When systems involve multiple talkers, a means of allocating the shared bus among talkers is required. This function is usually performed by a bus master using a wide variety of allocation algorithms.

Computer input/output involves a data exchange between a digital computer and an external device. Programmed I/O uses instructions to test the ability of the external device to accept or provide data. Handshakes and flags are used to provide status information to both the computer and the external device. Interrupt-driven I/O reduces the need to poll data flags and allows the computer to perform useful activities while waiting for slower I/O devices. Interrupt systems may be implemented in many ways. Multiple interrupting devices add complexity. Direct memory access allows an external device to request the use of the system expansion bus from the computer. When granted, very high speed data transfers are possible from memory devices on the expansion bus and the DMA controller.

This chapter concludes with a discussion of a serial expansion port, a Centronics printer parallel expansion port, an IEEE 488 instrumentation port, and a Multibus backplane expansion.

For additional information and references on the topics presented in this chapter, access the World Wide Web page provided at: `http://www.ece.orst.edu/~herzog/docs.html`.

Key Terms

address bus	interrupt facility
block DMA transfer	interrupt-driven I/O
bus master	interrupts
centralized protocol	listener
channel access protocol	master
control bus	memory-mapped I/O
cycle-stealing DMA	priority
data bus	polling
direct memory access (DMA)	programmed I/O
distributed protocol	semi-synchronous data transfer
fully synchronous data transfer	slave
handshaking	synchronizing
input/output processors	talker
interrupt enable	time-out

Readings and References

Herman H. Goldstine, *The Computer from Pascal to von Neumann*. Princeton, NJ: Princeton University Press, 1972.

Herman Goldstine was one of the insiders in the development of ENIAC. Goldstine was a collaborator with John von Neumann in the development of several early computers. His book offers a unique insight into the men and women responsible for the creation of computing devices.

Exercises

10.1 In a data-transfer handshake, why is the data presented before the TAKE_DATA control signal? Wouldn't higher performance be possible if both actions took place in the same clock cycle?

10.2 In a talker-initiated handshake, when is it possible to assert TAKE_DATA without using the DATA_ACCEPTED response?

10.3 In a listener-initiated handshake, when is it possible to assert the REQUEST_FOR_DATA without using the DATA_VALID response?

10.4 A talker-driven data-transfer handshake is to be used to transfer data to a location 10 kilometers distant. The rate of signal propagation is 3×10^8 meters per second. The clock cycle of the source is 100 nanoseconds. What is the maximum data-transfer rate possible?

10.5 In the listener-initiated handshake of Figure 10.14, what would happen if the last decision block in the task diagram for the listener, (DAV1* = 0), were eliminated?

10.6 In the listener-initiated handshake of Figure 10.14, what would happen if the last decision block in the task diagram for the talker, (RFD2* = 0), were eliminated?

10.7 A talker and a listener both have a clock rate of 10 MHz. They are not synchronized. Both use a synchronizing register to accept both data and handshake signals from the other. Each requires a setup time of one half their clock period to accurately accept information (data or handshake). The talker uses a one-period guard band on both sides of the data with its DAV signal. If a talker-driven handshake is used, what would the maximum expected data transmission rate be?

10.8 What happens in a talker-driven handshake if the listener fails to respond with DAC? How is the talker made aware of the malfunction?

10.9 In the listener-driven synchronous data transfer of Figure 10.6, what requirements are placed on the RFD signal?

10.10 In the semi-synchronous data transfer of Figure 10.7, the TAKE_DATA and DATA_ACCEPTED signals are removed simultaneously. Why is this different from the situation with the handshake in Figure 10.2?

10.11 Construct a microoperation diagram for sending data semi-synchronously. Allow a maximum of three clock cycles for a DAC response, then signal a transmission error.

10.12 What is the advantage of using a wait state rather then a traditional handshake using a DATA_VALID when the listener responds?

10.13 In a two-line handshake, how long should the master wait for a slave response?

10.14 In the talker-driven handshake of Figure 10.12,
  (a) is the loop on box Y necessary? Explain.
  (b) is the loop on box E necessary? Explain.

10.15 In the microoperation diagram for a three-line handshake, Figure 10.16, can boxes C and D be combined in a single box to improve performance? Why?

10.16 In a three-line handshake, can a fast listener leave the RFD line permanently asserted to speed up the data transfer operation?

10.17 How can bus logic be used to provide OR and AND logic functions?

10.18 Compare using open-collector circuitry and tri-state circuitry with respect to their use on a data bus.

10.19 In a time-share protocol, what is the relative efficiency and response time of:

    (a) Using a very short BUS GRANT of a few clock cycles?

    (b) Using a very long BUS GRANT of several hundred clock cycles?

10.20 In the round-robin protocol with BUS BUSY, as shown in Figure 10.22, what happens if:

    (a) The talker cannot assert BUS BUSY?

    (b) The talker cannot de-assert BUS BUSY?

10.21 When using a round-robin access protocol with BUS BUSY, how can hogging be eliminated?

10.22 Information exchange in asynchronous digital systems can involve (a) one talker and one listener, (b) one talker and many listeners, or (c) many talkers and many listeners. Provide an example of each. For each:

    (a) Why is the handshake needed?

    (b) Show a diagram of the essential hardware components for an example system.

    (c) Show a timing diagram of the information exchange. On your timing diagram indicate what information is conveyed by each transition.

10.23 Design a microoperation diagram for the following bus controllers:

    (a) Round-robin bus allocation.

    (b) Bus request access protocol of Figures 10.24 and 10.25.

    (c) Daisy chained bus allocation.

10.24 Develop a microoperation diagram for a talker-driven two-line handshake in which the handshake is aborted if the listener does not respond to the DAV signal within four clock cycles.

10.25 Compare the time-share, round-robin with BUS BUSY, and BUS REQUEST protocols.

    (a) Which is most/least efficient use of the data bus? Why?

    (b) Which gives the fastest/slowest response time when there is low bus traffic?

    (c) Which gives the fastest/slowest response time when there is heavy bus traffic?

    (d) Which has the least/most complex bus master? Why?

    (e) Which has the least/most complexity in the talker units? Why?

10.26 Often times in a bus-allocation environment it is desirable to allow a unit with higher priority to claim the bus from a unit of lower priority. Discuss the feasibility of implementing this type of algorithm with each of the four bus allocation methods discussed in Section 10.6.

10.27 Pin count is always a consideration when using integrated circuit devices. Show how it would be possible to use a separate memory and address space, yet use a single address bus and data bus.

10.28 The Intel 8085 microprocessor saves on the use of pins on the integrated circuit package by time-sharing 8 bits of the 16-bit address bus and the 8-bit data bus on a single set of pins. What are the advantages and disadvantages of this technique?

10.29 A Harvard computer architecture uses a separate memory for data and program code. Where would a memory-mapped I/O space be located? Why?

10.30 Assume that in BART the input register has address FE_{16}, and the output register has address FF_{16}. Assume that the code for testing the input flag bit is 110; the code for the output flag bit is 111. Write a sequence of assembly code to transfer the contents of memory locations 40_{16}–$4F_{16}$ to the output port.

10.31 Under the same assumptions as the previous problem, write a sequence of assembly code to read the next 16 bytes from the input register and store them in locations 50_{16}–$5F_{16}$.

10.32 Can anything useful be done while waiting for a status signal during polling?

10.33 Assume that it is desired to use a polled input port of a digital computer to read the value of a switch. It is known that the switch is manually operated (i.e. no more than 10 changes per second are physically possible). Following a switch transition, there may be a period of contact bounce which lasts up to 10 milliseconds. Show a software solution which will allow the switch value to be accurately read.

10.34 Can a dedicated controller manage data transfer functions faster than a computer with programmed I/O? Why?

10.35 When using I/O flags while performing programmed I/O:

(a) Must the output flag flip-flop in Figure 10.37 be synchronous? Why?

(b) Must the input flag flip-flop in Figure 10.38 be synchronous? Why?

10.36 In Figure 10.38, the input flag is set at the same time as the data input is loaded. Does this present a timing problem?

10.37 In interrupt-driven I/O, why does the controller wait until the end of the execute cycle before reacting to an interrupt?

10.38 Do any of the following cause program flow errors in a digital computer? Explain your response.

(a) A subroutine is called from within an interrupt.

(b) An interrupt is asserted from within a subroutine.

(c) In an interrupt-service routine, data is stored on the stack in preparation for returning from the interrupt.

10.39 Discuss two techniques for vectoring to an interrupt service routine.

10.40 What services are provided by a programmable interrupt controller (PIC)?

10.41 In what computational environment is DMA data transfer faster than programmed data transfer? When would programmed data transfer be faster? Why?

10.42 Some computers respond to an interrupt by storing registers on the stack under hardware control; others perform this action under software control. What is the advantage and disadvantage of each approach?

10.43 Why may a processor have several places in its execute cycle to respond to a DMA request but must wait until the end of the execute cycle to respond to an interrupt?

10.44 Assume that a computer requires 12 clock cycles of a 12 MHz clock to completely process an instruction. Data originates in the memory of the computer and is transferred to an I/O port. The memory can be read in one clock cycle. No handshakes are used. What is the maximum data rate that can be attained using programmed I/O?

10.45 Under the same circumstances as Exercise 10.12, assume that a DMA transfer is to be used. Using a 12 MHz clock in the DMA controller, what data-transfer rate can be attained?

10.46 The 8051 can be programmed to respond to either a level interrupt or an edge-transition interrupt. Under what circumstances is each appropriate?

10.47 Some computers such as the Motorola 6800 have an instruction called a software interrupt or SWI. When executed, the SWI instruction causes all the operations to be performed that are initiated from a hardware interrupt. Compare the SWI instruction to a subroutine call. Is there any advantage in using SWI?

10.48 The 6800 microprocessor has a wait-for-interrupt instruction. All registers are stored in advance of the interrupt signal. Of what use is this instruction?

10.49 Compare the operation of an interrupt and a subroutine call.

10.50 As part of a computer design, it is proposed to use a register called OLD_PC to store the program counter while an interrupt is being serviced. After completion of the service routine, PC←OLD_PC. Is this a useful approach? Explain.

10.51 Can a DMA data transfer be accomplished in 1 clock cycle with a pipeline approach? Explain your reasoning.

10.52 Assume that an asynchronous serial communication system is using a 9600 baud communication rate. What is the maximum percent change in clock frequency to allow error-free reception of characters?

10.53 Design a datapath and microoperation diagram for a device that will accept data from a Centronics port and store it in sequential memory locations.

10.54 Design a datapath and microoperation diagram for a printer data buffer. The device is to accept information from the Centronics port of a digital computer at a very high data rate. It is to supply data to a printer at the maximum rate the printer can accept the data.

10.55 In serial communication a framing error is created when a data bit is mistaken for a start bit. Is it possible to detect a framing error? How?

10.56 Rate the three types of interfaces (custom, expansion bus, standard) with respect to the following:

(a) Potential for high performance.

(b) Ease of designing compatible equipment.

(c) Economics of manufacturing.

11
Parallel and High-Performance Computing

John Atanasoff

An Iowa State prof. named Atanasoff
Used vacuum tubes to turn his bits on and off
His proponents proclaim
The modern computer belongs in his name
At Mauchly, Eckert, and von Neumann they still do scoff

Prologue

The electronic era of digital computation began in 1906 with Lee de Forest's invention of the triode vacuum tube. The triode permitted switching of electrical currents using a control signal applied to a grid structure between the cathode-emitting surface and a plate-collecting surface. By the mid-1920s electronic circuitry was used for radio sets and other purposes.

In the late 1930s John Atanasoff was on the faculty of mathematics and physics at Iowa State College. He had several graduate students working on techniques for solving sets of linear differential equations. By 1937 he was convinced that electronic techniques were best suited for the solution. He began designing, with colleagues, a special-purpose electronic computing device to expedite the computations.

Atanasoff's machine was designed to implement the Gaussian elimination technique for solving sets of linear equations. His electronic circuitry was capable of performing the addition, subtraction, and shifting operations required in the technique. All computations were performed in binary. Numerical coefficients were stored in capacitors on the surface of a rotating drum. The drum rotated at one revolution per second and allowed access to any of the data within a one-second period. Punch cards stored intermediate results.

Records indicate that Atanasoff's computer was constructed and functioned. It was undoubtedly the first electronic digital computer. Evidence suggests that John Mauchly visited Atanasoff and was impressed with the work in progress. Later, conflicting patent claims on computing devices led to legal battles. History seems to largely have neglected the contributions of Atanasoff in favor of the work of Mauchly, Eckert, and von Neumann on the ENIAC computer at the University of Pennsylvania.

11.1   Introduction

John Atanasoff at Iowa State University pioneered the era of high-performance computing. He was the first to use electronic vacuum tubes to perform the arithmetic and logic operations associated with digital computers.

Most electronic digital computers produced through the early 1960s used a processor organization similar to BART. Components were expensive and unreliable by today's standards. Early efforts resulted in computers costing in the range of $500,000 to $5,000,000. These computers were installed in air-conditioned environments and serviced by a priesthood of hardware and software technicians.

While the early electronic computers offered considerable improvements over mechanical calculators, rapid improvements in hardware and software have made further improvements possible. This chapter looks at several of the significant architectural changes that have improved processor performance.

11.2   Parallel Computing Structures

After initial development of the digital computer, many techniques that could be implemented at low or moderate cost were developed to improve machine performance. VLSI has allowed complex circuitry at very low cost, but physical and thermal constraints imposed by electronic materials make it difficult to further increase the density of components. Higher clock rates are limited by capacitive effects, propagation delays, and complexity associated with data skewing. RISC architectures are limited by the complexities introduced by increased length of the pipeline. Parallelism seems to be one of the few remaining techniques offering the possibility of continued performance increases at modest cost.

When discussing parallel systems, the level of parallelism must be clear. Parallel adders, for example, are parallel at the bit level. The term parallel is used here to contrast the operation to slower bit-serial methods of processing information. Within a processor, we can perform multiple register transfer operations simultaneously, or in parallel. In this case, the parallelism is at the word level.

The parallel operations presented in this chapter are parallel at the processor level. Parallel processing involves the simultaneous use of multiple processors. In **superscalar** computers, for example, the hardware attempts to begin executing multiple independent instructions in a single clock cycle. Other parallel structures may involve independent computational devices which cooperate to solve a single decomposed computational problem.

It should be noted that not all processing activities are suitable for parallel operation. As in the case of pipelined processing, there are some limitations.

- **Data dependency.** Some processing actions are sequential; that is, results produced by some processing devices are used later by other processors. It may be difficult or impossible to perform the actions in parallel.

- **Resource conflicts.** If data is available for processing, the required processing hardware must also be available. If resources are in use for other processing activities, parallel processing is not possible.

- **Jumps and branches.** Flow-control type instructions, such as jump and branch, present special problems. They can strongly impact the range of operations suitable for parallel processing. No more than one flow-control operation can be processed in parallel.

11.3   Performance Measurement in Parallel Systems

Two measurements of performance appropriate for parallel systems are the **speedup**, and **parallel efficiency**. The time to complete a task using n processor units is defined as T_n. Speedup, S_n, is defined as T_1/T_n, the ratio of the processing time using a single processor to the time required when n processors are used. The Efficiency, E_n, when using n processors, is defined as S_n / n. When the speedup is equal to the multiplicity of the processing elements, the efficiency is 1.0.

Consider a processing task requiring twelve instructions. Assume that all instructions, I_0–I_{11}, require the same length of time, one time unit, for execution. The processing can be accomplished using a single processor, P_0, sequentially as shown in Table 11.1.

Time Unit	Processing Action P_0
0	I_0
1	I_1
2	I_2
3	I_3
4	I_4
5	I_5
6	I_6
7	I_7
8	I_8
9	I_9
10	I_{10}
11	I_{11}

Table 11.1

In this example, the processing activity requires 12 time units and $T_1 = 12$.

Assume that some of the instructions have no data dependency or resource conflicts and can be executed simultaneously. The instructions can be grouped into five sets. All instructions in a set can be performed simultaneously.

Set	Instructions
1	$\{I_0, I_1, I_2\}$
2	$\{I_3\}$
3	$\{I_4, I_5\}$
4	$\{I_6, I_7, I_8, I_9\}$
5	$\{I_{10}, I_{11}\}$

Table 11.2

If two processors are used, the operations in Set 1 can be performed in two time units instead of three. No improvement can be achieved with Set 2 because of the limited concurrency possible. Table 11.3 shows the scheduling of processors P_0 and P_1 to execute the instructions.

Time Unit	Processing Action	
	P_0	P_1
0	I_0	I_1
1	I_2	
2	I_3	
3	I_4	I_5
4	I_6	I_7
5	I_8	I_9
6	I_{10}	I_{11}

Table 11.3

$T_2 = 7$ time units
$S_2 = T_1 / T_2 = 12 / 7 = 1.71$
$E_2 = 1.71 / 2 = 0.86$

Using the same analysis procedure with three, four, five, and six processing units:

$T_3 = 6$ time units
$S_3 = T_1 / T_3 = 12 / 6 = 2.00$
$E_3 = 2.00 / 7 = 0.67$

$T_4 = 5$ time units
$S_4 = T_1 / T_4 = 12 / 5 = 2.40$
$E_4 = 2.40 / 4 = 0.60$

$T_5 = 5$ time units
$S_5 = T_1 / T_5 = 12 / 5 = 2.40$
$E_5 = 2.40 / 5 = 0.48$

$T_6 = 5$ time units
$S_6 = T_1 / T_6 = 12 / 5 = 2.40$
$E_6 = 2.40 / 6 = 0.40$

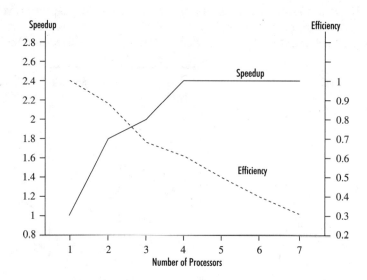

Figure 11.1   *Speedup and Efficiency vs. Number of Processors*

The resulting speedup and efficiency are plotted in Figure 11.1. Inherent sequential operations limit the ability to achieve a high speedup factor, even when many processors are available. In the above example, the speedup increases to a value of 2.4 when four processors are used. It is then insensitive to the availability of more processing units.

Using more processors also drops the efficiency because of the inability to use the extra processors for computations. They remain inactive.

11.4   Amdahl's Law

Gene Amdahl began his career as a computer designer for IBM and later formed his own company. Amdahl is credited with formulating Amdahl's law, an analysis of the effectiveness of parallel processing. Amdahl reasoned that in a computer system with n processor elements, a certain percentage, f, of the operations are inherently sequential. Under these conditions, Amdahl established a relationship which empirically captures the actual speedup observed in real systems.

$$S_n = \frac{n}{1 + (n-1)f}$$

In the above equation, if $f = 0$, there are no inherently sequential operations. This allows n operations to be performed simultaneously and $S_n = n$. At the other extreme, if $f = 1$, all operations are sequential and $S_n = 1$. Based on experimental measurement with real programs, Amdahl found f to be approximately 0.40. With this value of f,

$$S_n = \frac{n}{1 + 0.4\,(n-1)}$$

$$S_n = \frac{1}{\dfrac{0.6}{n} + 0.4}$$

The limiting value of S_n, as n is allowed to increase without limit, is 2.5. This appears to be a very modest improvement for the extensive amount of work that may be needed to establish a parallel computational environment. Critics are quick to point out some of the possibly false assumptions that lead to Amdahl's conclusion. Many massive computing problems, such as global weather prediction, involve algorithms that can effectively use parallel processing structures.

While the efficiency of parallel processing may decrease with additional processors, the cost of providing additional processors also decreases with quantity. Many processors, for example, may share a common power supply, mounting hardware, and input and output devices. The personnel and maintenance costs may also grow at less than a linear rate.

11.5 Flynn's Classification of Datapath Organization

Flynn's taxonomy is often used when discussing controller and processor organization. Flynn[1] classified systems according to their method of processing a data stream. A data stream is a sequential flow of related data originating from either an external source or the registers and memory of the datapath. The goal of the system is to process the entire data stream to produce desired results.

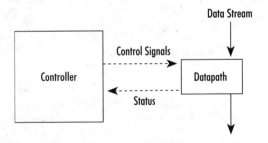

Figure 11.2   A Single Instruction Single Data Stream (SISD) Datapath

A single instruction single data stream (SISD, pronounced "sis dee") organization, as shown in Figure 11.2, is used in most conventional computers and processing systems. The controller

1.   Flynn, M.J. 1966. "Very High-Speed Computing Systems." *Proceedings of IEEE*, 54 (December); 1901–1909.

provides control signals for a single datapath structure, which sequentially processes the data stream.

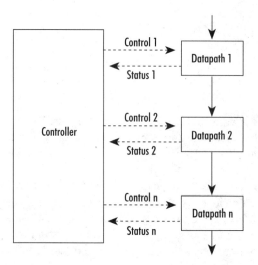

Figure 11.3   A Multiple Instruction Single Data Stream (MISD) Datapath

Pipelining the processing datapath, as shown in Figure 11.3, creates a multiple instruction single data stream (MISD, pronounced "mis dee") organization. In this organization multiple datapath elements are connected serially. The single data stream is subjected to different processing actions at different stages in the pipeline. The single controller is responsible for interpreting the status information from each of the pipelined stages and providing the proper control signals. Pipelining improves performance by using each processing datapath concurrently to work on different groups of data elements. Once the pipeline is filled, each clock cycle produces a new result.

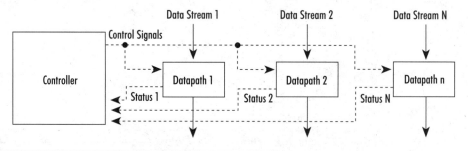

Figure 11.4   A Single Instruction Multiple Data Stream (SIMD) Datapath

Parallelism in the data stream is shown in Figure 11.4. If each datapath element performs the same processing action, a single instruction multiple data stream (SIMD, pronounced "sim dee") is created. During each clock cycle all processors perform the same operations, but on different data streams. The control signals generated by the controller are the same for each unit and only require replication. Each of the processor datapaths provides its unique status information to the controller. Using status information extensively may cause the controller to become quite complex. The complexity is somewhat limited, however, since each of the processor elements must perform the same operation.

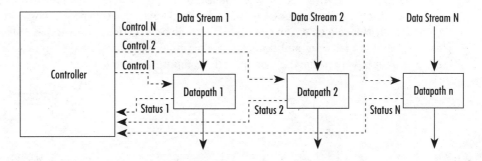

Figure 11.5   A Multiple Instruction Multiple Data Stream (MIMD) Datapath

The multiple instruction multiple data stream (MIMD, pronounced "mim dee") organization, shown in Figure 11.5, is the most complex of the parallel structures. Each of the processing architectures provides its unique set of status signals and receives a unique set of control signals. Each of the data streams can be subjected to different processing algorithms. Data can be exchanged between datapath elements. For a single controller to service many independent processor architectures in a timely manner is a complex task and quite difficult to achieve.

11.6   SIMD Parallel Structures

Two structural implementations of parallel architectures are possible. SIMD uses a single instruction stream with multiple data streams; MIMD allows different instructions for each processor.

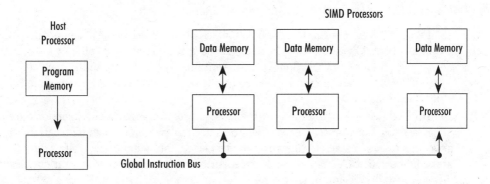

Figure 11.6   A SIMD Parallel Structure

Figure 11.6 shows a method of implementing the SIMD structure using multiple processors, each with their own data memory. This allows multiple data streams provided by the individual data memories to be processed. The host processor provides a single program memory and control unit. Each processor uses operands obtained from its own data memory to execute the identical instruction provided by the host.

The SIMD structure allows implementation of highly repetitive parallel operations such as adding vectors. Such machines are often referred to as **vector processors**. Each processor is associated with one element in the vector. A single instruction, for example ADD, can operate on each element of the vector simultaneously. This single operation would require an extensive number of sequential operations in a nonparallel computing structure.

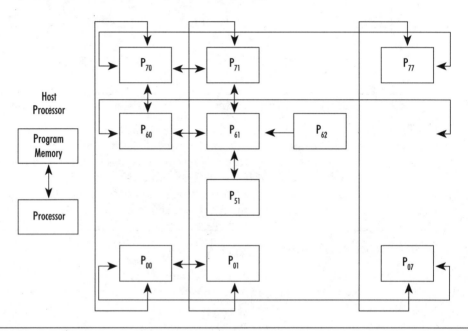

Figure 11.7   ILLIAC IV, A SIMD Parallel Machine

Figure 11.7 shows the interconnection structure of ILLIAC IV, one of the pioneering SIMD machines. ILLIAC was built in the early 1960s with a cooperative effort by the University of Illinois and Burroughs Corporation. ILLIAC contains 64 processing units, each connected with four of its neighbors. Processing units located on the boundary of the array are connected in a toroid, or wraparound fashion.

ILLIAC uses a single processor to access a program memory. Instructions are distributed to all processors. Each processor executes the same instruction in each instruction cycle. Each of the array processors has access to its own modest sized data memory as well as data from its immediate neighbors. ILLIAC is especially well suited for performing computations associated with a physical grid (row, column) structure.

Example 11.1:   Heat flow equation

Consider a two-dimensional slice of a plate of homogeneous material. The temperature readings along the outer boundaries are known. It is desired to calculate the temperature distribution for points internal to the plate.

Solution:

Problems of this type are often solved iteratively by dividing the region into small cells. The temperature readings of the cells along the boundary are known. The temperature of an interior cell is calculated as the mean of the temperature of the four neighbors located above, below, to the right, and to the left. The process is repeated until negligible changes are obtained by the calculations.

$$T_{i,j} = \frac{T_{i+1,j} + T_{i-1,j} + T_{i,j+1} + T_{i,j-1}}{4}$$

This form of calculation is ideally suited to structures such as ILLIAC. The cells correspond to ILLIAC processors. Each processor provides its temperature to its four neighbors. Processors on the boundary receive temperature data specified by the boundary conditions. All processors are able to update calculations simultaneously.

11.7   MIMD Parallel Structures

MIMD structures, with their independent instruction streams, offer more flexibility in implementing parallel processing algorithms than is possible with SIMD machines. For example, each of the processors can be unique and optimized for a highly specialized operation.

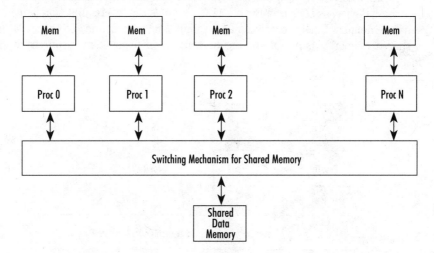

Figure 11.8   A Closely Coupled, Shared Data Memory MIMD Parallel System

With MIMD structures, the ability to interchange information between processors is quite important. Two techniques are used. **Closely coupled systems** exchange information using shared data memory, as shown in Figure 11.8. Each processor has its own dedicated program memory and may also have its own dedicated data memory. The shared memory may contain restricted regions as well as shared regions for each processor. Information processed and stored by one unit can be accessed by another unit. A method is needed to manage contention for the memory as multiple processors try to simultaneously access information.

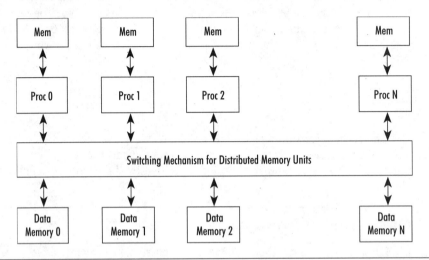

Figure 11.9   A Closely Coupled MIMD System with Segmented Shared Memory

In an alternative formulation, the data memory can be segmented into several units. A switching device, as shown in Figure 11.9, provides the interconnection between a processor and a selected memory. Each processor may be simultaneously connected to a different memory unit. In a shared memory system, assuming no contention for system resources, data exchange is very fast.

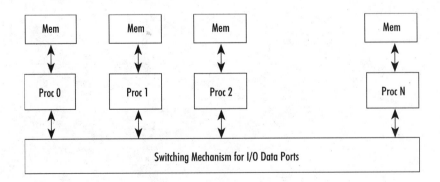

Figure 11.10 A Loosely Coupled MIMD System

Loosely coupled systems are characterized by relative independence of most components. Information is shared using I/O structures, which can send messages to other units, as shown in Figure 11.10. Because of the need to formulate message packets and use I/O facilities, data exchange in loosely coupled systems is normally quite slow compared to closely coupled systems. There is moderate similarity between a loosely coupled MIMD and a local area network (LAN), which is presented in Chapter 12.

Both the closely coupled and loosely coupled structures require a mechanism to interconnect the various system elements. In the case of a closely coupled system, the processors must be connected with the selected memory unit. In loosely coupled systems, the respective I/O ports of communicating processors must be interconnected.

Two general techniques exist for interconnecting devices: circuit switching and message switching. In **circuit switching** a dedicated communication path is established between the source and destination unit. Once established, information may traverse from source to destination with only a slight propagation delay attributed to logic elements.

Message switching transmits information in a hop-by-hop manner. The source node transmits the entire message directly to the destination node if a direct path exists. If no direct path is present, the source node transmits to an intermediate node, where the message is temporarily stored. If multiple hops are used, the time delay associated with message switching may be unacceptable. Most loosely coupled parallel systems use some form of circuit switching to minimize data transit delay.

In the following sections several possible topological configurations are presented for interconnecting parallel processing systems. In the case of closely coupled systems, the interconnection structure provides access to a common shared memory system. In loosely coupled systems, the interconnection structure provides a device-to-device path for interconnection of the respective I/O channels.

Bus Interconnection

The bus structure shown in Figure 11.11 offers a method to interconnect a moderate number of devices. Three areas of concern must be addressed:

- **Bus access control.** A bus is a shared communication medium. Rules must be established and administered to avoid simultaneous bus use by multiple devices. This may involve using a bus controller to accept bus request signals from the individual devices and grant the bus according to a predetermined set of rules. The concepts and handshakes in this type of management scheme were considered in Chapter 10.
- **Bus bottlenecks.** Because the bus must be shared, an individual unit may experience a delay in its processing algorithm while waiting for access to the bus. These delays could appreciably decrease the overall system performance.
- **Lack of parallelism in the communication path.** Although we are dealing with the topic of parallel systems, the bus interconnection structure precludes simultaneous communication.

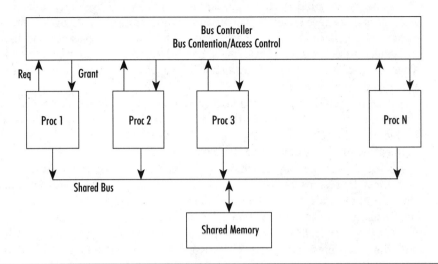

Figure 11.11 A Shared Memory System with Bus Contention Control

Fully Interconnected Processing Devices

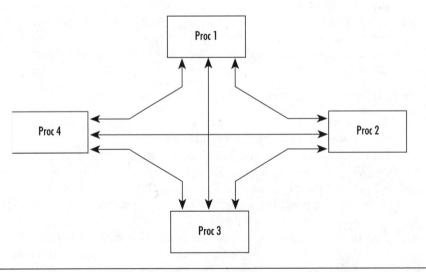

Figure 11.12 A Fully Connected MIMD Parallel System

A fully interconnected system, as shown in Figure 11.12, offers the potential of directly communicating between any two devices; no multi-hop routing is required. The many interconnections eliminate the possibility of communication link bottlenecks. Multiple concurrent communication activities are possible.

Fully interconnected systems require $N * (N - 1) / 2$ bidirectional data links to interconnect N nodes. This number grows as the square of N, and rapidly becomes unwieldy. Adding or deleting of a single node may require an enormous change in the number of data links. Fully interconnected structures are useful only where a few processor units are used.

Hypercube Connection

The bus structure and fully interconnected structure present two extremes in developing a method of interconnecting multiple computational nodes in a system. The bus offers too few data paths and presents a bottleneck when more than a few nodes must share the communication media. In the fully connected structure, the number of data links grows at a rate proportional to the square of the number of nodes, and rapidly becomes excessive.

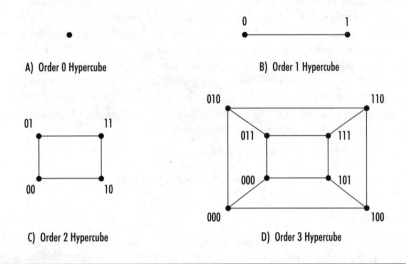

Figure 11.13 Hypercube Topologies

In many respects, the **hypercube** offers an interconnection structure that represents a good compromise between these two extremes. Figure 11.13 shows a progression of hypercubes of various orders. An n dimensional hypercube consists of 2^n nodes. A 0 dimensional hypercube is a single node. To create an $n + 1$ dimensional hypercube from an n dimensional hypercube, the n dimensional structure is first replicated. The corresponding nodes of the two are then joined with a bi-directional communication path. Each node can be assigned a binary numerical address based on its location. The algorithm is not complex even though the structures are difficult to draw for large values of n.

Assume that a hypercube involves a total of N nodes, where $N = 2^n$. Each element is connected to n other nodes through a bidirectional communication link. Thus for a system with $N = 2^n$ nodes, there are $N * n = N \log_2 N$ unidirectional connections or $N / 2 \log_2 N$ bidirectional connections. This value grows considerably more slowly than the $N * (N - 1)$ unidirectional connections encountered in the fully connected system.

Routing information from node to node must also be considered. In both the single bus and the fully connected structures, information may be passed between nodes without the need for interaction by an intermediate node. With hypercube, transmitting of information between nodes may require passing through intermediate nodes. Such action by nodes that are not destination nodes are called **hops**. Each node may communicate with every other node by a path comprised of a number of hops. The minimum number of hops for each pair of nodes can be determined. The maximum of these values is defined as the **spanning diameter** of the network. It is possible to communicate between any two nodes in the network with a number of hops less than or equal to the spanning diameter.

In an n-dimensional hypercube, the spanning diameter is n. Thus for a hypercube of order 10 with all $2^{10} = 1024$ nodes free, a message can be passed between any two nodes with 10 or fewer hops. If some of the nodes are busy and committed to other activities, an alternate path requiring more than 10 hops may exist.

Messages can be routed between hypercube nodes in several ways. One method is to formulate the messages with a field dedicated to the destination address, as expressed in binary. A node receiving a message first examines the destination address. The bits of the destination address and local address are compared bit-by-bit starting with the left-most bit location. When a bit location with no match is found, for example position q, the message is transmitted on the link associated with coordinate q. Thus with each hop there is one less coordinate that differs from the destination address. If the destination address is identical to the local node address, the message is accepted for interpretation by the local node.

Example 11.2:   Routing in a Hypercube

Assume that in a three-dimensional hypercube, Node 010 has a message for node 111. Find the routing the message will take.

Solution:

The source node compares the initial node address 010 with the destination node address 111. The first bit position is different. It is corrected by sending the message to node 110. At node 110, the first two bit positions are found to be in agreement. The third bit is corrected by sending the message to node 111.

The above algorithm guarantees that a message will always be routed using the fewest number of hops between the source and destination. It is possible, however, that intermediate nodes may be busy and unable to complete the routing in a timely manner. This phenomenon is called **blocking**.

NCUBE Corporation and Intel Corporation manufacture MIMD computing structures based on a hypercube topology. The NCUBE parallel processor is examined in Section 11.9.

11.8   Digital Switching Devices

Implementing a MIMD device requires a method of establishing the interconnection between processing nodes (in a loosely coupled systems) or processors and memory units (in a closely coupled system). This section considers the actual switching mechanisms for establishing such connections.

A **switching network** establishes a communication path between source and destination nodes, as shown in Figure 11.14. At any specific time period, multiple data paths may exist between source and destination nodes. There is no internal data storage; all internal components to the switching network consist of logic gates or contact switching devices. This is known as **spatial division multiplexing**.

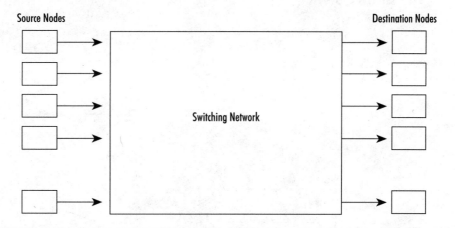

Figure 11.14 Spatial Division Multiplexing

Crossbar Switch

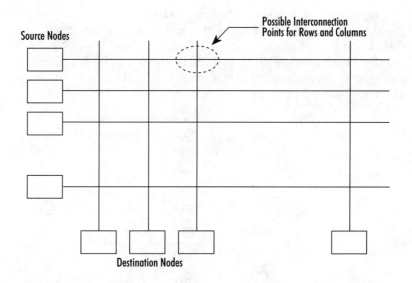

Figure 11.15 A Crossbar Switch

The crossbar switch shown in Figure 11.15 was studied in Chapter 1. Since a data path can be established between any source and any *free* destination, the device is nonblocking.

The Binary Switch

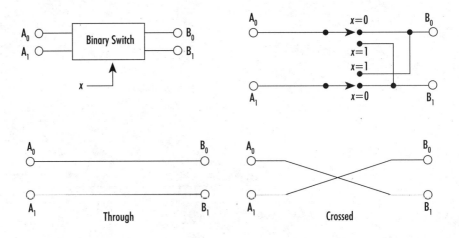

Figure 11.16 The Binary Switch

The binary switch, shown in Figure 11.16, has two inputs and two outputs. The internal switch connections are specified by a control input, x. The internal path is "through" if $x = 0$; it is "crossed" if $x = 1$. Binary switches can be interconnected as modular components to form a wide variety of useful switching components. In Figure 11.17 a multiplexer has been formed. The control input $x_0 x_1 x_2$ controls the connection pattern between the eight inputs, A_{000} through A_{111}, and the single output Y.

$x_0 x_1 x_2$	Input Selected
000	A_0
001	A_1
010	A_2
011	A_3
100	A_4
101	A_5
110	A_6
111	A_7

Table 11.4

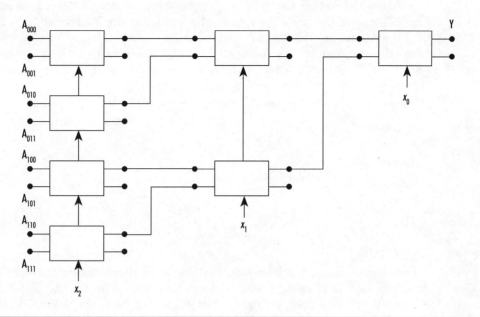

Figure 11.17 A Multiplexer Constructed from Binary Switches

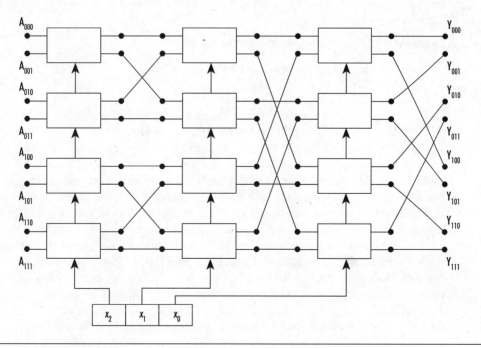

Figure 11.18 A 4 × 3 Butterfly Switch

In Figure 11.18 additional binary switches and interconnections have been added to the multiplexer. This structure, known as a **butterfly switch** or **shuffle switch**, uses both switch outputs. All switches in a column are controlled by the same single logic variable. Paths exist between the eight inputs, A_{000} through A_{111}, and the eight outputs, Y_{000} through Y_{111}, and are specified by the status of control inputs x_0, x_1, x_2. For the condition $x_0x_1x_2 = 000$, each input element is connected to the corresponding output element.

For the condition $x_0x_1x_2 = 001$, connections are made between the pairs:

$A_{000} \rightarrow Y_{001}$
$A_{001} \rightarrow Y_{000}$
$A_{010} \rightarrow Y_{011}$
$A_{011} \rightarrow Y_{010}$
$A_{100} \rightarrow Y_{101}$
$A_{101} \rightarrow Y_{100}$
$A_{110} \rightarrow Y_{111}$
$A_{111} \rightarrow Y_{110}$

Each input terminal has been connected with an output terminal in which the binary designator of the third bit has been inverted. A similar analysis shows that $x_0x_1x_2 = 010$ would connect input with the corresponding output in which the second bit in the binary designator bit pattern was inverted: A_{011} would be connected to Y_{001}. The pattern $x_0x_1x_2 = 111$ would connect A_{101} with Y_{010}. The routing algorithm requires a comparison of only the source node and the destination node. An exclusive OR operation on the binary designations provides the correct value for the X register.

Example 11.3:   Routing in a Shuffle Switch

Find the value of $x_0x_1x_2$ to connect node A_{110} to Y_{011}.

Solution:

Take the bit-by-bit exclusive OR of the respective binary designators for the nodes. $(110) \oplus (011) = 101 = x_0x_1x_2$. The validity of the solution can be checked with Figure 11.18.

The shuffle switch can be the basis for a structure to interconnect processors in a loosely coupled MIMD system, or processors and memory elements in a closely coupled MIMD system. The shuffle switch is a shared resource and must be allocated to a single processor. If circuit switching is used, the processor requiring a connection must gain access to the X register. This can be accomplished by time-shared access or a bus master to resolve contention. The processor then transmits its data. When finished, it releases the X register and the switch.

In a circuit-switching environment, the contents of the X register controls the setting of all of the individual switches. If one of the inputs is using the shuffle switch to send information, other free paths still exist through the switches. For example, if processor P_{011} requests an interconnection to processor P_{110}, then the X register requires $x_0x_1x_2 = 101$. This same value of X makes connections between processors pairs (P_{000}, P_{101}), (P_{001}, P_{100}), etc. Some of these connections may be useful for other processors, but not much flexibility exists and blocking is likely.

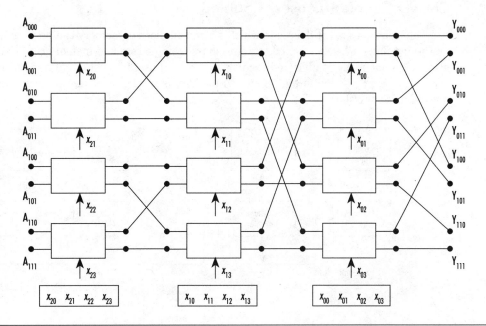

Figure 11.19 A Switching Network with Individually Controlled Binary Switching Elements

An alternate method for controlling an interconnection switching network is shown in Figure 11.19. If each binary switch is assigned a column/row address, then x_{ij} controls the switch located in column i and row j. For example, x_{00} would control a binary switching element attached to outputs Y_{000} and Y_{100}; x_{21} could control the element attached to inputs A_{010} and A_{011}.

The pattern of 12 control signals x_{00} through x_{23} can establish a connection pattern between the inputs and outputs, which allows greater connection flexibility than that previously described. Each processor needs to control only one of the switches in each column to establish a connection through the switch to the desired destination terminal. Since not all switching resources are used or specified, a second processor can establish a connection path if required switching resources are free. It can be shown that blocking is still possible.

11.9   Parallel Computation

Parallel computation involves both hardware and software. Of the two, hardware issues are much more easily addressed and solved. Computation problems, as perceived by humans, are usually engaged by sequential rather than parallel actions. Unless computations can be performed by concurrent actions, there is no hope that parallel computations will offer advantages over conventional approaches.

There are several issues to be addressed in implementing a parallel computing environment; some relate to software and some to hardware.

Closely Coupled/Loosely Coupled

The issue here is the degree to which a common block of memory can be accessed by multiple processing units. Closely coupled systems share a single common memory component. They may also have some memory allocated exclusively for local use. In most situations, the common memory is accessed by a shared bus. If shared memory is used extensively, performance deteriorates due to bus contention.

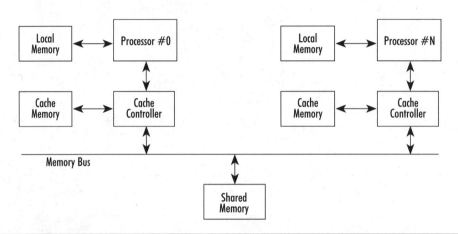

Figure 11.20 A Closely Coupled System Using a Cache Memory

A cache memory helps to avoid conflicts in accessing shared memory. A possible implementation structure is shown in Figure 11.20. Besides allowing the fast access of memory information, the cache decreases the need to access information from the shared memory. An intelligent cache controller automatically manages the shared memory and caches to assure data coherency.

Closely coupled systems are well suited for situations in which a single program can be decomposed into instruction blocks, or processes, which can be assigned to different processors. They are also well suited for multiple programs run under the control of a single operating system.

The major performance limits in closely coupled systems are the shared bus and inherently sequential aspects of the program. The shared bus, even with the use of cache memories, limits closely coupled systems to modest amounts of parallelism. Existing closely coupled systems, such as the Sequent system to be discussed later, involve fewer than 30 processors.

Loosely coupled systems avoid the bottleneck of a shared bus. Since each processor operates relatively independent of all others, the number of processors can be increased with few practical operational problems. Systems of 1024 processors, for example, are available from NCUBE Corporation. The NCUBE system will be discussed in a later section.

While avoiding a shared bus, loosely coupled systems must rely on inherently slower I/O ports to share information between processors. Shared facilities are also required for controlling processors, downloading programs and data, and providing shared I/O facilities.

Loosely coupled systems offer performance advantages in applications using separate data sets extensively, each operated upon by a separate instruction set. This situation may arise from processing independent or noninteracting data streams; it may also result from processing large numbers of small, independent programs.

Topology

Bus interconnection topologies dominate closely coupled systems. This allows any processor to easily gain access to a shared memory base. To avoid contention, a means of bus allocation must be imposed.

Loosely coupled systems, since they rely less on shared data, use more exotic interconnection schemes requiring multi-hop data communication. Because of the potentially large number of cooperating devices, grids and hypercube interconnections are common. Rings, trees, buses, and meshes may also provide appropriate interconnection.

Decomposition

Before a computation can be performed in a parallel hardware environment, it must be decomposed into multiple unique processes or tasks. Each task can then run on a separate processor using a shared or independent data set. When the data set can also be partitioned, less concern is needed for bus contention and communication activities.

Decomposition can be performed by either the programmer or by the operating system. Decomposition presents an additional layer of complexity to the programmer, who may not fully understand all of the intricate data interrelationships in the program being coded. In some environments, particularly those with large data sets and repetitive processing algorithms, the decomposition of the problem may not be difficult. In many situations, the advantages gained from parallel processing will adequately reward the extra programming effort.

Ideally, an operating system could examine programs and decompose them into modules suitable for use in a parallel processing environment. In such an environment, programs originally written for sequential processing could run on parallel machines.

Scheduling

Scheduling is the process of assigning tasks to processing elements. This includes making decisions concerning both the sequencing of processing a task and the processor to which the task should be assigned. Sequencing is concerned with the order of processing tasks. Single processor computers always execute tasks in a strict sequential order determined by the programmer when the code is written.

In parallel processing, a task cannot be processed until it has the necessary computational resources and data. The data resources can be assured if a prerequisite set of tasks is finished. The requirement for computational resources is least complex if all processing units are identical; a processor must be free. If the resources are varied and optimized for different operations, they may require a single resource or one of several devices.

At any time a task can be **idle** while it waits for data, it can be **ready**, which means it has the necessary data but is waiting for an available processor, it can be **running**, while executing instructions, or it is **finished** when it is completed. When a processor becomes available, a task from the pool of ready tasks is allowed to run. Scheduling parallel processing activities is complicated because it is often not known in advance how much time a given processing task will require. The time required for a program iteration loop, for example, may be data-dependent.

Scheduling may be performed by the programmer, at run-time by the operating system, or dynamically during execution of the program. Prescheduling by the programmer is most useful when the operation of the processing environment is well understood. Scheduling by the operating system before the program runs relieves the programmer of a difficult job that is prone to errors. If scheduling decisions are made while the program is running, data-dependent variations in the availability of resources can be considered.

Tuning

Tuning is the process of modifying the scheduling rules to attempt to optimize overall performance. In general, good performance requires balancing the overall operating load among the available processors. Any processor idle time will lower overall performance. In loosely coupled systems the ratio of communication time to computation time needs to be minimized. Communication is a relatively slow operation.

The Sequent Parallel Processor

Sequent Computer Systems offers an extensive line of tightly coupled parallel computing devices. A single shared bus interconnects up to 32 processors, a shared memory, and I/O facilities. All the processors are the same and are based on high-performance 32-bit microprocessors. The number of processors can easily be increased or decreased to match the computational needs of the application. The Sequent parallel processor is intended to run in a conventional computing environment. Programs need not be specifically written for a parallel machine.

The Sequent Symmetry system can function with from 2 to 30 processors. The shared memory ranges from 8 through 240 megabytes. A 64-kilobyte cache memory is maintained at each node. The cache's availability considerably reduces the contention for the single system memory. Load balancing is done dynamically during program execution.

Sequent provides the user with a Dynix operating system that offers the advantage of a standardized Unix environment. Fortran, C, Pascal, and Ada compilers are available. The Dynix operating system can service up to 256 users in a time-share environment. Parallel operation is achieved through **multiprogramming** and **multitasking**. In multiprogramming, multiple unrelated programs can be run on different processors. Since there is virtually no communication between programs except to share I/O facilities, multiprogramming is relatively straightforward on a multiprocessor system. In a multiprogramming environment, performance over conventional computers is improved by executing more than one program concurrently. An individual program requires about the same amount of execution time that would be required on a conventional machine.

In multitasking, a single application is decomposed into multiple processes. These processes (or tasks) are distributed to run on different processors. Running tasks in parallel reduces execution time for a single program. Multitasking is limited by some inherent limitations of parallel processing. This includes problems with inherently sequential coding blocks and the inability to continuously use a large number of processors.

The NCUBE Parallel Processor

NCUBE Corporation manufactures a range of loosely coupled MIMD computer systems. Each of the computational nodes is interconnected in a hypercube structure. Many designers view the hypercube as the most densely connected network scalable to thousands of processors.

The number of nodes in an NCUBE system can range from 16 to 1024 computing elements. Up to 64 processing elements can be included on one circuit board. A grouping of 16 boards can be assembled, with appropriate backplane wiring, to produce a full 1024-node hypercube.

Each node in the NCUBE system is an independent 32-bit processor with its own memory. Each node is connected to all other nodes that are within one hop on the hypercube. In the case of 1024 nodes, a tenth-order hypercube, ten bidirectional DMA communication channels are associated with each node. An additional bidirectional channel is available for system I/O. After a message transmission, a node can provide an interrupt to the receiving node to indicate that a message has arrived.

I/O is handled by special I/O boards. One bidirectional DMA channel from each of the processors is brought through the backplane to the I/O board. A variety of different I/O configurations are available. One, known as the host board, runs a Unix-like operating system and provides a user interface. It also controls the standard computer peripherals.

Since the NCUBE device is a loosely coupled system, its ability to achieve high performance is closely tied to the ability to keep each of the processors active for an appreciable percentage of the processing time. To effectively use the nodes, large hypercubes can be divided into smaller hypercubes to more closely match the needs of smaller problems.

The main application area of the NCUBE machine is inherently parallel computational problems. This includes signal processing using the fast Fourier transform. The hypercube's ability to function as a grid structure allows it to be used in the numerical solution of partial differential equations associated with electromagnetic fields and heat flow.

11.10   Pipeline Processing

Pipeline processing, while structurally different from parallel processing, can increase the effective throughput of the processing action. In an earlier chapter we studied pipeline processing at the register transfer level. Processor-level pipeline elements are also useful. Figure 11.21 illustrates a possible pipeline configuration for processors.

Multiple processors are used in pipeline processing. Each processor may be optimized to perform one stage of the pipeline operation. Results from a processor must be available to the succeeding unit. This is accomplished by placing the information in a memory element or, if the amount of information is relatively small, in a register. If a large amount of information must be placed in memory, storage time can significantly affect the performance. For modest amounts of information, this is a good application for a dual port memory.

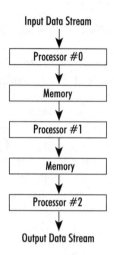

Figure 11.21 Pipeline Processing at the Processor Level

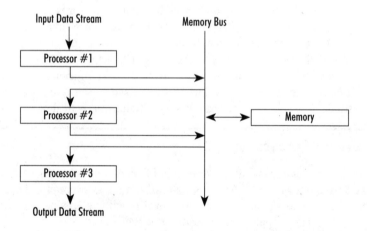

Figure 11.22 A Pipeline with a Single Shared Memory

Several implementations are possible. In Figure 11.22, all processors use a single shared memory. Each processing elements accesses the memory to get to information stored by the previous element in the pipeline. Each processing element requires time to store information intended for the next element in the pipeline.

Figure 11.23 shows a memory schedule for a four-processor pipeline. Each processor requires access to the shared memory to store results (S) from its current operation and to access data (A) for its next operation. After acquiring data, each processor can process information (P) and use its own memory for temporary storage and accessing instructions. While unit 1, for example, is processing data, unit 2 can perform its storage and access functions with the shared memory. The memory is shared among all four processors in the pipeline.

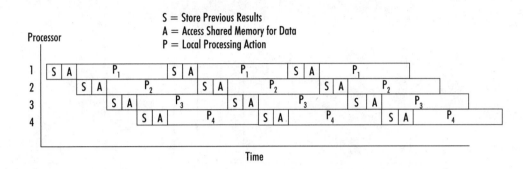

Figure 11.23 Scheduling a Memory for Use in Pipeline Processing

To fully benefit from pipeline processing, the following guidelines should be followed:

1. All stages of the pipeline should require about the same amount of processing time since pipeline processing requires that the time duration of all processing steps be equal. If this is not the case, timing must be adjusted to accommodate the slowest pipeline stage. Similarly, if one stage of the pipeline requires a data-dependent processing time, this could cause problems.

2. The amount of information passed between stages should be minimal since passing information between processing elements requires storage. It also may require considerable time to have one processor store the information in a memory and then have a second processor retrieve it.

3. The processing problem should require repetitive processing of a relatively long data stream. If the data stream is not sufficiently long, an excessive proportion of the processing time may be spent filling the pipeline and then emptying it.

4. The processing time of a single unit of data must not be extremely time-critical. In a pipeline, the processing time of a single data unit is longer than that required using a single processor. Extra delays are caused by passing information between the various processors. The overall processing rate of the pipeline exceeds that of a single processor because many processors simultaneously work on different parts of multiple data sets. Parallel systems perform different processing activities from the same data set simultaneously. This decreases the processing time for each data set.

11.11   Pipeline Scheduling

Scheduling a one-pass pipeline presents no conceptual problems. First, the time duration for each processing element is determined. The longest time duration determines the rate at which new information can be introduced into the pipeline. For example, consider a processing activity requiring five sequential processing actions. Figure 11.24 shows a **reservation table** for a five-stage pipeline. Each time slot for each processor may be allocated to a single data set. Three data sets, X, Y, and Z, are to be processed.

Time

	t_1	t_2	t_3	t_4	t_5	t_6	t_7	t_8
P_1	X	Y	Z	·				
P_2		X	Y	Z				
P_3			X	Y	Z			
P_4				X	Y	Z		
P_5					X	Y	Z	

Processors

Figure 11.24 A Reservation Table for Processing Three Data Sets in a Five-stage Pipeline

The speedup for using five processors to process the three data sets can be computed. With a single processor, the processing must be done sequentially, requiring 15 time slots.

$$T_1 = 3 * 5 = 15$$

If five processors are used, it takes five time slots to process data set X. It requires two additional time slots to produce the results from set Y and set Z.

$$T_5 = 5 + 2$$
$$S_5 = \frac{T_1}{T_5} = \frac{3 * 5}{5 + 2} = \frac{15}{7} = 2.14$$

For N data sets and five stages in the pipeline, the speedup becomes:

$$T_1 = N * 5$$
$$T_5 = N + 4$$
$$S_5 = \frac{T_1}{T_5} = \frac{N * 5}{N + 4} = \frac{5}{1 + \frac{4}{N}}$$

For N data sets, the speedup approaches n, the number of processors for large N.

$$S_n = \frac{n}{1 + \frac{n - 1}{N}} \approx n \quad \text{(for large } N\text{)}$$

In some computational environments, some processing elements may be used more than one time. This may decrease the cost of processing elements but increase processing time. When processing elements are used more than once, a reservation table can indicate the usage of processors as a function of the time unit in the processing algorithm.

Time

	t_1	t_2	t_3	t_4	t_5	t_6	t_7	t_8
P_1	X							
P_2		X		X		X		
P_3			X		X			
P_4							X	
P_5								X

(Processors)

Figure 11.25 A Reservation Table for Pipeline Processing

Consider, for example, the reservation table shown in Figure 11.25 for processing data set X. This processing operation requires eight stages in the pipeline, but uses only five processing elements. Processor P_2 is used three times and P_3 is used twice.

If data set X begins processing at t_1, when can an additional data set, Y, be introduced into pipe segment P_1? The answer can be determined by examining the reservation table. Clearly, a new processing activity cannot begin two time units later, at t_3, since there would be a collision in data from X and data from Y in both P_2 and P_3. The same reasoning precludes the initiation of a new processing activity at t_5; this would cause a collision in P_2.

The operation may be analyzed by determining the time delays, or latencies, which lead to collisions in the pipeline. In Figure 11.25, collisions result if time latencies have values of two or four time units. Figure 11.26 shows the reservation table if data set Y is initiated at time t_2, which corresponds to a latency of one time unit. There are no collisions. If a third data set, Z, is started one time unit after set Y, a collision would again result.

Time

	t_1	t_2	t_3	t_4	t_5	t_6	t_7	t_8	t_9
P_1	X	Y							
P_2		X	Y	X	Y	X	Y		
P_3			X	Y	X	Y			
P_4							X	Y	
P_5								X	Y

(Processors)

Figure 11.26 Pipeline Processing Two Data Sets with a Latency of One Time Unit

The **minimum constant latency** is defined as the smallest constant latency that can be used without causing collisions in the pipeline. If data sets are introduced at time delays corresponding to the minimum constant latency, the processing operations may continue without danger of data collision. Figure 11.27 shows this situation. Data sets X, Y, Z, A, and B are successively introduced with time intervals of 3. The integer 3 is the smallest integer that it does not cause a collision, nor does any of its integer multiples cause a collision.

Time

	t_1	t_2	t_3	t_4	t_5	t_6	t_7	t_8	t_9	t_{10}	t_{11}	t_{12}	t_{13}	t_{14}	t_{15}	t_{16}	t_{17}
P_1	X			Y			Z			A			B			C	
P_2		X		X	Y	X	Y	Z	Y	Z	A	Z	A	B	A	B	C
P_3			X		X	Y		Y	Z		Z	A		A	B		B
P_4							X			Y			Z			A	
P_5								X			Y			Z			A

(left margin label: Processors)

Figure 11.27 A Pipeline Using Minimum Constant Latency

Example 11.4:   Analyzing a Reservation Table for Speedup

What is the speedup attained in processing four data sets with the schedule table of Figure 11.27?

Solution:

If a single processor is used, eight time slots are required for each data set; a total of $4 \times 8 = 32$ time slots. Figure 11.27 shows that 17 time slots are required to fully process the four data sets X, Y, Z, and A. Therefore, the speedup is:

$$S_5 = \frac{T_1}{T_5} = \frac{32}{17} = 1.88$$

11.12   Data-flow Architectures

The goal of all parallel approaches to computation is to use multiple processing units to process different portions of the overall computational problem simultaneously. To work well, the various processors must be kept busy and avoid idleness. On the other hand, they must not be so over-worked that they create a bottleneck for other processing activities.

The conventional approach to managing a complex processing environment is to use a supervisory system to schedule the activity of the individual processors. This underlying principle of pipeline processing works satisfactorily if the processing algorithm is well understood and regular in its demands on processing elements. Performance may degrade if certain stages in the computation are data-dependent. Unpredictable events, such as interrupts announcing the presence of additional data, are difficult to include in the scheduling algorithm.

Centralized management can be compared to top-down management of a factory. Tasks are assigned, scheduled, and sequenced by management (the system scheduler). A work force (individual processors) attempts to carry out the assignment of tasks in a timely manner. The system can break down if management gets overloaded with complex scheduling algorithms and unforeseen events. Some of the highly scheduled and restricted production crew (processors) are forced into nonproductive idleness while others are hopelessly overworked.

An alternative management approach uses a decentralized organization of production personnel (processors). Workers (processors) don't follow a fixed schedule but perform whatever portion of the job can be done based on the availability of components (data). When finished with a portion of the job, they pass it to the next step in production. They attempt to keep busy at all times.

This latter approach is known as **data-flow processing**; processing activity is initiated based on the availability of data. The former technique is known as **control-flow processing**; processing activity is initiated by a control signal from a centralized scheduler.

Data-flow Token

A central element within a data-flow computer is a **token**, which includes a portion of the information needed to perform a computation. The **tag** portion of the token identifies the data and allows it to be grouped with other tokens associated with the same processing activity. The data portion of the token contains all or a portion of the data necessary to perform the computation.

Consider the activity of a worker in the printer room of a large corporation. Documents are continuously being printed on a large number of printers. A document may include text from one printer, graphs from another, and colored imagery from a third. The printed pages are the data; an identification tag allows all the pages from the same document to be identified. The worker is to assemble completed documents, place them in envelopes, and send them to the proper location in the organization. The worker (processor) simultaneously monitors the printed output (token) of all the printers until a complete document is available. The document is collected and placed in an envelope. Incomplete documents are ignored.

Data-flow Program

A data-flow program is often represented as a graph. In Figure 11.28, a processing operation involves the computation of:

$$Z = \frac{(X+Y)^2 - (X-Y)^2}{4}$$

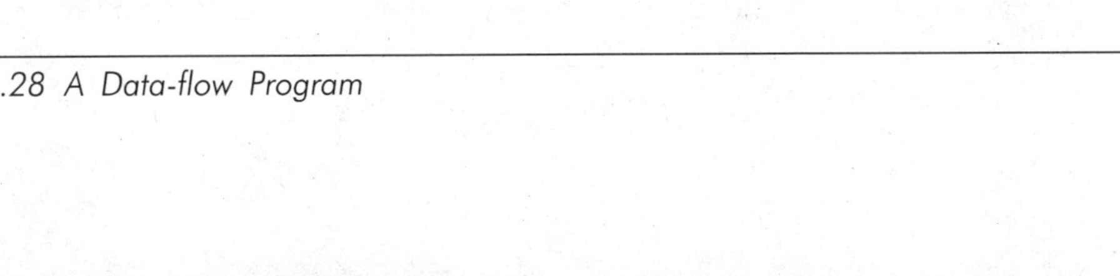

Figure 11.28 A Data-flow Program

This processing action is sometimes used to compute the product of X and Y. It has the advantage of not requiring a multiplier; only a squaring device, adders, subtracters, and a divider are required. The division is trivial if binary representation is used.

Each rectangle, P_1 through P_5, represents a processor. Each performs a set of operations to provide the function listed. The arcs represent paths taken by tokens. The data source of X and Y may be electronic instrumentation not synchronized to the data processor. Initially, the data for X and Y, along with an identifying tag, is placed in a token. When both tokens are present, P_1 and P_3 are able to perform a computation and create new tokens to present to P_2 and P_4. P_2 and P_4 require only a single token before performing their computation. P_5 requires tokens from P_2 and P_4 before computing a value for Z.

Data-flow Processors

Data-flow processors perform administrative work in addition to computations.

- **Token collection and storage.** Tokens are collected from all possible sources and stored in an internal memory.
- **Token matching.** As more tokens arrive, they are compared with tokens in storage to determine if enough information is present to perform a calculation. Associative memory is useful for performing these actions.
- **Algorithm execution.** When all tokens have arrived, the computational program begins execution.
- **Token generation**. Results of the computation are used to create a new token for passing to the next node in the processing operation.

Problem Areas

While data-flow computers offer a path toward high performance computation, there are several problem areas. There is not a good mapping between existing high-level languages and the graphs necessary to implement data-flow processing. Processing overhead in the individual processors is high. The need to manage tokens, perform calculations, create tokens, and pass tokens places a high computational burden on the processor. This overhead is wasted if significant portions of the program are inherently sequential and do not allow parallelism.

Data-flow computers also have performance problems when the size of the data field in the token is large and the required computation is relatively modest. It may be necessary to receive and transmit a large data array when only a single element requires an operation.

Summary

Parallel computing structures attempt to gain high performance using multiple simultaneous processing actions. SIMD processors are especially useful for solving vector operations on large data sets. MIMD structures require a means of exchanging data. Closely coupled systems use a shared memory; loosely coupled systems use an interconnection of I/O paths. Hypercube interconnections grow at a logarithmic rate and allow processor-to-processor communication with a modest number of hops.

Crossbar switches allow an interconnection between any two devices with no intermediate hops. The binary switch is a useful modular switching device that can implement a wide variety of interprocessor switching architectures.

Pipeline processing is a form of parallel processing. Data in the pipeline can be passed between processors using multiple dual port memories or a single time-shared memory. Complex pipelines require a reservation table to determine scheduling that will produce nonconflicting use of the pipeline resources.

Data-flow computers offer a different approach to parallelism. Rather than using the rigid scheduling of resources required by conventional approaches, data-flow processors are data driven. Processing action is initiated when all required data elements are simultaneously present as tokens in a processing unit.

For additional information and references on the topics presented in this chapter, access the World Wide Web page provided at: `http://www.ece.orst.edu/~herzog/docs.html`.

Key Terms

blocking	multitasking
butterfly switch	parallel efficiency
circuit switching	ready
closely coupled systems	reservation table
control-flow processing	running
data dependency	spanning diameter
data-flow processing	spatial division multiplexing
hops	speedup
hypercube	superscalar
idle	switching network
loosely coupled systems	tag
message switching	token
minimum constant latency	vector processors
multiprogramming	

Readings and References

Dennis, Jack B. 1980 "Data Flow Supercomputers," *Computer* IEEE Computer Society (November).
    Dennis gives a good presentation of the architecture of high performance data flow computers.
Lerner, Eric J. 1984 "Data-flow Architecture." *IEEE Spectrum* (April)
    This is a good tutorial presentation of the basics of data-flow architecture.

Exercises

11.1 A parallel computing system with 32 processors has 40% sequential instructions. What improvement in processing performance can be expected if the percentage of sequential actions is reduced to:

(a) 20%?

(b) 10%?

(c) 1%?

11.2 Show that for a MIMD system with n nodes, it requires $n(n-1)$ unidirectional data links to fully connect the nodes.

11.3 Consider a MIMD structure for n nodes in the topology of a ring. Unidirectional data paths are used. What is the spanning diameter of the structure? If communication between all nodes is equally likely, what is the mean number of hops that would be expected? If unidirectional communication is replaced by bi-directional communication, what is the spanning diameter and the mean number of hops?

11.4 What is the routing and address field in a eight-dimensional hypercube for a message originating at node 10010111 with a destination of 00111100? How many hops are required?

11.5 Develop an algorithm to allow a message to proceed in an alternative direction if some of the communication links on a hypercube are not available. How many possible minimum-length data routings are there between two nodes?

11.6 In an n-dimensional hypercube using bidirectional links between connected nodes, what is the mean number of hops for a message?

11.7 Compare the number of communication paths required to interconnect 64 elements using the following structures. What is the spanning diameter of each?

(a) Toroid (such as used by ILLIAC IV)

(b) Hypercube

(c) Bus

(d) Fully connected

11.8 Discuss the blocking characteristics of the butterfly switch of Figure 11.18.

11.9 Compare the blocking characteristics of the switching network in Figure 11.19 to that of the butterfly switch.

11.10 Show an algorithm for determining the contents of the 12-variable X register that will allow input terminal A_{ijk} to be connected to output terminal Y_{lmn} using 12 individually controlled binary switches. Use a register to indicate when a switching component is busy.

11.11 Determine if it is possible to simultaneously connect the switching network of Figure 11.18 between the following pairs of processors. Determine the required control vectors.

(a) (A_{000}, Y_{001}) and (A_{111}, Y_{101})

(b) (A_{001}, Y_{110}) and (A_{111}, Y_{000})

11.12 Which of the following sets of connections are possible using a 3×4 binary switch array of Figure 11.19 with each of the switches individually controlled?

(a) A_{000}, Y_{011}

A_{010}, Y_{111}

A_{100}, Y_{000}

(b) A_{000}, Y_{111}

A_{111}, Y_{000}

A_{001}, Y_{001}

(c) A_{000}, Y_{001}

A_{001}, Y_{010}

A_{010}, Y_{011}

A_{011}, Y_{110}

11.13 In what type of computing environment would you expect a loosely coupled parallel computing system to perform better than a tightly coupled system? Why?

11.14 Assume that it has been proposed to design a loosely coupled parallel computing system using a star architecture (all processors are connected only to a single hub node). What advantages and disadvantages would this have in comparison to a hypercube?

11.15 What is the minimum constant latency for the reservation table shown in the Figure 11.29? Show a possible scheduling algorithm for starting data sets through the pipeline.

	t_1	t_2	t_3	t_4	t_5	t_6	t_7	t_8
P_1	X							
P_2		X						
P_3			X		X	X		
P_4				X			X	
P_5								X

Figure 11.29

11.16 If an additional processor could be added to the reservation table in Figure 11.25, what would it be? Find a new scheduling algorithm for this condition.

11.17 What is the speedup for processing five data sets in the schedule table of Figure 11.29? What is the efficiency?

11.18 In Figure 11.29, what is the limit of speedup for very large number of data sets?

11.19 In Figure 11.25, what is the speedup obtained when processing two data sets?

11.20 A conventional computer is found to have inadequate processing power. Discuss three techniques that might be considered for redesigning to improve performance.

11.21 In what computational environment is a pipelined processor the best choice for high-speed computation? In what environment are parallel processors the best choice? Why?

12
Digital Communication and Networking

ENIAC

This war time project did attract
Some exceptional people with a knack
For designing a device
Which was very precise
And known to the world as ENIAC

Prologue

ENIAC, Electronic Numerical Integrator and Calculator, was a large-scale computing device intended to produce artillery tables for the United States Army in World War II. Under the direction of John Mauchly, J. Presper Eckert, and John von Neumann, ENIAC was assembled at the Moore School of Engineering at the University of Pennsylvania in 1946.

A large and clumsy machine by today's standards, ENIAC weighed over 30 tons and occupied a floor area of 1500 square feet; the size of mid-size house. ENIAC, with 18,000 vacuum tubes, 70,000 resistors, 10,000 capacitors, and 500,000 solder joints was the largest and most complex electronic device of its time.

ENIAC was credited with solving an atomic physics "wartime problem" in only two weeks. Conventional computation would have required over a year of calculations by 100 mathematicians.

Eckert and Mauchly later founded the Eckert-Mauchly Computer Corporation. Following acquisition by Sperry Rand, Eckert and Mauchly designed UNIVAC I, the first commercially feasible digital computer.

12.1   Introduction

Modern computer structures have experienced a remarkable evolution extending from single processors through closely coupled multiprocessors, loosely coupled multiprocessors, and distributed processors. Data communications and networking can be considered a logical extension of the concept of input/output structures. Both are concerned with exchanging information between a source digital computer and a device at an external destination.

Digital communication and networking is now a major technological component of today's society. Networks link offices, banks, factories, homes, and universities. In 1994 there were an estimated 100 million networked computing devices. Networked computers share information, databases, and remote computing resources. The lines of differentiation between the domain of the single computer and the network have continued to erode.

Digital communication has emerged as a unique discipline that has combined classical analog communication areas such as radio/television and telephone with digital computers. In this new field, computers are key elements in encoding, transmitting, routing, queuing, and receiving information. Computers also manage the complex rules (protocols) needed for media access, flow control, and error management.

The design structures and components necessary for digital communication are identical to those utilized in traditional computer design. It is now impossible to distinguish between computers for communication and communication for computers.

When considering networks of computers, some fundamental issues must be addressed.

- **Synchronization.** The source and destination do not share a common clock and may not necessarily operate at a common clock frequency. Techniques to synchronize the data units for reliable reception must be used.

- **Serial Data Transmission.** For reasons of economy, most data-communication techniques use a serial rather than parallel data-transmission channel. This requires serial-to-parallel and parallel-to-serial data conversion at the source and destination. Prior to transmission, the data is formed into packets and transmitted as a unit containing from eight to several thousand bits of information.

- **Non-Ownership of Transmission Channel.** In much long-range communication, the data channel is not owned or controlled by the same organization in control of the data source or destination. This limits the range of techniques for establishing and controlling the data transmission. Data security may also be a problem.

- **High Probability of Transmission Errors.** Long-distance communication is subjected to a wide range of atmospheric and man-made disturbances that can corrupt the transmitted data with errors. The probability of an error during long-distance communication may be hundreds or thousands of times more likely than with localized input/output. There is a strong need to incorporate techniques to detect errors and correct them by either mathematical techniques or data retransmission.

- **Time Delays.** Due to the finite velocity of information propagation, time delays are inevitable in long-distance communication. Propagation delays may range from a few nanoseconds on electrical or fiber-optic circuits to several hundreds of milliseconds when using satellite communication facilities to several minutes for space communication. Additional delays are possible if messages are stored and forwarded by intermediate components.

- **Data Modulation.** Because of the common use of radio, telephone, and fiber-optic transmission media, the transmission of binary voltages to represent digital data is not feasible. Instead, the information must be encoded, via modulation techniques, for passage through these components.

- **Communication Protocols.** In order to transmit information reliably between computing devices, a set of rules, or protocols, must be used by each. These rules control such things as a method to access a shared communication media, a method to limit the rate at which the source sends information (flow control), and methods to detect and correct errors.

The following sections will address these concerns.

12.2   Computer Networks

Figure 12.1 shows point-to-point data communications between two computers. In addition to the computers, both the source and destination need a **communication interface**. This device converts data from the computer to a form suitable for transmission. This might include parallel-to-serial data conversion, adjustment of voltage levels, or data modulation.

The **communication channel** links the two participants. Physically it may consist of dedicated media, such as a fiber-optics cable, or a mixture of components provided by common carrier organizations such as the telephone company. The channel is referred to as **full-duplex** if both units can simultaneously transmit to the other. A full-duplex channel may use two independent data paths or may encode information on a single path, perhaps through frequency multiplexing, to allow the simultaneous flow of information in both directions. The channel is **half-duplex** if it can support information flow in two directions, but not simultaneously. A **simplex** channel is restricted to a one-way flow of information.

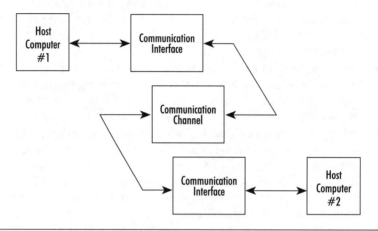

Figure 12.1   Point-to-point Data Communications between Host Computers

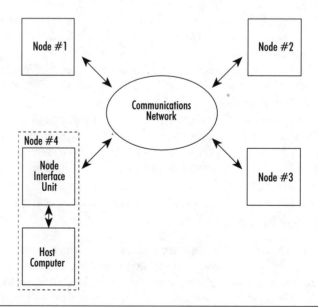

Figure 12.2   A Computer Network

Figure 12.2 shows a **computer network**. Multiple computers, called **hosts,** participate in exchanging information. The **communication network** includes the interconnection structure that distributes information to the hosts. This type of structure often uses a separate **network interface unit (NIU)** to relieve the host of many of the complexities associated with implementing network protocols used in information encoding, channel access, routing, and error control. The combination of the host and its NIU is a network **node**.

Networks are referred to as **local area networks (LANs)** if the inter-device distance is less than a few kilometers and the communication network is under the user's control. Networks involving greater distances are called **long haul networks** or **wide area networks (WANs)**.

Modern data communication systems use a wide variety of transmission media.

- **Twisted Pair.** Two metallic wires, twisted to cancel effects of crosstalk and electrical interference, are the basis of a twisted-pair communication channel. In many situations multiple pairs are bundled in a common cable. Because of their relative ease in connection and routing, twisted pairs are often used to provide communication cabling within a single building.

   Twisted pair is normally restricted to distances less than a few hundred feet and data transmission rates of less than 10 Megabits per second (Mbps). At very high frequencies, twisted-pair conductors act as antennas with a resulting loss of effective transmission power.

- **Coaxial Cable.** A coaxial cable also uses two conductors. A central signal-carrying conductor is completely surrounded by an insulating layer and an outer ground conductor. The complete enclosure of the signal by a grounded conductor minimizes radiation. Coaxial cables are suited for operation at frequencies as high as 500 Mbps.

   Coaxial cable is used in electronic instrumentation and television distribution systems. It is readily available. Splitters, amplifiers, and interconnection equipment developed for cable television are directly usable in data-communication systems.

- **Optical Fiber.** Optical fiber is formed by drawing optical glass into a single thin fiber that may be many kilometers long. Optical energy introduced in one end of the fiber travels, via internal reflections, to the other with very low loss of signal. Optical energy in a fiber-optic cable is insensitive to interference caused by electric and magnetic fields.

   Optical fiber is a medium of choice when a signal must be propagated long distances through an electrically noisy environment. It also offers the highest data rates. Disadvantages of fiber include the relative difficulty of terminating the fiber, the difficulty of splicing the fiber, and the complexity of converting the signal between optical and electronic form. Access to the signal is possible only at the end of the fiber. This offers some data security advantages but prohibits the use of bus type interconnection structures.

- **Microwaves.** Microwave transmission carries the information through a wireless media in the form of highly directional electromagnetic waves. Like fiber optics it can support a high data rate. Atmospheric disturbances, mountains, and structures can interfere with the line-of-sight requirements for microwave transmission. Microwave communication via orbiting satellites is now feasible and widely used.

- **Radio.** Radio, like microwaves, uses free space as a communication media. Unlike microwaves, radio is not directional. This permits a broadcast data transmission in which multiple communication nodes can simultaneously receive data. Radio has many of the characteristics of a data bus; a single medium is shared among all users.

12.3   Data Representation and Encoding

Digital data can be represented as a sequence of data words, each composed of a grouping of 1s and 0s. The 1s and 0s can be represented in a variety of ways.

Baseband and Broadband Data Encoding

In electronic digital systems, the values of logic 1 and logic 0 are usually mapped to voltages of 5 volts and 0 volts, respectively. The voltage remains constant throughout the bit period. This is known as **baseband** data encoding.

Baseband data representation is used exclusively within the system memory and data-processing elements associated with a digital system. A single data channel can transmit only one baseband signal at a time. If a source and destination were to simultaneously transmit baseband information on the same channel, the signals would interfere with each other and could not be reliably interpreted. This means that multiple data links must be used to achieve full-duplex (simultaneous two-directional) information exchange.

Within a communication system, however, there are several disadvantages to using baseband signals. The source and destination are often widely spatially separated. The ability to transmit simultaneous messages on a single channel has economic advantages. Since data elements often have long strings of 0s and 1s, baseband signals often have low frequency and constant voltage components. Amplifiers and signal splitters for such signals are very difficult to fabricate.

In **broadband** representation, the data is used to modulate a carrier frequency. This is comparable to data modulation in broadcast television. Multiple signals can be present on the same medium without interference. Cable television systems, for example, place hundreds of television signals on a single coaxial cable. Each signal has a different carrier frequency. Signals are separated with filters and tuners to provide a single channel of information.

Data transmission may be in parallel or serial form. With a parallel format, multiple data bits are sent simultaneously, each on a separate data line. An obvious advantage of parallel transmission is the high effective data rate that can be achieved. An overwhelming disadvantage is the need for multiple parallel-communication channels. Parallel transmission also may be subjected to different delays in each of the data channels, resulting in data skew. Such skews are very difficult to detect and correct.

Amplitude Shift Keying

In broadband data encoding the original binary signal can be used to control the amplitude of a sinusoidal signal. This is known as amplitude shift keying, or ASK. In binary systems, the two amplitudes are usually chosen to be a full amplitude and a minimum or zero amplitude, as shown in Figure 12.3. If two devices use different frequencies for their amplitude shift keying, both signals could exist simultaneously on the communication media without interference. Receiving devices have filters tuned to receive a single frequency and reject others. This allows full duplex transmission on a single line if the two communicating devices use different frequencies.

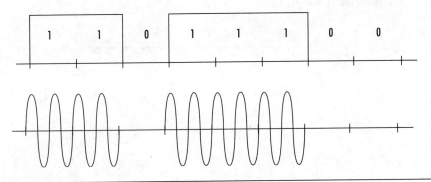

Figure 12.3   Amplitude Shift Keying Data Modulation

Frequency Shift Keying

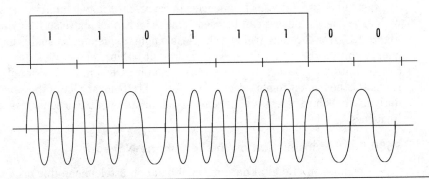

Figure 12.4   Frequency Shift Keying Data Modulation

With frequency shift keying, or FSK, as shown in Figure 12.4, two different frequencies represent the 1 and 0 data bits. With FSK the transmission medium will always carry a signal. This is in contrast to ASK, where a long string of 0s would result in no signal on the channel. The signal's presence provides the security of knowing that the transmission channel is functional, even if no data is currently being transmitted. FSK also supports full-duplex data transmission.

300 Baud Data Communication on the Telephone Network

The telephone system can transmit frequencies in the range from about 100 Hz through 4500 Hz, which is quite suitable for voice communication. When transmitting data, a modulator/demodulator, or modem, is used to encode the 1s and 0s of the digital data into tones.

Full-duplex transmission requires four different tones, two for each direction of transmission. The tones need to be relatively easy to detect with a minimum of errors. Based on these requirements, frequencies of 1070 and 1270 Hz are used for FSK data in one direction; 2025 and 2225 Hz are used for FSK transmission in the reverse direction. These frequencies are detected by band-pass filters within the modems.

Phase Shift Keying

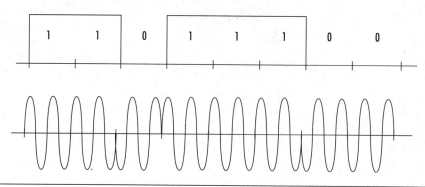

Figure 12.5   Phase Shift Keying Data Modulation

Instead of two different frequencies to represent the two binary elements, a single frequency with two different phases can be used. This is known as phase shift keying, or PSK, as shown in Figure 12.5. For encoding two logic levels, a single frequency sinusoid is used for a 1; the same frequency shifted 180 degrees is used to represent 0. More elaborate systems use multiple frequencies or multiple phase shifts to encode groupings of bits. For example, four frequencies can represent the two-bit combinations {00, 01, 10, 11}. Four phases of the same frequency can be similarly used. A third alternative is to use a combination of two frequencies with two phases of each frequency.

9600 Baud Transmission on the Telephone Network

Transmission of digital data at 9600 baud on a telephone line originally designed for voice with a maximum frequency component of about 4500 Hz presents a formidable design problem. Binary amplitude shift keying alone is not sufficient, since the rate of change of amplitudes would exceed the frequency capability of the line.

The solution used in a 9600 baud modem is a combination of ASK and PSK. Groups of 4 bits are encoded into a single signal element. Twelve phase angles are used. Each phase angle is referenced to the phase of the preceding character. Four of the twelve phase-modulated signals are also modulated in amplitude.

Bit Pattern	Encoding Phase Angle	Encoding Amplitude
0000	15°	high
0001	45°	high
0010	45°	low
0011	75°	high
0100	105°	high
0101	135°	low
0110	135°	high
0111	165°	high
1000	195°	high
1001	225°	high
1010	225°	low
1011	255°	high
1100	285°	high
1101	315°	low
1110	315°	high
1111	345°	high

Table 12.1

Data Encoding

Serial transmission uses data elements, called **data frames** or **packets**, which may contain hundreds or even thousands of bits of information. The baseband representation of a binary signal may offer some problems concerning the synchronization of the transmitted information. With a long string of 0s, for example, it is difficult for the receiving device to accurately sense the transition from one bit to the next. A long string of 1s causes a similar problem. The same problem exists in broadband representation where strings of 1s and 0s result in long segments of the same frequency or the same phase signal.

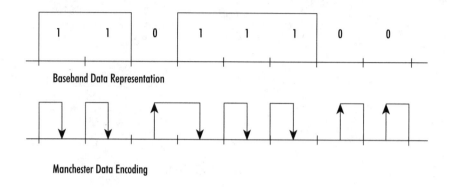

Figure 12.6   Manchester Data Encoding

This problem can be solved by transmitting a clock signal on a separate data line. A second, more efficient, option is to incorporate the clock signal into the data signal, as shown in Figure 12.6. In this data representation, known as **Manchester encoding**, a data bit of 1 has a transition of high-to-low in the middle of the data interval. A data bit of 0 has a low-to-high transition in the middle of the interval. Since there is a transition in the middle of every interval, transition interludes correspond to either one data period (for alternating data bits) or one-half data period (for strings of the same data bit). Circuitry can be devised to accurately extract the clock associated with the data.

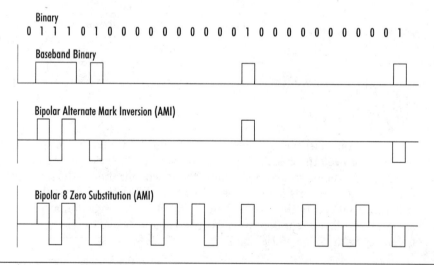

Figure 12.7   Bipolar Alternative Mark Inversion (AMI) and Bipolar with Eight Zeros
            Substitution (B8ZS) Coding

An alternative technique, used extensively in microwave communication, uses more than two signal levels. In bipolar-AMI (alternate mark inversion), as shown in Figure 12.7, a logic 0 is represented by a zero magnitude signal. A logic 1 is represented by a pulse. The polarity of the pulse alternates in direction. Long strings of logic 1s present no problem with this technique because of the transitions. Long strings of logic 0s, however, lack transitions necessary to maintain synchronization.

Bipolar with eight-zeros substitution (B8ZS) modifies bipolar AMI by breaking up long strings of 0s with a special string of bit transitions. In AMI, an alternating polarity pulse is used for each occurrence of a 1. In B8ZS, a substitution is made for a string of eight consecutive 0s. The pattern 00000000 is replaced by 000 + -0- + if the previous representation of 1 was a positive transition. It is replaced by 000- + 0 + - if the previous representation of 1 was a negative transition. In both cases the pattern introduced could not have been created by a combination of 1s since the polarities of the pulses are incorrect. This allows the bit pattern to be detected and replaced with 00000000 at the receiver.

In addition to synchronization properties, it is important that data frames have a net DC value of 0. Within a data frame, there should be equal portions at both of the data levels. Manchester encoding and both AMI techniques meet this requirement; standard baseband binary encoding does not.

12.4   Asynchronous Communication

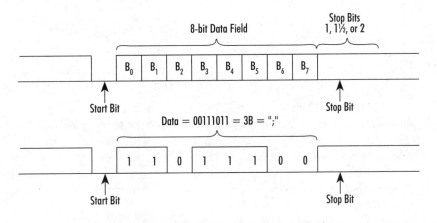

Figure 12.8   Asynchronous Serial Communication Data Format

Asynchronous serial communication, also known as RS-232, was studied with the topic of input/output systems in Chapter 10. It is characterized by the use of a start bit and stop bit transmitted with each byte of data, as shown in Figure 12.8. The start bit allows the receiving device to synchronize a local clock to acquire each of the eight data bits at the proper time period. Seven of the data bits are used for ASCII-encoded characters. Bit 8 is sometimes used for parity protection.

Most modern asynchronous systems use a bit rate, or baud rate, which is a multiple of 300 baud. Rates of 300, 1200, 9600, and 19200 are frequently encountered. In asynchronous systems the baud rate is not the same as the data rate. The baud rate is the number of signal elements transmitted per second. For example, a baud rate of 1200 bits per second will provide 120 data units, each of 10 bits. Since two of the bits are used for synchronization (start, stop) only 8 bits contain transmitted data. Therefore, 1200 signal elements per second provide 960 (120 * 8) data bits per second.

The need for constant synchronization is a major handicap of asynchronous transmission. The overhead of the start and stop bits consumes two of the ten available bit periods, a 20% penalty. The technique also requires multiple local clock periods within each data bit period. This limits the effective bit rate that can be transmitted.

RS-232-C Asynchronous Communication

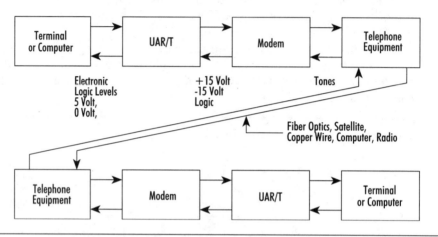

Figure 12.9   An RS-232-C Data Communication Network

Recommended Standard (RS) 232-C has been the most often used method of implementing serial data exchanges. It was originally developed for sending digital data via telephone lines, as shown in Figure 12.9. Terminals and computers, with their associated UART interfaces, are considered data terminal equipment, or DTEs. They are the source and destination of data.

A modulator/demodulator, or **modem**, is used to translate logic levels into audio tones for transmission through the voice-processing equipment of the telephone company. They are considered data communication equipment, or DCEs.

After coupling into the telephone equipment, the information may be transmitted by a wide variety of media to its final destination. Fiber optics, twisted pair, microwave, and satellite repeaters may all participate in data transmission. The reception process translates the information through a modem and back to baseband digital form where it can be used by the terminal or computer.

Many modern serial-communication activities with computers do not use telephone equipment or modems. Instead, they involve a direct connection to printers and instrumentation. The technology is robust enough that selective layers of hardware can be removed if not needed. The UARTs from two terminals may be directly interconnected, for example, so that information transmitted from one is displayed on the other.

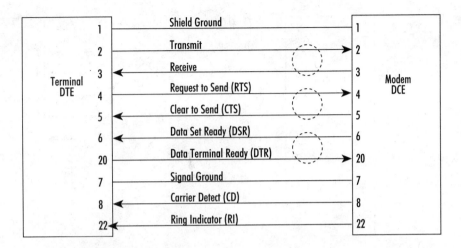

Figure 12.10 RS-232-C Interconnections between a Terminal and Modem

The RS-232-C interface involves more than just the data transmit and receive channel. A full 25 pins are assigned roles in the standard. Figure 12.10 shows the interconnection required to connect a terminal to a modem. The purposes of the transmit and receive data lines are obvious. Ring indicator and carrier detect are used only when telephone equipment is used. They provide status signals associated with the telephone ringing and the presence of a data carrier on the incoming line.

Request to send (RTS) and clear to send (CTS) are handshake signals that interconnect the terminal and modem. RTS requests permission to use the modem. The return handshake of CTS informs the terminal that it may begin transmission. Similarly, data terminal ready (DTR) and data set ready (DSR) involve an end-to-end handshake between the terminals serviced by the modem and telephone equipment.

When two serial devices are directly connected without a modem, as shown in Figure 12.11, some special concerns must be addressed. Since there is no modem, wires must be crossed so that the transmit of one device is connected to the receive of the other. Crossing is also necessary on request to send/clear to send and data set ready/data terminal ready. Although there is no carrier, since there is no telephone line, the carrier detect may need connection to a logic level or request to send to assure the terminal that all status signals of the communication link are in the proper state.

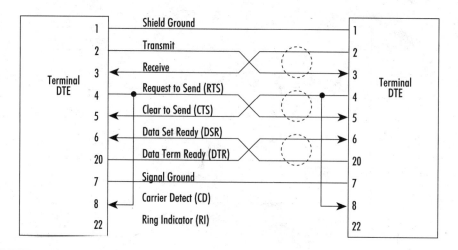

Figure 12.11 Serial Data Interconnection for Two RS-232-C Terminals

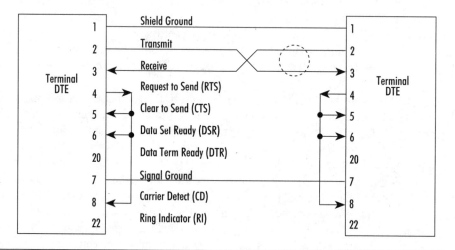

Figure 12.12 A "Null Modem" Interconnection

Problems often arise when using RS-232-C devices due to incompatibility of the handshake signals between the two communicating devices. Figure 12.12 shows a circuit that bypasses all of the handshake signals. A circuit of this type is called a **null modem**. It provides replacements for the handshake signals that would be provided by a modem. A device issuing a request to send to a (nonexisting) modem will immediately receive a clear to send, data set ready, and carrier detect.

As might be expected for a standard which was developed for use with telephone equipment and is used in a very different environment, many operational problems exist. RS-232-C is a one-to-one communication channel. It cannot directly be used to interconnect more than two modules. Difficult physical problems also exist in ground loops and noise immunity. A newer standard, RS-

485 has solved many of the physical problems while still maintaining a strong degree of logical compatibility with RS-232-C.

12.5   Synchronous Communication

In synchronous communication techniques, the data units are usually much longer than a single byte, as used in asynchronous techniques. A clock signal must be included, or derived, from the data signal to allow the destination to accurately determine the transitions between bit strings. Without clock information, it would be difficult to retain synchronization with a long string of 0s. Techniques such as Manchester encoding or B8ZS are well suited for synchronous data encoding.

Two or more hosts interchange information in the form of **messages** which are groupings of characters representing one continuous sequence of data. The arbitrary length of a message is inconvenient from the point of view of data transmission. A message may be divided into smaller units called packets. Each packet is handled individually by the communication network.

Synchronous techniques must address several issues:

- **Media access.** There must be a method for prospective data sources to share the communication media. The method may either be centralized within a single unit or distributed among all participants.

- **Synchronization.** Certain bit patterns must be used to allow the data source and data destination to recognize the start of a data packet and to synchronize the clock associated with the packet.

- **Field delineation.** A data packet will have fields assigned for addresses, data, error protection, and other uses. There must be a method for the destination to determine the boundaries between adjacent fields.

- **Error control.** Long data packets are more likely to have multiple errors than short data units. A more sophisticated error-protection scheme is required for error detection.

- **Error recovery.** Data packets with errors must either be corrected by the destination or retransmitted by the source. There must be a set of rules to allow error recovery to work smoothly and efficiently.

The data packets used in synchronous data communication have a format similar to that shown in Table 12.2. The exact format varies slightly depending on the communication technique used.

Start Flag	Control	Data	FCS Error Control	End Flag

Table 12.2

The start and ending flag are special 8-bit character strings, usually 01111110. All other fields in the packet are in fixed positions with respect to the start flag. Therefore, delineators between fields are not necessary. Precautions are made to assure that the flag string does not occur in any of the other fields of the packet. For example, a flag pattern appearing in the middle of a data string might mistakenly be interpreted as the end of a packet. Through a process called

bit stuffing, any strings containing 011111, other than a flag, have a 0 inserted after the fifth 1. This prevents a flag character from being placed in a data packet. The stuffed 0 is removed when the data is recovered.

The control field is multiple bytes in length and is used for a wide variety of purposes depending on the application of the packet. It may contain several subfields. One field, for example, may identify the type and purpose of the packet. The control field may also identify the source and destination address for the packet. A sequence number is also common. This allows duplicate packets to be identified. A subfield in the control field may also indicate the number of data elements that are included in the data field.

The data field contains the numeric or text data associated with the packet. The length of the data field may be fixed or variable.

The frame check sequence, or FCS field, is used for error detection. The FCS is usually based on a **cyclic redundancy check (CRC)**. To generate the CRC value, the control field and data field are processed by special hardware or software to produce a CRC field of 16-bits. The error-protection process will be discussed in more detail in the next section.

High Level Data Link Control (HDLC)

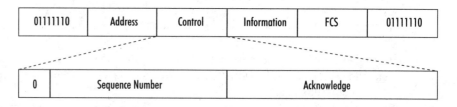

Figure 12.13 An HDLC Data Packet

HDLC is one of several standardized data protocols established by the International Standards Organization. Its packet format is similar to several others in common use. Figure 12.13 illustrates the main features in the data packet.

- **Flag.** Both the beginning and end of the packet have a flag with the bit pattern 01111110. This pattern is used to synchronize the receiver with the beginning and end of the packet. Bit stuffing is used to assure that a flag character does not appear in any other field of the packet.
- **Address.** Following the start flag is a one or more byte-address fields. The address field identifies the destination that is to receive the packet. Usually the address 11111111 is reserved as a universal address for broadcasting information to all possible destinations.
- **Control field.** The one-byte control field identifies certain characteristics of the packet. In addition to information packets, other frame formats exist for acknowledgments and other supervisory functions.
- **Information field.** The n-byte data field contains the information in the packet. The number of bits is variable but usually limited to a maximum value for an application.
- **Frame check sequence field.** The FCS field is a two-byte error-check field used for error-control purposes.
- **Flag.** An end flag terminates the packet.

12.6   Error Control

Data corruption of packets due to faulty equipment or environmental disturbances is possible. There are two types of errors: **random errors**, which are infrequent and result in an erroneous bit appearing in the data stream, and **burst errors**, which often result from electrical interference and are characterized by a high percentage of errors occurring in a narrow block of characters.

The integrity of a data packet is maintained by use of a cyclic redundancy check, CRC. A CRC sequence is a bit pattern derived from the message bits in the packet. It is appended to the message in the FCS field of the data packet. The device receiving the packet examines the message and FCS field and makes a judgment concerning the presence of an error in the packet. The technique works well and can be implemented with software or hardware.

Although the technique is usually applied to packets of hundreds of bits, the following example uses a much smaller message. Assume it is desired to transmit the message {1101} and protect it using a 2-bit error-correcting code.

A generator is chosen (the term **generator polynomial** is often used). A generator polynomial of $x^2 + 1$ is equivalent to the bit sequence {101}. The 1s and 0s relate to the binary coefficients for the terms in x.

$$a_n x^n + a_{n-1} x^{n-1} + ... + a_1 x^1 + a_0 x^0 \rightarrow \{a_n \, a_{n-1} \, ... \, a_1 \, a_0\}$$

The choice of the generator polynomial is not random. Different generators will produce different error-protection characteristics in the code. A few standardized polynomials are normally used. The standardization allows much of the encoding and decoding circuitry to be created in hardware.

The encoding process proceeds as follows:

1. If the generator polynomial is of kth order, then the message is appended with k 0s. The generator polynomial $x^2 + 1$ requires two 0s to be appended to the message.

    message = 1101
    message with two appended 0s= 110100

2. The binary representation of the message is then divided by the generator, 101. The process uses modulo 2 arithmetic instead of addition and subtraction. The addition (or subtraction) of two bits is performed by an exclusive OR gate. For the generator and message chosen this becomes as follows (note that in the division operation, a quotient bit is 1 whenever the leading bit of the partial remainder is 1. There is no magnitude comparison as in numeric division):

```
         1110
     101 )110100
         101
          11100
          101
           1000
            101
             010
             000
              10 = remainder
```

3. The remainder, 10, from the division is the desired CRC protection. Though it may seem strange, the quotient is of no interest and can be discarded.

4. The remainder is appended to the message to create {110110}.

After transmission, the receiver performs the same division operation on the received message. No additional bits are added.

```
         1110
     101 )110110
         101
          11110
          101
           1010
            101
             000
             000
              00 = remainder
```

Following the division, the remainder is 00. This indicates that the message bits are consistent with the CRC bits. Assume that in transmission, an error occurs in the third bit position of the transmitted message. The received message is then {111110} instead of {110110}. Performing the division operation would produce:

```
         1100
     101 )111110
         101
          10110
          101
           1010
            100
             010
             000
              10 = remainder
```

The presence of a remainder indicates an error but does not indicate the position of the error. In most communication systems, errors are corrected by retransmitting the message.

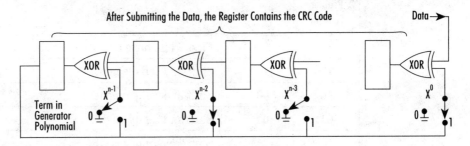

Figure 12.14 Generation of a Cyclic Redundancy Check

The modulo 2 division operation can be implemented with a shift register and exclusive OR gates as shown in Figure 12.14. There is one exclusive OR gate associated with each term of the polynomial except the highest order. The fewer the number of terms in the polynomial, the simpler the circuitry to perform the CRC.

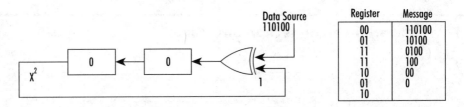

Figure 12.15 Circuitry for Encoding with Polynomial $X^2 + 1$

The circuitry for the $X^2 + 1$ polynomial used in the example is shown in Figure 12.15. The single exclusive OR gate receives its feedback input from the last flip-flop of the circuitry. After the complete message has been entered into the shift register circuitry, the remaining bits in the shift register are the CRC bits for the packet.

The algorithm can be explained by reference to Figure 12.15.

1. If the most significant bit in the shift register is a 0, the combined contents of the polynomial shift register and message are shifted one position to the left.

2. If the most significant bit in the shift register is a 1, the combined contents of the polynomial shift register and message are shifted one position to the left. During the shift, the incoming data bit is complemented by the exclusive OR circuit. The positions to be complemented are a function of the generator polynomial.

3. The process continues until the number of bits remaining is the same as the order of the generating polynomial.

Table 12.3 shows the way the algorithm would proceed.

Register Contents	Message	Comment
11	0100	First 2 bits have already been shifted.
11	100	Shift and complement.
10	00	Shift and complement.
01	0	Shift and complement.
10		Shift; remainder is CRC.

Table 12.3

CRC codes in common use are:

$$\text{CRC-12} = X^{12} + X^{11} + X^3 + X^2 + X + 1$$
$$\text{CRC-16} = X^{16} + X^{15} + X^2 + 1$$

12.7   Network Interconnection Structures

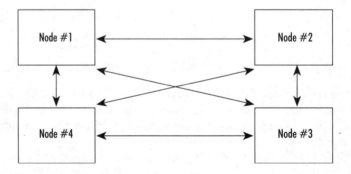

Figure 12.16 A Fully Connected Network

A network involving only a single source and destination represents the least complex form of network structure. When multiple hosts are available, many methods of interconnection are possible. Figure 12.16 illustrates a **fully connected** structure; the NIU (Network Interface Unit) of each host has a dedicated connection to each of the other hosts. No routing or use of shared network components is required. A major disadvantage of the fully connected structure is the difficulty and expense of expanding the network to include additional nodes. Multiple interfaces must also be monitored by each node.

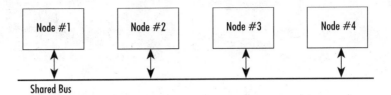

Figure 12.17 A Network Using a Data Bus

In Figure 12.17, all the NIUs are interconnected on a common bus. The shared bus necessitates the use of a communication protocol to allow the sharing of the bus among all of the participants. Each unit on the bus must also be able to recognize messages for which it is the destination and to ignore all others.

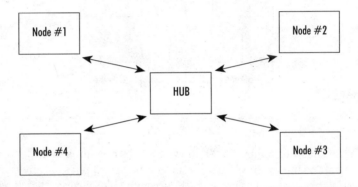

Figure 12.18 A Star Topology Network

Figure 12.18 shows a **star** interconnection structure. All messages must pass through the hub of the star, where they are routed for delivery to the proper destination. This technique requires the hub to have most of the responsibility for assuring that messages are correctly routed to their destination. The individual NIUs can perform similar to having a single source and destination. Data routing is relatively easy. Expansion is relatively easy until the hub reaches capacity. In some situations, the hub may present a bottleneck and limit the ability of the network to achieve a high data rate.

The **ring** of Figure 12.19 offers a structure which is easy to expand. Data packets are introduced to the ring by the source node. They circulate until reaching the destination node. Since packets must be temporarily stored at each node in order to investigate the destination address, the ring may introduce unacceptable data-transmission delays.

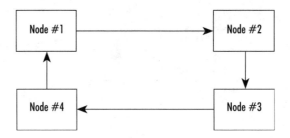

Figure 12.19 A Ring Topology Network

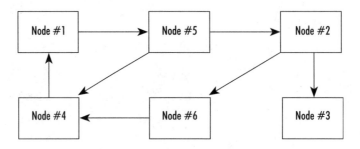

Figure 12.20 A Mesh Topology Network

The **mesh** of Figure 12.20 allows an arbitrary interconnection of nodes. Its use allows fine tuning of the interconnection structure to the actual physical environment. For example, it allows direct source-to-destination paths to accommodate devices that have frequent and high-volume communication needs.

The bus and the ring have been incorporated into IEEE standardized network methodologies. These IEEE standards create an environment in which device designers can more easily interface spatially distributed systems. Later sections will present the token bus, token ring, and CSMA/CD (Ethernet).

12.8   Communication Protocols

For reliable communication between two or more computing devices, a set of rules, or protocols, must be established. These rules cover many situations that must be consistently interpreted by both the source and destination.

Circuit Switching

There are two methods of transmitting information through a communication network. In **circuit switching** a dedicated data path from source to destination is established before data communication begins. The communication path may pass through several network nodes. Once the path is established, the data stream passes via a hardwired path from source to destination. Message transmission delay is very small. Old-time telephone service, in which a copper path was patched between source and destination, provides a good model of a circuit-switched network. The nodes of the network are the various central switching offices that provide trunk lines between various geographic areas.

Packet Switching

The second method, **packet switching**, allows data transmission to begin without establishing an end-to-end path. Packets are routed from one node to the next according to a **routing algorithm**. At each node the packets are stored, analyzed, routed and queued for transmission to the next node. Error control is required for each link of the journey, called a **hop**. The store-and-forward nature of packet switching may introduce appreciable delay. If a large message is broken into several packets, there is no assurance that each of the packets will take the same route in reaching the destination. The situation is analogous to the postal system. Individual envelopes may traverse different delivery routes and may arrive at the destination out of sequence.

Stop-and-Wait Protocol

Automatic repeat request (ARQ) protocols involve rules to assure the reliable delivery of packets. In these protocols the source sends the destination a packet. The destination responds with either an ACK or NAK data-acknowledgment packet. ACKs are used in communication systems in much the same way that handshakes are used for the reliable exchange of localized information. An ACK is a packet sent by the destination node to acknowledge a correctly received message. A NAK is a response to an incorrectly received message. If the source receives neither an ACK or a NAK, this indicates a lost packet.

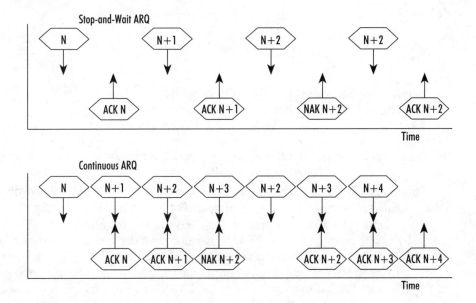

Figure 12.21 Data-transfer Protocols

The least complex form of ARQ protocol is known as **stop-and-wait ARQ** as shown in Figure 12.21. After each packet is transmitted, the source waits to receive confirmation from the destination. If the destination does not detect an error, a positive acknowledgment, ACK, is transmitted to the source. The source then sends the next packet in sequence. If the destination discovers an error by means of the CRC evaluation, it transmits a negative acknowledgment, NAK. When the source receives an NAK it retransmits the packet.

While the stop-and-wait protocol is uncomplicated, substantial time may be wasted on the data channel waiting for ACK or NAK signals. Since errors are normally very infrequent, a **continuous ARQ** offers higher channel efficiency. In continuous ARQ, as shown in Figure 12.21, data packets are continuously transmitted. The responding ACK or NAK for one packet overlaps the transmission of the next. This requires a full-duplex channel to accommodate simultaneous data transmission in both directions. A NAK requires the source to retransmit the damaged packet. After retransmitting the packet, the source continues transmitting packets sequentially.

In some situations, for example spacecraft and satellite communication, the transmission time delay may be significantly longer than the time required to transmit a packet. With stop-and-wait ARQ there may be an intolerable delay between the time the packet is transmitted and when an ACK/NAK response is received. This may cause a significant portion of the available data capacity of the channel to go unused.

Example 12.1:   Satellite Communication

The transit time for a packet using a satellite repeater is 270 msec. One thousand bit packets are transmitted at 1 Mbps using a stop-and-wait protocol. How many packets per second can be transmitted?

Solution:

$$\text{Transmit time} = \frac{1 * 10^3 \text{ bits}}{1 * 10^6 \text{ bits/second}} = 1 \cdot 10^{-3} \text{ seconds}$$

Transmit time for message = $270 * 10^{-3}$ seconds

Transit time for ACK/NAK = $270 * 10^{-3}$ seconds

Total time for transit and acknowledgment = $541 * 10^{-3}$ seconds

$$\text{Maximum transmission rate} = \frac{1}{541 * 10^{-3}} = 1.8 \text{ packets per second}$$

This transmission rate is very slow. The transmitter is capable of transmitting 1000 packets per second, but the stop-and-wait protocol restricts the rate to 1.8 packets per second.

For situations in which the transit time exceeds the transmission time, efficiency can be gained by having many packets simultaneously in transit. In the situation shown in the example, 540 packets could be simultaneously in transit while awaiting for an acknowledgment for the first packet. ACKs or NAKs are not received until multiple additional packets have been transmitted. In this case a **go-back-*n* protocol** may be used. If an error in a packet is detected, the source is required to retransmit the last *n* packets.

The efficiency of the go-back-*n* protocol can be improved by using selective retransmission. Only the damaged packet is retransmitted. The destination maintains all data packets that have been correctly received within a fixed time window in order to place the retransmitted data packet in its proper sequence in the received message.

Flow Control

Even when information is flowing reliably from the source to the destination, it is possible that the destination cannot accept and process the information as fast as the source can send it. A method is required to allow the destination to inform the source to temporarily stop sending data until the destination is sufficiently able to accept more data.

Several solutions are possible. The destination can refuse to accept information by purposefully failing to send an ACK/NAK. In this case the source would continue to resend packets. In another approach, a destination sends a WACK (wait acknowledge) instead of an ACK when it wishes the source to stop sending packets. If a future query from the source is acknowledged with an ACK, the transmission can begin.

With a **sliding window protocol** a limit is established for the number of packets that can be sent without receiving an acknowledgment. If this limit, for example seven, is exceeded, the source terminates transmission. With this protocol the destination can withhold the ACK to limit packet transmission.

12.9   Local Area Networks (LANs)

Local area networks, LANs, are characterized by relatively high data-transmission rates of data packets over dedicated data channels. Most LANs are owned by a single organization with control over both the hardware and software components of the system. A university campus or industrial corporation are examples of places where a LAN can provide data-communication services. Using bridges and gateways, local area networks can be interconnected to achieve communication over inter-city or global distances.

LANs are characterized by their interconnection topology, their transmission medium, and the method used to access shared communication components. Three forms of LANs, token bus, token ring, and CSMA/CD (Ethernet) have been standardized by IEEE. They have dominated most current applications.

Centralized and Distributed Control

The control function of a LAN can be either centralized in a single location or distributed among the various nodes. In centralized control, individual nodes may request to use the network. Access to the network is granted by a single centralized controller. The centralized controller has ultimate authority and responsibility for the correct operation of the network.

Distributed control minimizes the role of a central authority. Allocating network resources and implementing communication protocols are the responsibility of the individual nodes. A set of rules specify how each node is to cooperate with others. By distributing the control function, the network may perform adequately even if one or more components are nonfunctional. In some cases, especially if the network is lightly loaded, distributed control may improve delay times and overall performance.

Performance of Local Area Networks

There are several possible ways of measuring performance in LANs. **Throughput**, S, measures the amount of data successfully transmitted between nodes in the network. The **offered load**, G, includes not only the node-to-node data, but also any acknowledged packets, packets retransmitted due to errors, and packets associated with network management. The offered load is always greater than or equal to the throughput.

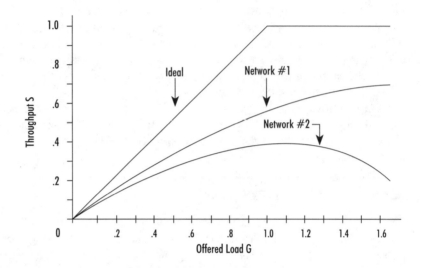

Figure 12.22 Throughput as a Performance Measure of a LAN

Ideally, the network throughput increases as the demand for network service, the offered load, increases. The maximum depends on the data-handling capacity of the network. After reaching capacity, further offered load produces no increase in throughput. Figure 12.22 shows example plots of throughput versus offered load. Both throughput and offered load are normalized with respect to the data capacity of the network.

For real networks, network inefficiency and errors result in the offered load exceeding the throughput. In Figure 12.22, the throughput is always less then the offered load. The maximum throughput is often considerably less than the capacity of the network. Under some conditions, as shown in network 2, throughput can decrease as offered load increases. This can happen if heavy usage leads to excessive errors and decreases the quantity of error-free messages.

Delay is the time between the presentation of the packet to the network and its successful delivery at the destination node. Delay is very strongly influenced by the method of accessing the media and the manner in which data is transmitted through the network. Delay can be broken into two parts. The network-acquisition delay, D_a, is the time it takes to get permission to transmit a packet. This is strongly related to the media-access mechanism. Transit delay, D_t, is the time the data is in transit through the network. If data is transmitted in a hop-by-hop fashion, with each intermediate node receiving and storing the message before retransmission, the transit delay can be quite long. Communication satellites can also introduce significant delays.

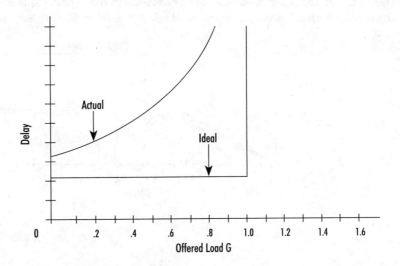

Figure 12.23 Delay as a Function of Offered Load in a LAN

Figure 12.23 shows the relationship between delay and the offered load. Ideally the delay is constant until the capacity of the network is reached. At this point it grows without limit. For actual networks, the delay increases as offered load increases. Since transmission is serial, the delay will, at minimum, include the transmission time for the packet. For two packet lengths, one short and the other long, the longer packet experiences the greater transmission time delay. Delay will increase as offered load increases. This is because there is less free time on the network, and the overhead associated with allocating the resource requires time.

12.10   Bus Networks

Bus networks are characterized by a single transmission media shared by all nodes. A message sent by any node is received by all nodes on the bus. If two nodes attempt to simultaneously transmit data, a **collision** occurs. The electrically jumbled data sequences will have extensive errors.

Token Bus

Token bus uses a single bus as a shared transmission media. Access control of the media is gained using a special data packet called a **token**. Each node on the network has an address; each also knows the address of a successor node. The token circulates from node to node on the bus. Each node always transmits the token to its successor node. A logical loop is formed among all the nodes.

A node in possession of the token may use the full bus resources to transmit data packets. Data is broadcast on the bus directly from the source node to the destination node; there is no store-and-forward action by intermediate nodes. The concept of a successor node is used only for transmission of a token and allocation of the bus, not for data transmission. When a node ends its transmission, the token is passed to the next node in the logical sequence of nodes. A node receiving a token without messages to transmit relays the token its successor node.

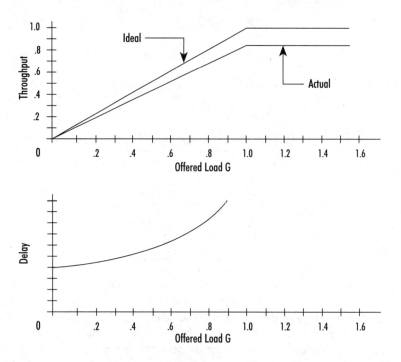

Figure 12.24 *Performance Measures of Token Bus*

Token bus is highly efficient in allocating the bus to useful data transmission. The main overhead is in transmitting the token. If the token length is small compared to the length of the data packets, the throughput versus offered load curve closely approximates the ideal case, as shown in Figure 12.24. The delay in token bus is primarily due to the time it takes for a token to circulate among all nodes. For a large number of nodes this can be significant. As the offered load increases, more time is dedicated to data transmission and the token does not circulate as fast. As the offered load approaches the capacity of the network, delays grow without limit.

Despite the simplicity of the token bus concept, there are many issues that make its implementation fairly complex.

- **Initialization.** Before the network can begin data transmission, it needs to establish the sequence in which the token will be passed. This requires that participants in the network be known. Each node must know its successor in the logical ring. One network participant is responsible for initializing the token.
- **Addition or deletion of nodes.** If new nodes are added, the new node must receive the token from an existing node. This requires a change in the successor node for one or more nodes. Likewise, deleting a node requires a modification of one or more nodes.
- **Lost token.** Since possession of a token is necessary for any node to initiate a data communication, loss of a token is serious. The token can be lost by sending it to a nonexistent or nonre-

sponding node. The token can also be damaged in such a way that it is not identifiable as a token. One network node is usually charged with the responsibility of detecting a lost token. A token may be considered as lost if a given node fails to receive it in a reasonable length of time. Generating a replacement token when the original token was not lost is also a very serious situation.

CSMA/CD (Ethernet)

CSMA/CD (carrier sense media access/collision detection), known as **Ethernet**, also uses a bus as the transmission media. The main difference from token bus is in the method used to access the transmission media. CSMA/CD is modeled after the protocols used in polite interpersonal communications within a group.

A node wishing to initiate a communication monitors the transmission medium to detect a period of inactivity. This is the carrier sense, multiple access, or CSMA portion of the CSMA/CD representation. When the medium is free, packet transmission begins.

While transmitting, the node continues to monitor the medium for a collision. A collision results when another node, sensing the same period of media inactivity, simultaneously begins to transmit information. The two messages interact to produce a composite message with many errors. When a sending node, by monitoring the bus, senses a collision, it terminates its transmission and transmits a brief jamming signal to assure that all other nodes are aware of the collision. This is the collision detection, or CD, portion of the name.

Following a collision, the node waits a random period of time before again monitoring the medium for a gap in activity. The random period of time assures that two nodes causing a collision will not cause a second collision.

Despite the seemingly complex rules for media access and collision detection, CSMA/CD is fairly easy to implement. Nodes may be added or deleted from the network without affecting network operation. In contrast to token bus, there are no complex rules for generating a token or adding nodes.

CSMA/CD has very different performance characteristics from token bus. For a lightly loaded system, with low offered load, the throughput closely agrees with the offered load, as shown in Figure 12.25. As loading increases, the probability of a collision increases. Collisions result in higher offered load but no additional throughput. As the collision rate gets very high, the throughput decreases as the offered load increases. The maximum throughput for CSMA/CD is only 0.18, a low figure. The throughput performance of CSMA/CD is poorer than that of the efficiently managed token bus. By slight changes in the access rules, the maximum throughput for CSMA/CD could increase, but at the expense of increased delays.

A main advantage of CSMA/CD is the short delay time needed to access a lightly loaded network. The delay increases as network loading increases. With token bus, even if the network is lightly loaded, a node is required to wait until a circulating token arrives.

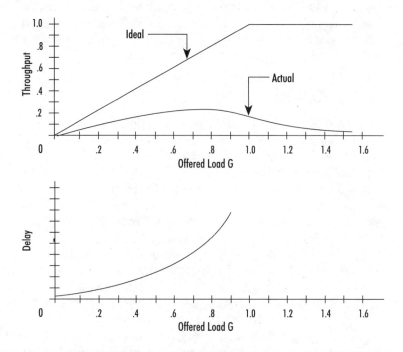

Figure 12.25 Performance Measure of CSMA/CD

12.11   Ring Networks

A ring LAN is a set of nodes arranged in a circular chain. Each node receives information from only one node; each node transmits information to only one node. A node must be able to receive and store at least a portion of the data packet. Figure 12.26 shows a possible organization for a ring node.

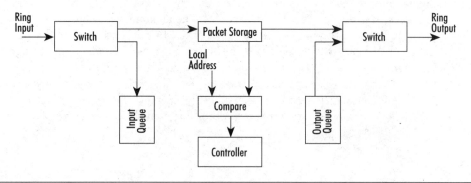

Figure 12.26 Structure of a Node for a Ring Network

Information arriving from the previous node in the ring is stored temporarily in a packet-storage shift register. At least one bit must be stored to allow a buffer for reception and retransmission. As the address portion of the packet passes through the node, the destination address can be examined. If the address agrees with the local node address, the remainder of the packet is stored in a local queue for further processing. If the addresses do not match, the packet is allowed to proceed through the packet storage shift register and is transmitted to the successor node. This causes a delay corresponding to the number of bits temporarily stored. Data originating in the node may be introduced to the ring through a switch.

Token Ring

The same media-access technique used in token bus can also be used in a ring structure. The token is passed sequentially among the devices connected on the ring.

A node wishing to transmit a packet waits for a token to arrive. It converts the token into a data packet by modifying a bit in the packet header and appending appropriate addresses, control fields, data field, FCS, and flag. The packet is propagated from node to node. Each node accepts the information packet bit-by-bit. Only after the address field is received can a node determine if the packet is addressed to itself. If the packet address agrees with the node address, the information field for the packet is routed to the local host. If the addresses do not agree, the information field and the remainder of the packet are relayed to the next node in the ring.

In ring networks, the data packet may be removed from the ring by either the source or destination node. In source-removal protocols, the packet is allowed to loop around the ring to the source, where data fields are removed and the packet is converted back to a token. Since data packets are usually longer than the delays on the ring, the source usually sends one portion of a packet while receiving an earlier portion of the packet which has returned after traversing the ring.

In destination-removal protocols, the destination node stores the packet locally until it can determine the contents of the address field. It can then either be removed from the ring and a token inserted, or the packet may be transmitted to the next node. The total time delay is increased by the additional processing at the node.

12.12   Mesh Networks

Mesh networks are interconnections of nodes not following a regular and predictable pattern. Routing information between nodes in the mesh requires intervention of other nodes along the transmission path. Either circuit switching or packet switching can be used. Figure 12.27 shows an example packet-switching mesh network of five nodes.

Routing information is required to direct each packet to its final destination. In Figure 12.27, each node has a routing table. As packets are received from other nodes, the destination address field is examined. Messages destined for the local node are removed and sent to local memory. Messages destined for other nodes are placed in queues for transmission to the address indicated in the routing table. For example, a message from node A with an address of node D would be routed to node B. Node B, when processing the packet, would route it to node D. The entire process may require several hops.

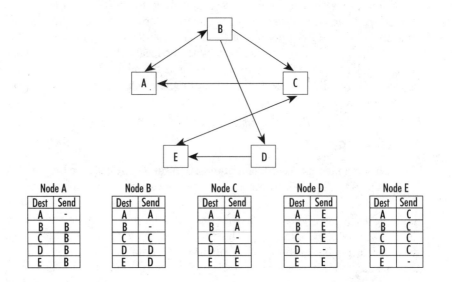

Node A	
Dest	Send
A	-
B	B
C	B
D	B
E	B

Node B	
Dest	Send
A	A
B	-
C	C
D	D
E	D

Node C	
Dest	Send
A	A
B	A
C	-
D	A
E	E

Node D	
Dest	Send
A	E
B	E
C	E
D	-
E	E

Node E	
Dest	Send
A	C
B	C
C	C
D	C
E	-

Figure 12.27 Packet Routing in a Mesh Network

In some situations, alternate routing may be available. For example, routing from node A to node E could follow the path ABDE or the path ABCE. If, however, network traffic were such that node C was overburdened with communication activity, the path ABDE might provide higher performance.

Although the routing function in mesh networks is more complicated than routing in rings, the mesh provides redundancy. Some node and link faults may be tolerated by using alternative transmission paths.

For maximum reliability, multiple paths may be simultaneously used to route a message. In a technique called **flood routing**, a node wishing to send a message transmits the message to all its outgoing data links (with the exception of the link which was the source of the packet). Each node receiving a message also retransmits the message to all of its outgoing links. If a path exists, flood routing will find it and deliver the packet in the shortest possible time. Flood routing may deliver multiple copies of the message, each delivered by a different path. The destination node would require sufficient intelligence to reject duplicate packets.

Another potential problem with flood routing is the presence of message loops. For example, a message flooded from node A might circulate endlessly in the loop ABC. This can be avoided by attaching a counter field to the data packet. Each time the message is retransmitted, the counter is incremented. When the counter exceeds a threshold, a node receiving the message destroys it.

Example 12.2: Network Hop Threshold

Determine the value of the counter threshold for the network of Figure 12.27.

Solution:

The longest possible valid data path in the network is ABDEC, a total of four hops. Therefore if a node, other than the destination node, receives a message with a counter value of four, the message should be destroyed. Note that the value of the threshold is always less than or equal to the number of nodes.

In this example it can also be noted that there exists a path of no more than 3 hops which will connect any source with any destination. Possible pathways between node A and node C include ABC and ABDEC. By allowing four hops, the message will be successfully transmitted even if the link between nodes B and C is not functional.

12.13   Layered Model of Communication Networks

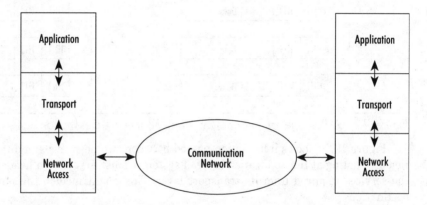

Figure 12.28 A Layered Network Architecture

Network operations can be decomposed into several network functions, called layers. In Figure 12.28, three layers are shown; each is responsible for one of the major network functions.

- **Network-access Layer.** This layer is responsible for all operations concerned with accessing the network and exchanging information, in the form of data packets, with other nodes in the network.

- **Transport layer.** The transport layer provides management functions to break large messages into smaller packets. It also appends information to assure that packets can be spliced together in the proper order to produce the information frame needed by the destination application. The transport layer also provides error-control services.

- **Application layer.** This layer provides the actual services to the user. It uses data provided via the communication structure.

Layered structures decompose the overall network into subfunctions to be independently implemented. The structure is top-down. The application layer does not need to be concerned with the implementation details of lower levels in the hierarchy. Each layer provides services that can be requested by higher levels.

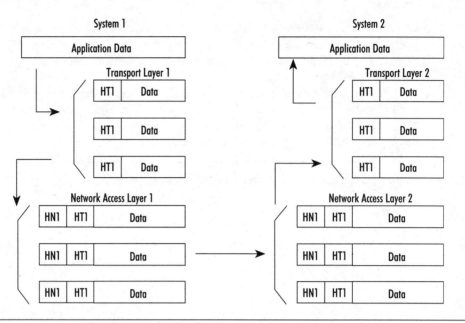

Figure 12.29 Passage of Information from Source to Destination

Figure 12.29 shows information passed between two peer levels in different nodes. Information originating at the application layer of system 1 must first have a header attached to indicate the services required from its transport layer. This might include the name or address of the destination.

Within the transport layer, the original data frame is broken down into packets. Each packet is provided with header information indicating the address of the packet, error-control information, a sequence number for the packet, and perhaps additional information needed for packet transmission and delivery.

The packets are submitted to the network-access layer. This layer must gain access to network resources. It adds an additional header needed to verify arrival of packets at the network-access layer of the destination node. It then transmits the packets and awaits confirmation of correct delivery.

At the destination, the packets are received by the network layer. The network header is removed, interpreted, and a confirmation is sent if necessary. The remaining packet material is then sent to the transport layer, where the transport header performs additional error control and assembles the packets in the correct format for the application. The information unit delivered to the application level of system 2 is identical to that produced by the application level of system 1.

The International Organization for Standardization has standardized a seven-layer model which further decomposes the actions of the network-access level and the transport level. This model is shown in Figure 12.30.

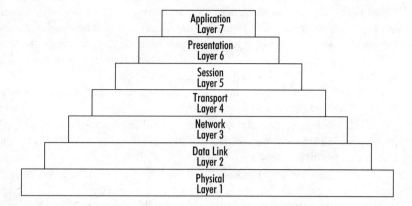

Figure 12.30 The Open System Interconnection Model of the International Organization for Standardization

- **Application layer.** This layer provides the user interface and coding for the application.
- **Presentation layer.** Provides such services as data encryption, data compression, and code conversion.
- **Session layer.** Provides services related to load sharing and access rights to resources.
- **Transport layer.** Breaks longer files into message packets for lower layers. Assembles message packets into user files destined for higher layers.
- **Network layer.** This layer is concerned with routing functions associated with the network.
- **Data link layer.** This layer is responsible for sending packets, error free, over the network.
- **Physical layer.** This layer delivers a sequence of bits from the source to the destination.

The seven-layer model is recognized as a standard and is often used as a reference for explaining communication networks. Standards allow manufacturers to create an apparatus suitable for many uses. Equipment from a large variety of sources can be assembled into a functional system.

Summary

This chapter is concerned with transmitting digital data among multiple cooperating computing devices. Encoding the binary data is necessary to assure synchronization and the ability to use communication networks such as a telephone. A wide variety of encoding based on amplitude, frequency, and phase are possible.

Asynchronous serial communication is widely used in personal computers and instrumentation. It is fairly inefficient because of low baud rates and the need for synchronization characters after every byte of data.

In synchronous communication the data is formed into packets and transmitted serially. Packets include information for synchronization purposes, an address field for the destination (and perhaps the source), data fields, and a field for error control.

Communication protocols are rules for the efficient and effective interchange of data packets. These protocols cover such issues as error control and lost packets. Communication networks can be structured as centralized or distributed. They can also be classified based on their topology and method for accessing shared network resources. Buses and rings are the most common forms of network topology. Tokens and CSMA/CD provide two methods for nodes to access the network resources. The IEEE has adopted IEEE 802.3 (CSMA/CD), IEEE 802.4 (token bus) and IEEE 802.5 (token ring) as standards for three of the more common implementations of the lower three levels of the ISO communication hierarchy.

Mesh interconnection structures are useful in situations in which greater redundancy and/or highly specialized interconnections are needed. Routing algorithms for meshes are more complex than those required for buses or rings.

Communication networks are often viewed as hierarchical or layered structures. Each level provides a support service to higher layers in the structure. The ISO seven-layer model provides a mechanism for decomposing network activity. This provides a mechanism for implementing hardware and software services that can be easily integrated to perform network functions.

For additional information and references on the topics presented in this chapter, access the World Wide Web page provided at: http://www.ece.orst.edu/~herzog/docs.html.

Key Terms

asynchronous serial communication	ethernet	null modem
automatic repeat request (ARQ)	flood routing	offered load
baseband	full-duplex	packets
bit stuffing	fully connected	packet switching
broadband	generator polynomial	random errors
burst errors	go-back-*n* protocol	ring
circuit switching	half-duplex	routing algorithm
collision	hop	simplex
communication channel	host	sliding window protocol
communication interface	local area network (LAN)	star
communication network	long haul network	stop-and-wait ARQ
computer network	Manchester encoding	throughput
continuous ARQ	messages	token
CSMA/CD	mesh	token bus
cyclic redundancy check (CRC)	modem	token ring
data frames	network interface unit (NIU)	wide area network (WAN)
delay	node	

Readings and References

Stallings, William. *Data And Computer Communications*, Third Edition. New York: Macmillan Publishing Co., 1991.

This book provides a good overview of data communication and local area networks. It also includes interesting information regarding the higher levels of the communication hierarchy.

Exercises

12.1 Encode the following using Manchester data encoding (the spaces are for ease of representation only).

(a) 10010100 11110000

(b) 00000000 11111111

(c) 10101010 10101010

(d) 00000000 00000000

(e) 11111111 11111111

12.2 Encode the following using Bipolar AMI data encoding.

(a) 10010100 11110000

(b) 00000000 11111111

(c) 10101010 10101010

(d) 00000000 00000000

(e) 11111111 11111111

12.3 Encode the following using B8ZS data encoding.

(a) 10010100 11110000

(b) 00000000 11111111

(c) 10101010 10101010

(d) 00000000 00000000

(e) 11111111 11111111

12.4 Why is frequency shift keying rather than baseband encoding used when sending digital data on the telephone system?

12.5 For very low cost storage, cassette tapes have been used for data storage in small computer systems. The data formats used Manchester data encoding. Why was this choice made?

12.6 Why is it desirable that the encoding scheme have a net DC value of 0?

12.7 Assume that a digitized image is to be transmitted via digital communication techniques. The image has 512×512 picture elements (pixels). Each pixel may be represented with one byte of information to provide the gray scale. How long will it take to send data for one image under the following conditions:

(a) Asynchronous communication using 9600 baud?

(b) A 64 kbps digital communication channel?

12.8 Assume that synchronous serial data communication relies on two clocks, one at the source and the other at the destination. Each can have an accumulative error of up to one minute per year. How long a sequence can be sent before clock drift becomes a problem under the following conditions:

(a) Asynchronous communication using 9600 baud?

(b) A 64 kbps digital communication channel?

12.9 Which is more efficient, asynchronous serial transmission or synchronous serial communication? Why?

12.10 What is a null modem? Why are they often necessary in implementing RS232 serial communication?

12.11 Assume that you have been hired as a consultant to determine why an instrumentation device is unable to communicate via the serial port of a digital computer. What are possible problem areas? How would you test them?

12.12 Use bit stuffing in the following sequences (spaces are for ease of representation only).

    (a)   11000000 00001111 11111100

    (b)   11111111 11111111 11111111

    (c)   01111110 01111110 01111110

12.13 The following data sequences have been bit stuffed. Recover the original data (spaces are for ease of representation only).

    (a)   00001111 10111110 01111100

    (b)   01110011 11001111 11011110

    (c)   01111110 01111110 01111110

12.14 Using the generator polynomial $x + 1$, encode the eight binary codes between 000 and 111. Show shift register circuitry to compute the CRC. What is a characteristic of the code?

12.15 Using the generator polynomial $x^2 + 1$, encode the messages $M_1 = \{001\}$, $M_2 = \{010\}$, and $M_4 = \{100\}$. Is it possible to create codes for the other possible 3-bit data using superposition? Demonstrate.

12.16 Encode the message 11001011 using the generator function 110011.

12.17 For a generator function of 110011, determine if the following are error free.

    (a)   1010101010

    (b)   10011110

    (c)   10000001

12.18 A data-communication system transmits 2000 bit packets at a data rate of 5 mbps. Propagation time is negligible. A stop-and-wait protocol is used for each packet. The ACK/NAK packet length is 100 bits. Find the data throughput (correct data) under the following conditions:

    (a)   There are no errors.

    (b)   1% of the packets have errors and require retransmission.

12.19 Assume that data is being transmitted to the moon, a distance of 240,000 miles. If the data rate is 30 Mbps, how many bits are in transit between the earth and the moon? The velocity of propagation is 186,000 miles per second.

12.20 Assume that data is being transmitted to a geosynchronous satellite at an altitude of 23,500 miles above the earth. 1K bit packets are transmitted at a rate of 50,000 bps.

    (a)   What average packet rate is possible if stop-and-wait ARQ is used?

    (b)   If a go-back-*n* protocol is used, what should be the value of *n*?

12.21 If "*a*" is defined as the ratio of the propagation time of a packet to the transmission time of a packet, and acknowledge packets are assumed to be very short (negligible), develop an expression for the utilization of the channel. Utilization is defined as the ratio of the transmission time of a packet to the total time to transmit a packet and receive an acknowledgment.

12.22 In Example 12.1, how many packets should be in transit, without confirmation of previous packets, to achieve a utilization of the channel of more than 90%?

12.23 What are some of the advantages and disadvantages in each of the following interconnection topologies?

(a) Bus

(b) Ring

(c) Star

(d) Fully connected

(e) Hypercube

(f) Mesh

12.24 In CSMA/CD (Ethernet), what happens when there is a collision?

12.25 In comparing CSMA/CD and token bus, which has the capability of higher throughput? Why?

12.26 Compare CSMA/CD and token bus for performance under light loading and heavy loading conditions.

12.27 In token ring, what is the minimum delay that can be experienced by each packet as it passes through a node? Show a register structure to support this minimum delay.

12.28 A packet in a token ring includes address fields for both the source and destination. What would be the advantage of removing the packet at the destination end, rather than allowing it to recirculate back to the source? How might this be accomplished?

12.29 For long rings, a short packet with an error in its address field might circulate continuously, since its destination would never be found. What is a possible remedy for this condition?

12.30 In a CSMA/CD system, there are too many collisions. How could the number of collisions be reduced?

12.31 Discuss the ability of token ring, token bus, and CSMA/CD to function if one or more network nodes are nonfunctioning.

12.32 How might a unit on a token ring determine if the token has been lost? How might it determine if there are multiple tokens? Develop an algorithm that would remove multiple tokens.

12.33 The response time on a token bus is too slow. How could it be improved?

Appendix A
Foundations In Digital Logic

George Boole

An Englishman by the name of Boole
Was definitely nobody's fool
Instead of thinking archaic
He treated logic as algebraic
And created a very useful tool

Prologue

A contemporary of Abraham Lincoln, George Boole was born in England in 1815. Because of his modest middle-class background, he began employment at the age of 16 rather than attend the university. Working as an assistant teacher in a high school, he taught himself French, German, and Italian in preparation for a career as a clergyman. He was convinced by associates that his personality was not suitable for working in the church.

At the age of 20 he opened his own school. After teaching himself higher mathematics, he made a major contribution to the area of numerical analysis by developing the calculus of finite differences. Two books, *The Mathematical Analysis of Logic*, 1848, and *The Laws of Thought*, 1854, established a sound mathematical basis for the field of logic. His contributions to the science of logic are compared to the contributions of Euclid in the study of geometry. Boole demonstrated that logic can be reduced to a simple algebraic representation known today as Boolean algebra.

Boole's work was largely unknown until it was presented again in *Principia Mathematica* by Bertrand Russell and Alfred North Whitehead in 1910. Boole's work undoubtedly influenced the first binary digital computing devices produced in the 1930s.

A.1   Introduction

This appendix examines some techniques used in digital logic design. It provides a minimal coverage of logic gates for the benefit of those without a formal logic design course. This appendix may be skipped or used as review material for those with previous logic design experience. Since this book emphasizes system design, most discussion will use AND-OR logic. AND-OR logic tends to be much more friendly with humans than NAND-NAND or NOR-NOR implementations. Any of these representations can, of course, be converted to the other.

From the time of Aristotle, the science of logic has been applied to truth or falsehood of statements. George Boole established that logic relationships could be manipulated algebraically. His work led to Boolean algebra and switching algebra. These concepts have led to representation and manipulation of logical equations describing complex relationships encountered while performing digital arithmetic and digital control operations.

In the past, logic design was closely related to the goal of logic minimization. Good designs were designs that performed the desired function with a minimum number of (expensive) logic gates. This situation has changed dramatically with the availability of inexpensive and highly complex integrated circuits. Many designers now adopt the strategy that "hardware is free" when evaluating the overall cost of a system. While not really free, it can certainly be argued that increasing importance must be placed on other aspects of the design, such as performance, design time, and reliability.

Another radical change in design techniques has evolved from computer aided design (CAD) workstations and their supporting software. It is now possible to specify, design, test, simulate, and fabricate complex logic circuitry without ever interconnecting devices at the logic gate level.

This appendix emphasizes design techniques that work, are easy to apply, are easy to document, use currently available components, and illustrate proven design strategies.

A.2   Digital Design

Design is the process of creating a working device to meet a previously defined set of design specifications. In digital logic design, many different kinds of components and design approaches are possible.

Digital logic involves using signals with two values: 0 and 1. Digital logic gates combine groups of input logic values to produce output logic values. A truth table can specify the relationship between the inputs and outputs. The truth table lists all the possible combinations of the input variables with the corresponding outputs produced. The truth table in Table A.1 has three input variables, X_2, X_1, and X_0, and two output variables, Y_1 and Y_0.

Input Variables			Output Variables	
X_2	X_1	X_0	Y_1	Y_0
0	0	0	0	1
0	0	1	1	1
0	1	0	1	0
0	1	1	1	0
1	0	0	0	0
1	0	1	1	0
1	1	0	0	1
1	1	1	1	0

Table A.1

Digital **analysis** starts with a known grouping of logic gates and determines the logic relationship between inputs and outputs. The truth table is one way of expressing this relationship. Digital **synthesis** begins with a truth table and determines a grouping of logic devices that will provide the specified logic function.

A.3   Logic Gates

While a large and diverse selection of logic gates is available, primary attention will be restricted to three types: AND, OR, and NOT. NAND and NOR gates will be considered because of their ability to economically implement logic circuitry. The tri-state buffer/driver will be included for its extensive use in bused data transmission. The exclusive OR gate is included because of its general usefulness in several situations. The role of programmable logic devices (PLDs) and read-only memories (ROMs) will also be considered.

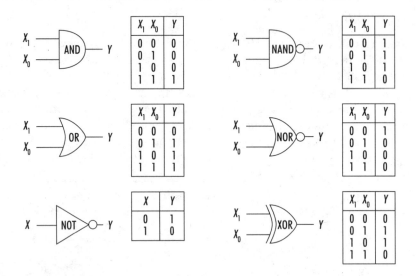

Figure A.1   Logic Symbols and Truth Tables for AND, OR, NOT, NAND, NOR, and XOR Logic Functions

Figure A.1 shows the graphical representation and truth table for AND, OR, NOT, NAND, NOR, and exclusive OR (XOR) logic functions. The AND output is 1 only when all of the gate input variables are 1; otherwise it is 0. This is shown in truth-table form for the case of two input variables. The number of input variables to the device may be increased to suit the requirements of the specific situation. Unused gate inputs are forced to a 1 state.

The output of OR is 1 if one or more of the input variables is a 1. Otherwise, the output variable is 0. Unused inputs to the gate are forced to a 0 state. The NAND (Not AND) function is the inverse of the AND function. The XOR has a value of 1 whenever one of the input variables, but not both, have a value of 1.

Note that with the AND, OR, XOR, NOR, and NAND gates, the order in which the variables is presented to the gate has no effect on the output. Each of the input lines to the gate is treated alike. This characteristic is known as **commutativity**.

AND, OR, and NOT Logic

It is important that a set of gates be able to implement any logic function represented by a truth table. This characteristic is called **functional completeness**. The set AND, OR, and NOT is functionally complete and has the advantage of being intuitive for the designer—two-level AND-OR representations are compatible with human logic. Most of the designs discussed in this book will use the AND-OR-NOT set with the full understanding that the logic can easily be optimized when converted to other realization forms.

NAND and NOR Logic

The NAND gate and NOR gate are functionally complete by themselves. This allows any logic function to be implemented using only one type of gate logic. A two-level NAND-NAND gate circuit is logically equivalent to a two-level AND-OR circuit, and a two-level NOR-NOR gate circuit is logically equivalent to a two-level OR-AND circuit.

The Tri-State Buffer/Driver

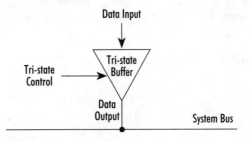

Figure A.2   A Tri-State Buffer/Driver

Figure A.2 shows the tri state buffer/driver. Using the control input to the buffer, the output is either the logic value provided by the input (two possible states) or a high impedance (the third state). When active, or **enabled**, the output circuitry can drive the data bus and all attached components to a voltage associated with a logic 1 or logic 0. When inactive, the device presents a high impedance, which allows another device to drive the bus without interference. The main purpose of the tri-state buffer is to allow logic components to share a common logic bus. The output of most logic devices cannot be electrically connected, or wire OR-ed together, without producing indeterminate logic levels and possibly damaging one or more of the devices.

Tri-state buffer/drivers are available in both inverting and noninverting output configurations. Many devices, such as multiplexers and memory devices, include tri-state drive capability in their output circuitry.

The Multiplexer

Several devices are not logic devices in the classical sense but have characteristics useful in implementing designs. The multiplexer (MUX), as shown in Figure A.3, is one such device. It selects one of several possible input lines and directs it to the single output line. Input selection is based on an n-bit code provided by a control input called SELECT. When used in this manner the multiplexer is a form of logical switch which selects one of 2^n data inputs and routes it to the data output. A multiplexer may also have an OUTPUT ENABLE control line to provide tri-state output-control. The SELECT and OUTPUT ENABLE pins are control points for the device.

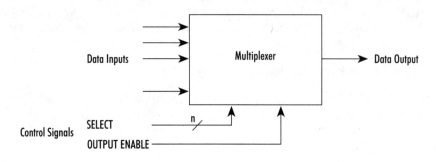

Figure A.3   A Multiplexer

Multiplexers are available as integrated circuit chips with up to 16 multiplexer input lines and 4 select lines. They can effectively be expanded to include more input variables if used with multiple levels of multiplexers.

The Demultiplexer/Decoder

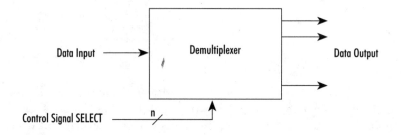

Figure A.4   A Demultiplexer

The demultiplexer, as shown in Figure A.4, is the inverse of the multiplexer. A single data input is routed to one of several data outputs depending on the select code provided. With an input of logic 1, the selected output receives a value of 1. Outputs other than the one selected receive a logical value of 0. With an input of logic 0, all outputs receive a logical value of 0.

The decoder, as shown in Figure A.5, and the demultiplexer are logically equivalent. The select lines are considered data input lines which are to be decoded. The output pin, corresponding to the code of the input data, is asserted. In the Figure A.5B, a series of decoders decodes the data word, 7405_{16}.

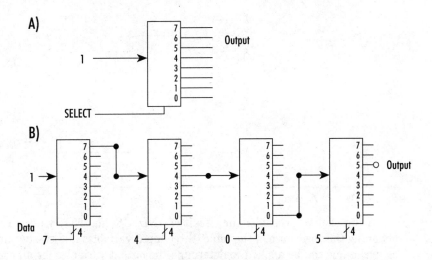

Figure A.5 A Decoder

A.4 Truth Table Analysis of Gate Circuits

Analysis involves determining and producing a description of a circuit's characteristics. Analysis is the inverse of design. Several methods exist for specifying characteristics of a gate circuit. In the most general form, these circuits will have several input variables and several output variables, as shown in Figure A.6.

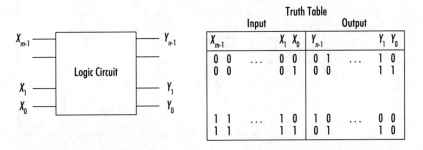

Figure A.6 A Multiple Input/Multiple Output Logic Circuit and Its Truth Table

The truth table is a tabular representation that enumerates all possible conditions of the input variables and the associated values of the output variables. For the condition shown in Figure A.6, the truth table would require 2^m rows to express all possible input conditions for m input variables.

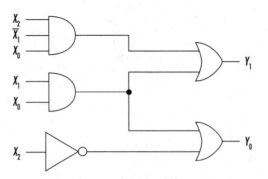

Figure A.7   A Logic Circuit

As an example, consider the circuit of AND, OR, and NOT gates shown in Figure A.7. After obtaining the basic form of the table (three input variables and two output variables), a methodical technique can be applied to determine the output variables for all combinations of the input variables. First, the input conditions for all first-level AND gates are determined. Using the gate definitions, the output variables of these gates may be determined and indicated in the table. Finally, the definitions of the OR gate are used to determine the circuit output variables.

X_2	X_1	X_0	$\bar{X}_1$	$X_2\bar{X}_1X_0$	X_1X_0	$\bar{X}_2$	Y_1	Y_0
0	0	0	1	0	0	1	0	1
0	0	1	1	0	0	1	0	1
0	1	0	0	0	0	1	0	1
0	1	1	0	0	1	1	1	1
1	0	0	1	0	0	0	0	0
1	0	1	1	1	0	0	1	0
1	1	0	0	0	0	0	0	0
1	1	1	0	0	1	0	1	1

Table A.2

A shorthand notation is an alternative to presenting the complete truth table. The truth table for Y_1 has a value of logic 1 for input configurations (X_2, X_1, X_0) of {011, 101, 111}. This three-variable example could be represented by indicating the numeric (decimal) value of the terms for which the output is a 1.

$$Y_1(X_2,X_1,X_0) = \Sigma\ (3,5,7) = \bar{X}_2X_1X_0 + X_2\bar{X}_1X_0 + X_2X_1X_0$$
$$Y_0(X_2,X_1,X_0) = \Sigma\ (0,1,2,3,7) = \bar{X}_2\,\bar{X}_1\bar{X}_0 + \bar{X}_2\bar{X}_1X_0 + \bar{X}_2X_1\bar{X}_0 + \bar{X}_2X_1X_0 + X_2X_1X_0$$

A.5   Algebraic Analysis of Gate Circuits

The characteristics of a logic network may also be expressed in an algebraic form commonly known as switching algebra or Boolean algebra. This form of analysis is reserved almost exclusively for circuits using AND, OR, and NOT gates.

The symbols + and OR represents the OR operation. The symbol + will be used when there is no confusion with the addition operation. The symbol $\bullet$ is used for the AND operation. When the intent is clear, the absence of a symbol indicates the AND operation. The symbol $\overline{X}$ is used to indicate the logical complement of variable X. Other notations for the complement operation include $/X$, X' and $!X$. The circuit of Figure A.7 may be represented by the logic expression:

$$Y_1 = X_2\overline{X}_1X_0 + X_1X_0$$
$$Y_0 = X_1X_0 + \overline{X}_2$$

Because the AND operation is commutative, the order of presentation of the variables is not important.

The main advantage of using logic expressions is they allow manipulation of the expressions using Boolean algebra or switching algebra. A truth table can verify that the two logic expressions

$$Y_1 = X_2X_0 + X_1X_0$$

and

$$Y_0 = X_2\overline{X}_1X_0 + X_1X_0$$

are equivalent because they have identical truth tables. The first expression has fewer appearances of the logic variable X_1 and is less costly. It is useful to perform some operations on logic expressions. Table A.3 summarizes the identities associated with Boolean algebra.

Identity	Comment
(1) $X + 0 = X$	Property of element 0.
(2) $X \bullet 0 = 0$	Property of element 0.
(3) $X + 1 = 1$	Property of element 1.
(4) $X \bullet 1 = X$	Property of element 1.
(5) $X + X = X$	Idempotency.
(6) $X \bullet X = X$	Idempotency.
(7) $X + \overline{X} = 1$	Complement.
(8) $X \bullet \overline{X} = 0$	Complement.
(9) $X + Y = Y + X$	Distributivity.
(10) $X \bullet Y = Y \bullet X$	Distributivity.
(11) $X + (Y + Z) = (X + Y) + Z$	Associativity.
(12) $X \bullet (Y \bullet Z) = (X \bullet Y) \bullet Z$	Associativity.
(13) $\overline{(X + Y)} = X \bullet Y$	DeMorgan's theorem.
(14) $\overline{(X \bullet Y)} = \overline{X} + \overline{Y}$	DeMorgan's theorem.
(15) $X \bullet (Y + Z) = (X \bullet Y) + (X \bullet Z)$	Commutativity.
(16) $X + (Y \bullet Z) = (X + Y) \bullet (X + Z)$	Commutativity.

Table A.3

Example A.1:   Boolean Equality

Verify the following equality:

$$X_1 X_0 + X_1 \overline{X}_0 = X_1$$

Solution:

The expressions for X_1 and $X_1 X_0 + \overline{X}_1 X_0$ have identical truth tables and are therefore equivalent.

X_1	X_0	$\overline{X}_0$	$X_1 X_0$	$X_1 \overline{X}_0$	$X_1 X_0 + X_1 \overline{X}_0$	X_1
0	0	1	0	0	0	0
0	1	0	0	0	0	0
1	0	1	0	1	1	1
1	1	0	1	0	1	1

Table A.4

Or, using identities 15 and 7 in Table A.4:

$$X_1 X_0 + X_1 \overline{X}_0 = X_1 (X_0 + \overline{X}_0) = X_1 (1) = X_1$$

Example A.2:   Boolean Equality

Verify the following equality:

$$X_1 + \overline{X}_1 X_0 = X_1 + X_0$$

Solution:

The two expressions have identical truth tables.

X_1	X_0	$\overline{X}_1$	$\overline{X}_1 X_0$	$X_1 + \overline{X}_1 X_0$	$X_1 + X_0$
0	0	1	0	0	0
0	1	1	1	1	1
1	0	0	0	1	1
1	1	0	0	1	1

Table A.5

Judicious use of the above identities will allow many logic expressions to be manipulated to the form most appropriate for the circuit realization.

A.6   Design Implementation Using AND, OR, and NOT Gates

The design techniques of this section assume that the design has been specified either in the form of a truth table or in the form of AND/OR logic expressions. Both forms allow a straightforward implementation using AND and OR logic. This implementation is called **two-level logic** because it requires logic signals to propagate sequentially through AND gates and then OR gates. Complementing variables in advance of the AND gate, if necessary, is performed by NOT gates. The NOT gates are not considered a level of logic.

Design Using the Truth Table

A truth table lists the desired output for each possible input conditions. Each output condition is considered separately.

1. Include an OR gate to produce each output variable in the truth table.
2. Include an AND gate connected to the OR gate for each appearance of a 1 in the truth table for that output variable.
3. For each AND gate, provide inputs for each of the input logic variables. The variable appears complemented if the input variable has a value of 0 in its truth table assignment; it appears uncomplemented if the input variable has a value of 1.
4. All complemented variables may be generated by NOT gates.

Example A.3:   Logic Gate Implementation

Design a logic gate realization for the following truth table:

X_1	X_0	Y_1	Y_0
0	0	1	0
0	1	1	1
1	0	0	1
1	1	1	0

Table A.6

Solution:

Figure A.8 shows the solution obtained by the above algorithm. Two OR gates are required: one each for Y_1 and Y_0. Five AND gates are required; three for Y_1 and two for Y_0 It should be noted that two of the AND gates produce identical logic expressions of $\overline{X}_1 X_0$. The output of one of these gates could have been used as a source for both of the OR gates. This would save one gate. Single gates could also be used to generate the complemented variables.

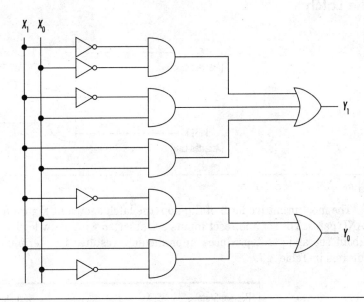

Figure A.8   AND OR Gate Circuit

Design Using a Logic Expression

When using a logic expression as the source of the design specification, two approaches are possible:

1. The truth table for the logic expression may be determined and the design may proceed using the algorithm discussed previously.
2. The logic expression (or the logic expression manipulated through the use of Boolean algebra) may be used directly.

When using the logic expression directly, the procedure is as follows. As in the previous case, each output variable is considered separately:

1. Include an OR gate for each output variable.
2. Include one AND gate for each appearance of an AND grouping in the logic expression for each of the output variables.
3. For each AND gate, provide inputs for the variables designated in the corresponding AND grouping of the logic expression. Inputs to the AND gate are inverted if the corresponding variable is a 0; they are not inverted if the variable is 1.
4. All complemented variables may be generated by NOT gates.

A.7   Memory Devices—The Flip-Flop

The flip-flop is the basic memory element for digital devices. It appears in a variety of different forms and with different control points.

The Latch

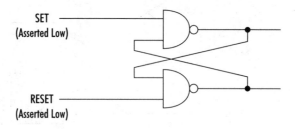

SET
(Asserted Low)

RESET
(Asserted Low)

Figure A.9   The Latch

The most primitive form of flip-flop, the latch shown in Figure A.9, is a cross-coupled pair of NAND gates. The set and reset inputs are asserted with a low logic level. Simultaneous assertion of both set and reset produces unpredictable results. The terminal behavior of the device is indicated in Table A.7.

Present State	SET	RESET	Next State
0	0	0	—
0	0	1	1
0	1	0	0
0	1	1	0
1	0	0	—
1	0	1	1
1	1	0	0
1	1	1	1

Table A.7

The latch is an asynchronous device; its output changes are initiated by changes in the input. The time of transition between the present state and the next state is not based on the system clock's established time base. Logic design using latches and other asynchronous devices is considerably more complex than design using synchronous devices such as the type D flip-flop of the following section.

The Type D Flip-Flop

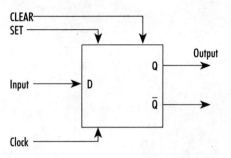

Figure A.10   The Type D Flip-Flop

The type D flip-flop, as shown in Figure A.10, is designed to function synchronously. Changes of state are synchronized to a transition of the system clock. All transitions occur very rapidly. The flip-flop is designed so that the window for the input conditions is very narrow. This prevents newly changed output conditions from affecting the input to the same flip-flop.

The design characteristics of the D flip-flop are shown in the following table. The next state is always the same as the D input. The transition to the next state is initiated by the positive transition of the system clock.

Present State	D	Next State (after next clock)
0	0	0
0	1	1
1	0	0
1	1	1

Table A.8

The D flip-flop may also have SET and CLEAR terminals. These are asynchronous control points used for establishing initial conditions on the flip-flop.

The J-K Flip-Flop

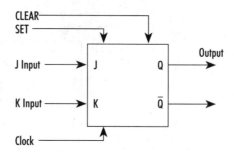

Figure A.11   The J-K Flip-Flop

The J-K flip-flop, as shown in Figure A.11, is also designed for application in synchronous systems. The device has the following characteristics:

Present State	J	K	Next State
0	0	0	0
0	0	1	0
0	1	0	1
0	1	1	1
1	0	0	1
1	0	1	0
1	1	0	1
1	1	1	0

Table A.9

Both the type D flip-flop and J-K flip-flop are extensively used. They are included as memory components in sophisticated PLDs and state machine components. Because of their versatility, they are easily configured into counters and registers.

The Monostable Multivibrator

The monostable multivibrator, shown in Figure A.12, is also known as a one-shot. It offers an attractive method of obtaining time-sensitive signals. When triggered by an input, the one-shot produces an output with a duration dependent on resistor and capacitor circuit elements. Using high-quality components, the time delay can be controlled to reasonable limits. Uncontrolled environmental variables, such as temperature, can cause considerable variation in the circuit element values and affect the time duration.

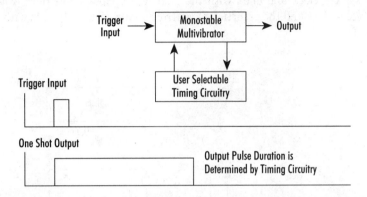

Figure A.12   The Monostable Multivibrator

A commercially available one-shot, the SN74121, can produce an output pulse with a duration from 35 nanoseconds through 28 seconds by varying the combination of resistance and capacitance. The device is also designed with internal compensation for temperature independence. The one-shot is not a synchronous device. The duration of its output is independent of the system clock. Debugging and troubleshooting circuits with one-shots is difficult. With synchronous devices, the clock can be stopped for diagnosis using a technique called single stepping. One-shots create signals independent of the system clock and cannot be fully controlled.

For many reasons, using one-shots is generally discouraged where other techniques can be used. Historically one-shots have been used to mask timing problems rather than to solve them with good design.

A.8   Registers and Register Files

Constructed from an array of flip-flops, a **register** is a memory device wide enough to store a data word. A logical group of registers with provisions for selecting one register in the file for reading or writing is a **register file**. **Asynchronous registers** (often called latches) change their content whenever input signals change. **Synchronous registers** are restricted so that their contents change only in synchronism with the assertion of a clock signal. Design, analysis, and debugging techniques are much simpler when synchronous registers are used. Asynchronous techniques are potentially more high-performance since it is not necessary to wait for the arrival of a clock to complete a processing action. For most applications, the added performance possible with asynchronous systems seldom justifies the added design complexity. This text will use clocked synchronous registers. Since a clock signal is always required for a register, its presence is implied and will not be specifically shown unless additional emphasis is needed.

Control signals to registers, such as LOAD and CLEAR, cause actions synchronized by the system clock. Be aware that some commercially available synchronous registers have a CLEAR and a PRESET control inputs that are asynchronous with the system clock.

The **clock period** is the time between successive active transitions of the system clock. The clock period is the reciprocal of the clock rate. A **clock cycle** begins *after* the active transition of the system clock and ends with the next active clock transition. A clock cycle begins with the arrival of data and the assertion of control signals (if any). It ends with the modification of data in registers.

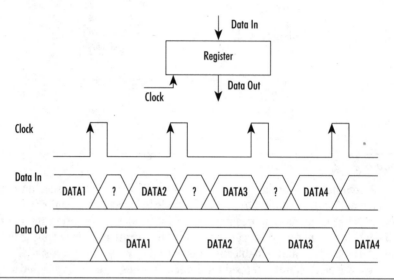

Figure A.13   Timing of Register-Transfer Operations

Figure A.13 shows the timing associated with data acceptance in a simple synchronous register. Data change in the register is initiated by the active transition of the system clock. This is indicated by the arrow in the clock signal. An up arrow indicates that new information is accepted in synchronism with the positive transition of the clock pulse. Stable data must be present at the input of the register in advance of the active clock transition. The minimum time data must be available in advance of the clock pulse is called the register **setup time**. In Figure A.13, a portion of the time interval between clock pulses is shown as indeterminate, and indicated by a ?. In this period the input data may undergo transitions to assume its new value. The data must then remain stable until the clock makes its next active transition. The downward (inactive) transition of the clock signal is usually not significant to the designer.

Clocked Synchronous Registers

Clocked synchronous flip-flops and registers allow the reliable transfer of information between registers. Their characteristics are such that newly accepted data in the output of a register cannot find a loop path back to the input of the register to cause unpredictable performance. Two different techniques are used—both use a simple clocked latch flip-flop device as shown in Figure A.14.

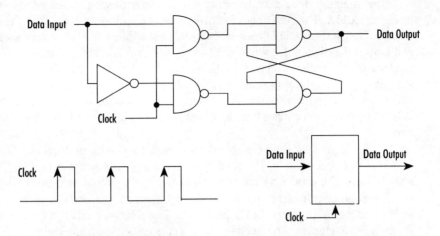

Figure A.14   The Clocked Latch

The clocked latch uses a set/reset flip-flop as its basic storage element. Additional NAND gates allow the input data to be presented to the set/reset flip-flop only when the clock is in the asserted (logic 1) condition. When the clock is not asserted, the flip-flop retains its contents. The clock pulse effectively creates a window during which data can be accepted. The clocked latch can be used as a register if the width of the clock pulse is closely controlled. It must be wide enough to allow the data input signals to be captured by the latch. It must be narrow enough so that data accepted into the register does not have sufficient time to loop back and affect the input while the data window is still open.

A) Edge-triggered Flip-Flop

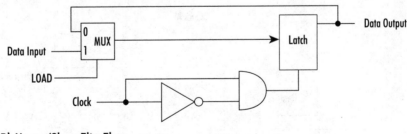

B) Master/Slave Flip-Flop

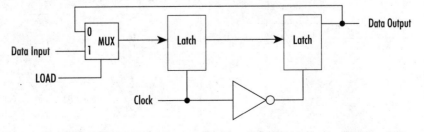

Figure A.15   The Edge-triggered and Master/Slave Flip-Flop Registers

The simplest clocked synchronous flip-flop, circuit wise, is the edge-triggered circuit shown in Figure A.15A. The AND gate in this example circuitry assures that the data window is very narrow. The NOT gate is fabricated to have a time delay of a few nanoseconds between its input and output. Consequently, when the input clock makes a transition from 0 to 1, there is a brief time period (the gate delay) when both the input and output of the NOT gate are logic 1. The AND gate detects this situation and generates a very short width pulse. Data acceptance is allowed only during a very narrow time slot during the rising edge of the system clock. Logic propagation delays prevent data changes from looping back to the input until after the access window has been closed.

Data selection for the load operation is performed by a multiplexer. When not performing a LOAD of new information, the register is reloading its current contents. The register performs either a load of new information or a reload of the current contents during every clock cycle.

A second circuit technique for designing a reliable clocked synchronous register, the master/slave flip-flop, is shown in Figure A.15B. Two clocked latches, a master and a slave, are used for each bit of storage. The master unit accepts data only while the clock is high; the slave unit accepts data only while the clock is low. Only one of the two registers can change during a portion of the clock cycle. The source of information for the master is specified by an input multiplexer. When the clock resumes its low value, the slave section accepts information passed from the master. Since at any instant only one of the flip-flops is being clocked it is impossible for a loop-back of information to alter the register contents. With each clock pulse the device performs a load of new information or a reload of current information.

The edge-triggered flip-flop is popular for integrated-circuit fabrication since the circuitry has fewer components than the master/slave flip-flop. Edge-triggered registers and master/slave registers involve different timing. Both should not be used in the same system without thoroughly analyzing the implications.

Clock Skew

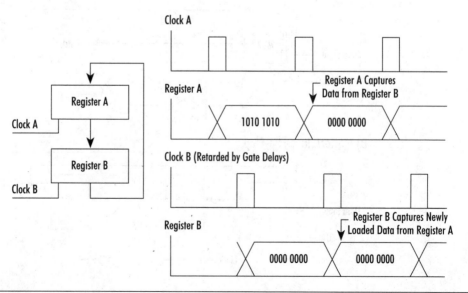

Figure A.16   Effect of Clock Skew on Data Transfers between Registers

As digital logic circuitry processes logic signals, time delays are introduced. These delays can be especially troublesome when they affect clock signals. Figure A.16 shows an extreme case of clock skew. Register A and register B have initial values of 10101010_2 and 00000000_2. Clock B is retarded in time with respect to clock A, perhaps due to the cumulative effect of digital logic delays or path propagation delay. This delay is assumed to be greater than the setup time of the register. Register A reliably accepts information (00000000_2) from register B and places it on its output, replacing its initial value (10101010_2). Register B, with a retarded clock, does not accept new information until after register A has provided its new data (00000000_2). The initial value of register A (10101010_2) is lost. If, in this example, the delay of clock B exceeds the set-up requirement of some, but not all, of the flip-flops, the result will be unpredictable.

Problems of clock skew become more severe as the speed of digital circuitry increases. Clock skew can be avoided by prudent use of logic in clock distribution circuitry. In common Transistor-Transistor Logic (TTL) logic families, each logic level contributes approximately 2–5 nanoseconds of delay. If buffers and inverters are necessary to adequately fan out the clock, all versions of the clock need to be subjected to equivalent buffering and delay. Propagation delays are approximately one nanosecond for each foot of wiring. A difference of 10 feet in cabling distance could result in a clock skew of 10 nanoseconds. In TTL logic, data must be available and stable at the input of a register for 5–10 nanoseconds (setup time) before the active clock transition.

With a careful choice of components and good design practices in distributing the system clock, clocked synchronous registers provide a very reliable design component. For the design techniques presented, they will be considered ideal components. For very high performance devices, we must study gate delays and propagation delays in detail when designing the device.

Performance characteristics for registers used in data-transfer operations are easy to specify: they must be synchronous, they must have control inputs to allow a choice of load and do-nothing operations, and they should be capable of being cleared, preferably synchronously. Most computer-aided design packages have provisions for a large number of suitable register configurations.

A.9   Counters and Shift Registers

Because of their general usefulness, many of the following devices are available from manufactures as MSI devices. In most cases they may also easily be fabricated using common flip-flops. In some cases idealized devices are described. Standard components can closely approximate their behavior.

Counters

Counters are characterized in several ways. Of primary importance is whether the device is a parallel counter or a ripple counter. In a parallel device a control signal initiates the change in count. All bits change simultaneously.

In a ripple counter the prior stage initiates the count action of each stage of the counter. The bits assume their correct value in a ripple fashion from the least significant bit to the most significant bit. This characteristic is undesirable. Delays in getting the correct count can cause delays in other parts of the circuitry, especially if decisions need to be made based on the count.

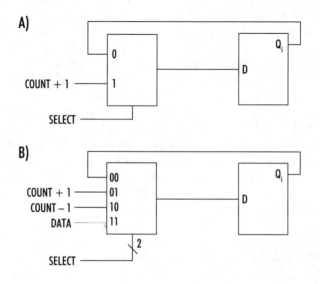

Figure A.17   The Counter

Most counters have only a few control points. In many cases the only control input is the one affecting the advance of the count. Figure A.17A shows a single flip-flop from a simple counter which loads the new count value when the control input is asserted. It maintains its current value when the control is not asserted. Clock signals are not shown. The device in Figure A.17B has additional multiplexer inputs that allow an increment, a decrement, and a parallel load of external data. For some applications, a synchronous clear would be useful.

Shift Registers

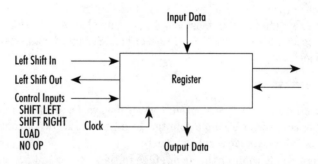

Figure A.18   The Shift Register

The shift register is useful in many data processing activities. It allows a bit pattern to be shifted one bit location in either direction. Figure A.18 shows an idealized implementation of a shift register. All actions are synchronous to the system clock. Useful modes include the following:

- **Logical shift right.** The contents of the register are shifted one bit position to the right. The right-most bit is discarded. The left-most position is loaded with a logic 0.
- **Rotate right.** The contents of the register are shifted one bit position to the right. The right-most bit is loaded into the left-most bit position.
- **Algebraic shift right.** The contents of the register are shifted one bit to the right. The right-most bit is discarded. The left-most bit position retains its previous value. This operation is useful when performing arithmetic on signed numbers represented in two's complement.
- **Logical shift left.** The contents of the register are shifted one bit position to the left. The left-most bit is discarded. The right-most position is loaded with a logic 0.
- **Rotate left.** The contents of the register are shifted one bit position to the left. The left-most bit is loaded into the right-most bit position.
- **Load.** Parallel load the register with the input data on the next clock pulse.

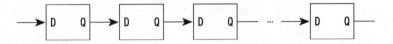

Figure A.19   A Shift Register with Type D Flip-Flops

Shift registers are easily implemented with type D flip-flops as shown in Figure A.19.

Shift Register Generators

A shift register generator has many of the desirable features of a parallel counter with some significant reductions of logic complexity. The shift register generator is fabricated from a shift register and additional logic that produces a bit to be shifted into the position vacated during the shift. This bit is usually generated by placing the two least significant bits into an exclusive OR logic gate.

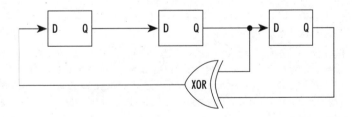

Figure A.20   A Shift Register Generator

Figure A.20 shows a shift register generator for three flip-flops. If the register originally contains the value 100, the contents will advance with each clock pulse as shown in Table A.10.

Clock Cycle	Register Contents	Comment
t_0	100	Initial contents.
t_1	010	
t_2	101	
t_3	110	
t_4	111	
t_5	011	
t_6	001	
t_7	100	Sequence repeats.

Table A.10

Barrel Shifter

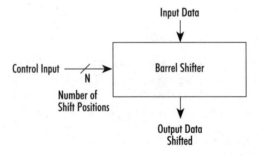

Figure A.21   The Barrel Shifter

The barrel shifter, as shown in Figure A.21, can implement many shifting functions associated with arithmetic operations. Its main advantage over a shift register is the ability to perform a variety of shifting operations within one clock cycle. A shift code is presented to the device by means of the control input. The code specifies the number of bit positions to be shifted. Shifts are performed circularly.

A.10   State Machines

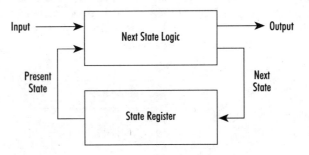

Figure A.22   A State Machine

The counter, shift register, and shift register generator are all examples of a finite state machine (or state machine). The state machine shown in Figure A.22 contains a state register synchronized to the system clock. The content of the register is the **state** of the state machine. The state of the system changes as a function of both the system inputs and the current value of the state. The output produced is also a function of the current state and the input.

State Diagram

The operation of a state machine may be described by a **state diagram**. Each of the states is represented by an ellipse. Each state is named by indicating the **state variable** values associated with the state. This is the value of the flip-flops of the circuitry. In the **Moore** form of the state machine, there is an output for each state; this is indicated following the state variables and separated by a "/." Inputs to the state machine may cause a transition of states. Arrows with an indication of the input values show state transitions. Figure A.23 shows a flip-flop circuit and its corresponding state diagram.

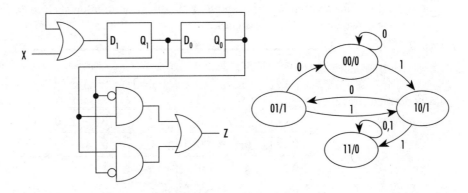

Figure A.23   A Flip-Flop Circuit and Its State Diagram

Example A.4:   State Machine Design

Design a state machine that will accept two input variables X_1 and X_0. The output of the machine is to be a two-bit binary number, Z_1 and Z_0. The output is to remain constant if $X_1 = 0$. It is to change at each clock pulse if $X_1 = 1$. It is to be incremented if $X_0 = 1$, and it is to be decremented if $X_0 = 0$. All counting is to be modulo 4 (i.e., the next count after 11 is 00). X_1 and X_0 change synchronously with the system clock.

Solution:

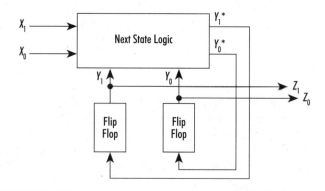

Figure A.24   A State Machine

A state machine as shown in Figure A.24 is selected. The output variables Z_1 and Z_0 are chosen to be identical to the state variables Y_1 and Y_0. This eliminates the need for an output logic block. X_1, X_0 and the current state of the state register (Y_1, Y_0) provide input variables to the next-state logic. Output from the logic is passed directly to the state register, which is composed of two type D flip-flops. The initial state following power-up is 00. Every clock pulse will cause the contents of the flip-flops to accept the value applied to the D input.

The truth table for the next-state logic is shown in Table A.11. An asterisk indicates a next-state condition for a variable. For example, Y_1 is the present value of state variable Y_1. Y_1^* is the value Y_1 will assume after the next clock pulse.

Present State		Input Variables		Next State	
Y_1	Y_0	X_1	X_0	Y_1^*	Y_0^*
0	0	0	0	0	0
0	0	0	1	0	0
0	0	1	0	1	1
0	0	1	1	0	1
0	1	0	0	0	1
0	1	0	1	0	1

Present State		Input Variables		Next State	
Y_1	Y_0	X_1	X_0	Y_1^*	Y_0^*
0	1	1	0	0	0
0	1	1	1	1	0
1	0	0	0	1	0
1	0	0	1	1	0
1	0	1	0	0	1
1	0	1	1	1	1
1	1	0	0	1	1
1	1	0	1	1	1
1	1	1	0	0	1
1	1	1	1	0	0

Table A.11

The detailed design of the logic block could proceed using the logic gate techniques discussed in an earlier section.

A.11   Positive and Negative Logic

Formal logic design begins with the premise that there are two logic values, 0 and 1. In implementation, it seems reasonable to assign the lower voltage to the logic value of 0 and the higher voltage to the logic value of 1. This assignment is arbitrary. It is often convenient to assign the voltages in the exact opposite manner.

In the case of control signals, assertion rather than polarity of a signal is important. When asserted, something is to happen; when passive, no action is to happen. The asserted value of the control signal can be either high or low.

The choice of assertion polarity is often dictated by physical characteristics of the devices. For example, a data bus is usually designed to be in the high condition when no devices are driving it. Therefore, to be recognized, a signal must be asserted low. Many logic components are designed so that when control points are not used, they are in the high condition. This would require a control signal to be asserted low.

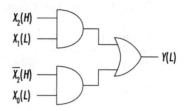

Figure A.25   Using Positive and Negative Logic

Figure A.25 shows a logic device in which some of the input variables are asserted high and others are asserted low. The output logic is asserted low. When necessary for clarity, the polarity of the logic is identified by an H or L in parentheses, following the variable name.

From a logic design point of view, there is no problem designing logic for mixed logic systems. A mixture of logic assertion levels, however, obscures the action of the circuit from the designer.

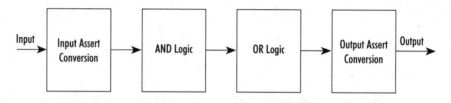

Figure A.26   A Technique for Working with Mixed Logic Systems

Figure A.26 shows an alternative method of implementing a logic block for mixed logic systems. The logic is functionally divided into four blocks. The first block converts all logic variables to a common form, either positive or negative logic. The second and third blocks perform the AND/OR logic function. The last block converts the outputs to the logic polarity required at the output terminals. As a reminder, a bubble is usually placed on the input or output of a logic gate to indicate that negative polarity logic is used.

Logic levels can be converted using only invertor gates.

$$\overline{X}(L) = X(H)$$
$$\overline{X}(H) = X(L)$$

This technique has the advantage that the designer can focus attention on the underlying AND/OR logic rather than the complexities introduced by multiple logic polarities. Because of the ease of converting between positive and negative logic representations, most of the discussion related to logic will assume only positive logic. Unless necessary for clarity, the concept of assertion allows us to often avoid discussions of positive and negative logic.

A.12   Programmable Logic Devices

Programmable logic devices, PLDs, have recently become very important design components. These devices, which come in several different forms, allow the user to implement relatively extensive and complex logic functions using only a small number of parts. Some devices allow programming at the user's location; others require operations during the final phase of the manufacturing process to customize the design. PLDs are usually designed with the aid of special design software. Using a programming device, the user is able to implement the final design using a small assortment of parts.

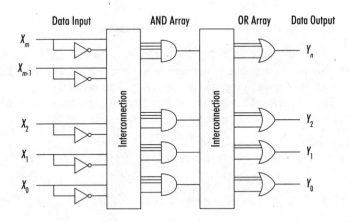

Figure A.27 Generalized Implementation of a Logic Circuit

In PLDs, all logic implementations involve interconnecting an array of AND gates and an array of OR gates, as shown in Figure A.27. These gates are interconnected to produce the desired logic expression. When using gate-level logic devices, interconnections are made by wiring together the inputs and outputs of gates on a printed circuit card. When using PLDs, the connections are made on-chip using crossbar switches.

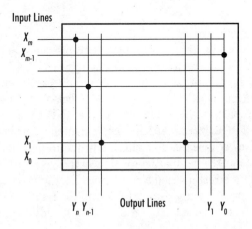

Figure A.28 A Crossbar Switch Interconnection Block

Figure A.28 is a schematic representation of a crossbar switch. By proper connection of internal rows and columns, any input can be connected to any output. In PLDs, the outputs are directed to inputs of logic gates. Two types of crossbar switches are used in PLD devices. Programmable crossbar switches can be specified by the user. Fixed crossbar switches are created as a part of the manufacturing process and cannot be changed by the user. In Figure A.28, input X_m is connected to output Y_n; input X_1 is connected to both outputs Y_{n-2} and Y_3.

ROM Replacements for Random Logic

ROM devices offer an attractive alternative to the use of gate type logic. A ROM with m address lines and n outputs, is the equivalent of a two-level AND/OR circuit. The first level has 2^m AND gates, each with m inputs. There is a gate to generate each of the possible combinations of the m input variables. The second-level equivalent logic has n OR gates, each corresponding to one of the outputs. Each OR gate has inputs for each of the 2^m first level AND gates.

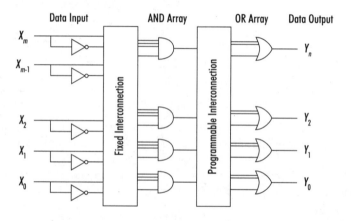

Figure A.29   Equivalent Logic of a ROM as a PLD

Figure A.29 shows the equivalent logic representation of a ROM. The blocks labeled AND gate array and OR gate array consist of the first-level AND gates and second-level OR gates. The crossbar switches contain the pattern of interconnections which exist between the input lines and output lines of the block. The first block is labeled as **fixed** because the user cannot alter the interconnection pattern. Logically, each of the AND gates is connected to one of the 2^m different combinations of input variables and complemented input variables. The second crossbar block is **programmable** because the user effectively determines the interconnection pattern of the crossbar switch when programming the ROM device.

ROMs offer a direct method for transforming the truth table of the desired logic device into a single component. They are especially useful for implementing multiple output logic functions. The ROM size is selected such that the number of address lines on the ROM is equal to (or greater than) the number of input variables. The word width of the ROM is selected to be equal to (or greater than) the number of output variables. The ROM is then programmed such that the input lines are interpreted as an address and the output lines are interpreted as the memory contents. Every combination of input variables produces an output. This is very much like a truth table.

Example A.5:   State Machine Design

Design a state machine that will have four states. The states will be designated both by a mnemonic name and a binary state variable designation. The first state is to be START(00); the last is to be a STOP(11). It is to begin in START and progress sequentially through ONE(01) and TWO(10) to STOP where it is to remain. In the START and STOP states there are to be no outputs. In state ONE the output Z_1 is to be asserted. In state TWO output Z_0 is to be asserted.

Solution:

Two logic functions are needed: the next state logic and the output logic. This could be provided by two ROMs. Since both ROMs would have the same input variables, it is possible to implement both logic functions with a single ROM with two fields; one for the next state and the other for the output. There are no input variables in this example.

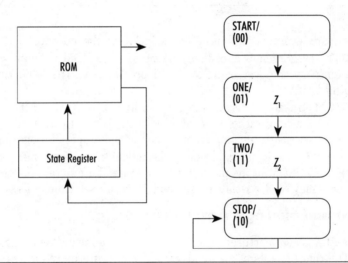

Figure A.30   State Diagram

The state register is initialized in the 00 state when energized. Figure A.30 shows the state diagram for this state machine. The content of the single ROM is as follows:

ROM Input (Address)		ROM Output (Contents of memory cell)			
		Next State		Output	
Y_1	Y_0	Y_1^*	Y_0^*	Z_1	Z_0
0	0	0	1	0	0
0	1	1	1	1	0
1	0	1	0	0	0
1	1	1	0	0	1

Table A.12

There are several variations of ROMs, including the following:

- **PROM** (programmable read-only memory). This ROM is blank when purchased. It is programmed by the user using a special programmer. The ROM cannot be erased and used again.
- **EPROM** (erasable programmable read-only memory). This ROM differs from PROM in that the user can return it to its blank condition. This is accomplished by exposure to ultraviolet light. After erasure, it can be reprogrammed using special programming devices.
- **EAPROM** (electrically alterable programmable read-only memory). The EAPROM differs from the EPROM in the method used to erase and program the device. In the EAPROM the memory may be electrically erased and programmed while in use.

Programmable Logic Array (PLA)

The PLA is manufactured as a form of generalized two-level interconnection of AND/OR logic gates. The complements of all input variables are internally created and are available. Figure A.31 shows its logic structure. A device with m inputs and n outputs has a fixed number of first level AND gates, which is generally much less than 2^m. There are n OR gates.

By a variety of different techniques, the user can customize the device by programming the internal interconnection structure between the inputs and the first-level AND gates and the connections between the outputs of the AND gates and the inputs to the OR gates. This means that both the AND array and the OR array are programmable. One implementation uses the same fabrication technology as EPROM memory chips. The connection pattern is programmed electronically using the same process to program EPROMs. The patterns can be erased and reprogrammed.

A typical PLA allows the user to construct logic devices involving 16 input variables and 8 output variables. This is subject to the restriction that the equivalent two-level (AND/OR) logic functions require no more than 48 product terms (AND gates). The product terms may be used by more than one second-level OR gate.

PLA approaches are attractive for situations in which there are a large number of input variables but fairly simple logic functions. A ROM to handle 16 input variables would require a memory space with 64K elements (K = 1024).

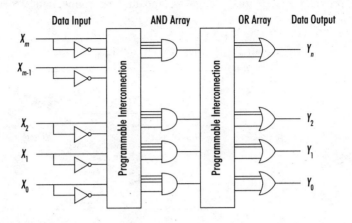

Figure A.31   Equivalent Logic of a PLA

Generic Array Logic (GAL)

GAL is an alternate name for a device that incorporates a programmable AND array and a fixed OR array. (Calling these devices GALs is an attempt to avoid use of their common name, which is a copyrighted trademark). The AND array and OR array of a GAL, as shown in Figure A.32, is similar to a PLA. There are programmable connections between the inputs and the first-level AND gates. The OR array connections are fixed and not user programmable.

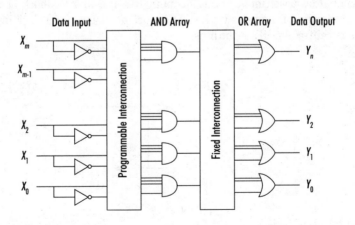

Figure A.32   Equivalent Logic of a GAL

PLDs With Macrocells

The most general purpose form of PLDs are devices that expand versatility by including programmable macrocells. The general structure is shown in Figure A.33. The logic section of the

PLD provides AND gates, OR gates, and an interconnection structure for implementing logic functions.

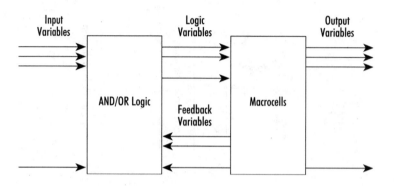

Figure A.33   PLDs Using Programmable Macrocells

The macrocell array provides flip-flops, a means of selecting the polarity of the logic (positive or negative), a method of providing feedback to the logic section, and flexibility in assigning pins as either inputs or outputs. Figure A.34 shows a more detailed presentation of one of the macrocells from an Intel 85C220. Thirteen lines are available for inputs; eight lines can be designated as either inputs or outputs. Eight GAL type logic units provide eight AND gates feeding each of the eight OR gates. The output from the logic may be directed to the output pins or directed to a flip-flop. Feedback into the gate array is available from either the output of the flip-flop or one of the dual-purpose (input or output) pins. Register and feedback usage is controlled by specifying the condition of the two switches. There are provisions for a clock input and an output-enable (tri-state) control for all output pins.

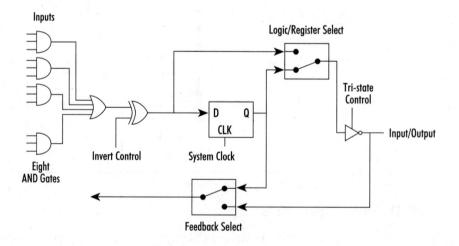

Figure A.34   I/O Macrocell of an Intel 85C220 PLD

Summary

This chapter begins with a description of the logic gates used to implement digital logic, emphasizing the use of two-level AND/OR logic. Both the truth table and Boolean algebra techniques may be used to analyze digital logic circuitry. Given either a truth table or a logic expression, a two-level AND/OR gate can be easily realized. Tri-state buffers, multiplexers, and demultiplexers are also useful design components.

Truth tables provide a unique means for expressing the characteristics of logic circuits. Equality of truth tables implies equivalence of logic expressions and logic circuit implementations. Logic expressions can also be manipulated algebraically using Boolean algebra. Logic designs can be implemented directly from the truth table or logic expression using AND/OR/NOT gates.

The flip-flop is the fundamental memory component in digital systems. Synchronous (clocked) flip-flops are the component of choice in designing logic circuitry that uses memory. Registers composed of flip-flops are basic building blocks for designing more complex digital systems. Clocked synchronous circuitry assures predictable performance with a minimum of concern for the intricacies of timing and electronic circuitry. Logic circuits with memory are finite state machines. The characteristics of a state machine can be described with a state diagram.

Programmable logic devices allow complex logic circuitry to be implemented with very few logic packages. In PLDs large arrays of AND and OR gates can be interconnected (programmed) to produce the desired logic characteristics. With macrocell devices, complex logic circuitry including state machines can be implemented in a single package. Sophisticated design software allows the designer to specify device characteristics and simulate logic designs.

For additional information and references on the topics presented in this appendix, access the World Wide Web page provided at: http://www.ece.orst.edu/~herzog/docs.html.

Key Terms

analysis

asynchronous registers

clock cycle

clock period

commutativity

enabled

fixed

functional completeness

Moore state machine

programmable logic device

register

register file

setup time

state

state diagram

state variable

synchronous registers

synthesis

truth table

two-level logic

Exercises

A.1 Determine if the following sets of digital operations can generate all possible logic functions:

(a) {AND, NOT}

(b) {OR, NOT}

(c) {AND, OR}

(d) {XOR, AND}

(e) {NOR}

A.2 Show how it is possible to make the logical equivalent of an AND, OR, and NOT gate using only a NAND gate.

A.3 Show how it is possible to make the logical equivalent of an AND, OR, and NOT gate using only a NOR gate.

A.4 Verify that any two-level AND/OR logic circuit is equivalent to the corresponding two-level NAND/NAND logic circuit.

A.5 Verify that any two-level OR/AND logic circuit is equivalent to the corresponding two-level NOR/NOR logic circuit.

A.6 Show how a 16-input multiplexer can be used to generate the logic function

$$F = X_1\overline{X}_2 + X_2X_3X_4 + \overline{X}_2\overline{X}_4$$

A.7 Show how four multiplexers with four inputs each can be connected to make a 16-input multiplexer.

A.8 Use a truth table to find the equivalent AND/OR logic for the following:

(a) $(X_1 \text{ NAND } \overline{X}_0) \text{ NAND } (\overline{X}_1 \text{ NAND } X_0)$

(b) $((X_2 \text{ NAND } \overline{X}_1) + X_0)$

(c) $(X_2 + \overline{X}_1X_0) \text{ XOR } (1)$

(d) $X_1 \text{ NOR } (\overline{X}_1 + X_2)$

(e) $X_1 \text{ XOR } X_0$

A.9 Find a (nonsimplified) AND/OR representation for the following.

(a) $Y_0(X_2, X_1, X_0) = \Sigma (0,2,3,4,6)$

(b) $Y_1(X_2, X_1, X_0) = \Sigma (4,5,6,7)$

(c) $Y_2(X_2, X_1, X_0) = \Sigma (0,2,4,5,6,7)$

A.10 Find a (nonsimplified) AND/OR representation for the complement of each function in Exercise A.9.

A.11 Use a truth table to verify the following:

(a) $(X_3 \text{ NAND } X_2) \text{ NAND } (X_1 \text{ NAND } X_0) = (X_3 \bullet X_2) + (X_1 \bullet X_0)$

(b) $(X_3 \text{ NOR } X_2) \text{ NOR } (X_1 \text{ NOR } X_0) = (X_3 + X_2) \bullet (X_1 + X_0)$

A.12 Find the Σ representation for each function in Exercise A.8.

A.13 Use a truth table to verify DeMorgan's theorems.

A.14 Use the Boolean identities from Table A.3 to simplify the following to two-level AND/OR logic with the fewest number of terms.

(a) $\overline{X}_2\overline{X}_0 + \overline{X}_2\overline{X}_1X_0 + \overline{X}_2X_1X_0$

(b) $X_2\overline{X}_0 + X_2X_1 + \overline{X}_2X_1X_0$

(c) $X_2X_1 + X_1X_0 + X_0\overline{X}_2$

A.15 Use Boolean identities to simplify the following:

(a) $Y_0(X_2, X_1, X_0) = \Sigma\,(0,2,3,4,7)$

(b) $Y_1(X_2, X_1, X_0) = \Sigma\,(0,1,4,5)$

(c) $Y_2(X_2, X_1, X_0) = \Sigma\,(0,1,2,3,4,5)$

A.16 Use Boolean identities to simplify the following:

(a) $Y_0(X_3, X_2, X_1, X_0) = \Sigma\,(5,7,13,15)$

(b) $Y_1(X_3, X_2, X_1, X_0) = \Sigma\,(0,4,5,10,11,13,15)$

(c) $Y_2(X_3, X_2, X_1, X_0) = \Sigma\,(1,4,5,6,7,9,13)$

A.17 Use Boolean identities to verify that $X_1X_2 + X_2X_3 + X_3\overline{X}_1 = X_1X_2 + X_3\overline{X}_1$.

A.18 Use Boolean identities to simplify the logic expressions from Exercise A.16. Complement the resulting logic expression using DeMorgan's theorems. Show the resulting OR/AND logic circuit. Comment on the relative complexity of the AND/OR and OR/AND logic representation.

A.19 Assume that the 3-input logic device in Figure A.35 is available in large quantities at a very low price. Is it useful for general-purpose logic design? Explain and justify your conclusion.

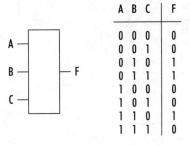

A	B	C	F
0	0	0	0
0	0	1	0
0	1	0	1
0	1	1	1
1	0	0	0
1	0	1	0
1	1	0	1
1	1	1	0

Figure A.35

A.20 The circuit in Figure A.36 is to perform as a synchronous modulo 5 counter. It is to count from 0 through 5 and then repeat back to 0. The count is to advance with each clock pulse. What contents are required for the ROM?

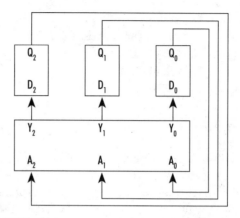

Figure A.36

A.21 Figure A.37 indicate two methods of fabricating a modulo 16 counter. Circuit 1 uses J-K flip-flops that change state on the $1 \rightarrow 0$ transition of the clock. The clock input is obtained from the output of the previous flip-flop in the counter. In circuit 2 additional logic is used to create the inputs to the J and K terminals. Sketch the output of the AND gate as the clock produces eight pulses. Pay special attention to the transient phenomena during state transitions. What conclusions can be drawn?

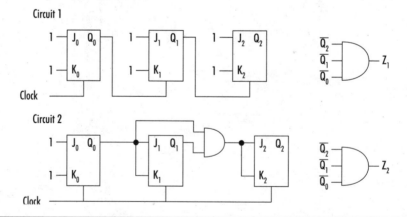

Figure A.37

A.22 The velocity of electromagnetic propagation is approximately 186,300 miles per second. What length of wire can be traversed in one nanosecond (10^9 nanoseconds = 1 second)?

A.23 The flip-flop circuit in Figure A.38 is fabricated from synchronous J-K flip-flops. Analyze the circuit and prepare a state diagram.

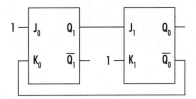

Figure A.38

A.24 A programmable logic array is represented by the diagram shown in Figure A.39. What is the logic function for F_1, F_2, and F_3?

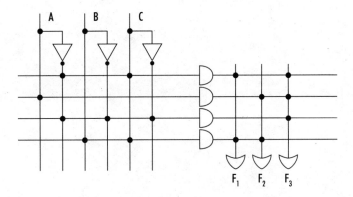

Figure A.39

A.25 In a crossbar switch used in programmable logic:

(a) can an input be connected to more than one output? Why?

(b) can an output be connected to more than one input? Why?

A.26 Design a modulo 4 counter using the macrocell structure of the Intel 85C220.

A.27 Design a modulo 5 counter using the macrocell structure of the Intel 85C220. The device is to have a logic 1 output whenever the count is 1 or 3.

Appendix B
Design For Test

Testing

Devices which one proposes
Seldom turn out as one supposes
At times it turns out
Without any doubt
They badly need fault diagnosis

B.1   Introduction

The complexity of digital systems is increasing at a rapid rate. Automated design tools now permit novice designers to develop complex systems and application-specific integrated circuits (ASICs). The designers are effectively isolated from the details of low-end circuit design and fabrication method. Multiple independent designs are often combined on a single chip.

This enormous increase in the number of active elements in a single package has been accompanied by a corresponding decrease in the accessibility of the devices. The number of access pins for an integrated circuit package has grown at a relatively slow rate. Probing internal circuit elements to determine their viability is no longer a feasible solution.

The ability to test devices and circuits composed of complex devices is now important. It has often been observed that on some products the work required to test a device greatly exceeds the effort expended on the product's initial design.

The cost associated with faulty products is real and substantial. A rule-of-thumb used by designers states that the cost to detect a fault at the chip level is in the range of $0.50; at the board level $5.00; at the system level $50; and in the field $500. These costs are associated with the detection and repair or replacement of faulty components. If the added costs associated with safety issues are considered, the importance of eliminating faulty devices increases dramatically.

The role of testing, once the domain of test engineers or quality-control engineers involved in manufacturing, is now increasingly the responsibility of the design engineer. The concept of device testing must be considered as an integral part of the design process. The device must be designed to allow testing.

This chapter introduces the techniques used in testing digital systems and designing digital systems in a manner in which they can be tested. It covers the relationship of the testing to digital logic and provides an overview of the IEEE 1149 Design for Test standards.

B.2   Faults in Digital Systems

Faults occur in digital systems for many reasons:

- **Design faults.** Design faults are attributable to errors of human judgment made in the design process. They may also be caused by imperfections in computer-aided design, CAD, and automated design environments. Design faults include races, logic hazards, and logic errors. It is also possible that improper components were selected with inappropriate speed or drive capability. Gross logic errors should be detected early in the design process, perhaps using logic simulation.

- **Fabrication faults.** In fabricating the system, errors may exist in the circuit board due to open traces, solder bridges, or cold solder joints. The wrong components may have been used or inserted incorrectly in their solder pads. Fabrication faults are often difficult to detect by optical or manual inspection.

- **Birth defects.** One or more of the components in the system may have been fabricated incorrectly. Although they may perform correctly during initial functional testing, they may fail after a relatively short lifetime. Some of these defects are detected by subjecting components to a "burn-in" period followed by functional testing before assembling devices into larger systems.

- **Component failures.** As a result of aging, exposure to environmental stresses, or random failures, a device which worked at "birth" may fail after a longer period of service in the system. With modern solid-state components, aging and use of devices has only a slight impact on reliability.

Faults may be permanent, intermittent, or transient. Intermittent faults are characterized by periods of proper operation interspersed with periods of improper operation. Transient faults are present only for short periods of time. Intermittent and transient faults present extremely complex obstacles to methodical detection techniques.

There are four approaches to handling faults in digital devices:

- **Fault detection using test vectors.** Existing components often are plagued with faults. These faults may be detected by analyzing the circuitry and determining test signals (vectors) capable of testing for a suspected fault.

- **Design for test.** Design for test (DFT) involves the conscious effort of the system designer to design the system such that it can be tested by conventional means. Such design modifications may involve dramatic increases in the number of components to accommodate testing.

- **Built in self-test.** In some situations a complete testing algorithm can be built into a system. This reduces the need for external components, access points, and connectors. A complete self-diagnosis of the device can be accomplished at regular intervals or at device startup.

- **Fault tolerance.** Fault-tolerant systems are designed in such a way that the performance degrades only slightly in the presence of faults. These systems are useful in situations in which repair or replacement of components is impractical or impossible. Digital systems configured for spacecraft travel lasting many years use fault-tolerant techniques.

B.3 Fault Models

A wide variety of circuit materials and fabrication techniques are used for devices, circuit boards, and interconnecting signal paths. While the specific cause of a fault may have a rather technical description, such as an incorrect time exposure of a layer of electronic material to an etching agent, the effect of the fault is usually to change electronic circuitry from its desired configuration.

A choice needs to be made concerning the level of fault to be analyzed. At the circuit level, the range of possible faults is very large. Voltages may be too high or too low. Resistors and capacitors may have incorrect values. Transistors may have characteristics that are out of tolerance. Fault detection at the circuit level is very difficult and usually not attempted for quality control or production test purposes.

One step up from the circuit level is the gate level. Circuit faults affect the performance of logic gates by introducing a 1 or a 0 at an inappropriate location in the circuitry. From the system and test perspective, it is sufficient to consider that the effect of the fault is to produce errors in logic values. Since functional elements in logic circuitry are gates, faults result in incorrect logic values at the input or output of logic gates or logic circuits. Three types of faults are considered:

- **Stuck-at-zero (s-a-0) faults.** S-a-0 faults may be caused by a variety of conditions such as a "shorted" diode or nonfunctional transistor. The effect of the fault is that the input (or output) is permanently stuck at a value of logic 0.

- **Stuck-at-one (s-a-1) faults.** S-a-1 faults are caused by such events as an open input line to a logic gate which has an internal pull-up resistor. The input (or output) is permanently stuck at a value of logic 1.

- **Bridge faults.** A bridge fault is caused by an unintended connection of two signal conductors, forcing both to have the same logic value. The dominating logic value shared by the two lines depends on the nature of the connected components. With standard TTL logic outputs, for example, two crossed outputs of different polarity will be dominated by the "low" value.

Although there are many potential faults in a device, most analyses are based on a single fault assumption. A fault, if it exists, will be a single fault of the s-a-0, s-a-1, or bridge type. This assumption is justified for several reasons. It is generally not necessary to know how many faults are present; one or more faults require corrective action. Multiple faults will usually be detected by tests designed for single faults. In some cases, however, multiple faults will produce the same test result as a no-fault condition and the test will fail. These situations are normally quite rare.

B.4   Fault Detection in Combinational Logic

This section will cover some of the techniques for detecting faults in combinational logic circuits. Under certain circumstances, some of the techniques may also be useful as a basis of the more complex problem of testing sequential circuitry.

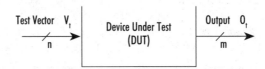

Figure B.1   Testing a Logic Circuit

Figure B.1 shows an input/output model for a logic device under test (DUT) with n inputs and m outputs. Testing is accomplished by applying an n-bit test vector V_t and observing the m-bit output vector, O_t.

Three generalized approaches to fault testing can be used:

- **Exhaustive testing.** The device can be tested exhaustively by applying all possible test vectors to the device input. The corresponding outputs can be compared with the "correct values" as stored in a table. The exhaustive test is straightforward, but may require an unreasonably large number of test vectors to be applied. For example, a device with 64 bits of input, if tested at the rate of one million tests per second, would require over 100,000 years to use all possible test vectors.

- **Functionality testing.** Rather than testing all possible input conditions, a subset of conditions related to the device's intended operation may be tested. For example, if the device is an ALU, a set of operands may be presented as test vectors for each of the available computation functions. The results of these sample computations would be compared with a table of correct results. This approach has the advantage of greatly minimizing the set of test vectors. It is possible, however, that the test will not be sensitive to enough potential fault conditions to be satisfactory.
- **Hardware testing.** Hardware testing involves probing for specific potential hardware faults, such as "stuck at" faults at predetermined locations. Hardware testing allows the test engineer to specifically design the testing procedure to test known problem areas. For example, if detailed investigations have shown that a particular gate inside an ASIC is prone to failure, a test to specifically test that gate can be formulated.

B.5   Logic Techniques for Fault Detection, Boolean Differences

In order for testing to be effective for a given fault, there must be a difference between the output elicited by the test vector for the faulty and fault-free conditions. This observation can be used to generate a set of test vectors for a given situation. The following analysis assumes a single fault. Multiple faults may create situations which are more complex to diagnose.

Example B.1:   *Using a Truth Table to Identify Test Vectors*

Consider the AND/OR circuit shown in Figure B.2. Find test vectors to test for s-a-0 and s-a-1 faults at points A and B.

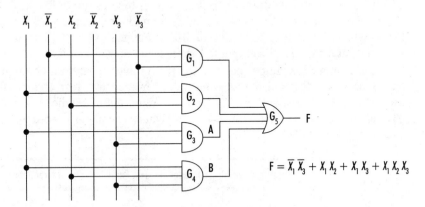

Figure B.2   *An AND/OR Circuit with Potential Faults*

$$F = \overline{X_1}\,\overline{X_3} + X_1 X_2 + X_1 X_3 + X_1 X_2 X_3$$

Solution:

The following truth table includes entries for F, the fault-free logic output. F_{A-0} and F_{A-1} are the logic outputs which would be seen if point A is stuck-at-0 or stuck-at-1 respectively. F_{B-0} and F_{B-1} are the corresponding s-a-0 and s-a-1 outputs associated with point B.

$X_1X_2X_3$	F	F_{A-0}	F_{A-1}	F_{B-0}	F_{B-1}
000	1	1	1	1	1
001	0	0	1	0	1
010	1	1	1	1	1
011	0	0	1	0	1
100	0	0	1	0	1
101	1	0	1	1	1
110	1	1	1	1	1
111	1	1	1	1	1

Table B.1

Consider first the problem of determining if an s-a-0 or s-a-1 fault exists at point A. To detect a fault, there must be a difference in the response to a test input for the faulty and fault-free conditions. For an s-a-0 fault the outputs for F_{A-0} and F are compared. They differ only for the condition in which $X_1X_2X_3 = 101$. Test vector 101 is applied to the DUT. If the output is 0 there is an s-a-0 fault at point A. If the output is 1 there is not an s-a-0 fault at point A.

To test for a s-a-1 fault at point A, the truth table is searched for situations in which an s-a-1 fault produces a different output from a fault-free device. Test vectors 001, 011, or 100 can be used. For example, test vector 001 is applied. An output of 1 indicates an s-a-1 fault; an output of 0 indicates there is no s-a-1 fault. The device can be completely tested for faults at point A by submitting test vectors 101 and 001.

Fault Test	Test Vector	Test Result	Interpretation
A stuck-at-zero	101	0 1	Possible s-a-0 No s-a-0
A stuck-at-one	001	0 1	No s-a-1 Possible s-a-1

Table B.2

A different result is obtained when trying to test for an s-a-0 fault for point B. The truth table for the faulty condition is identical to the truth table for the s-a-0 condition. There is no test vector that will detect this condition. It may be easily verified that the AND gate associated with point B is redundant; it can be eliminated without changing the logic function of the circuit. By similar reasoning, test vectors 001, 011, and 100 can be used to test for an s-a-1 condition at point B. Note that an s-a-1 condition at point A produces an identical result to an s-a-1 condition at point B. It is not possible to distinguish between them without establishing greater observability of the test points.

Fault Test	Test Vector	Test Result	Interpretation
B stuck-at-zero	None	— —	No Test No Test
B stuck-at-one	001, 011, 100	0 1	No s-a-1 Possible s-a-1

Table B.3

This technique, in which there is a search for all conditions in which the logic output for the faulty and fault-free conditions are different, can be expressed mathematically as the **Boolean difference** technique. The set of test vectors may be found algebraically by writing logic functions for both F, the fault-free condition, and F_f, the logic equation produced by the fault being considered. The EXCLUSIVE OR of F and F_f can then be evaluated to produce a list of test vectors. Each is associated with a condition in which a difference exists between the faulty and fault-free devices.

Example B.2:   Boolean Differences

Use the method of Boolean differences to find the set of test vectors associated with detecting an s-a-0 fault at point A in Figure B.2.

Solution:

First, determine the logic equations describing the system in the fault-free case and the case with point A s-a-0. These are F and F_{A-0} respectively. With point A s-a-0, one of the AND gates is effectively removed from the logic.

$$F = \overline{X_1}\overline{X_3} + X_1 X_2 + X_1 X_3 = \overline{X_1}\overline{X_3} + X_1 X_2 + X_1 X_3$$

$$F_{A-0} = \overline{X_1}\overline{X_3} + X_1 X_2 + X_1 X_2 X_3 = \overline{X_1}\overline{X_3} + X_1 X_2$$

The two logic functions are then XORed to identify the required test vectors.

$$F \oplus F_{A-0} = \overline{F} F_{A-0} + F \overline{F_{A-0}} = 1$$

$$(\overline{\overline{X_1}\overline{X_3} + X_1 X_2 + X_1 X_3}) \, (\overline{X_1}\overline{X_3} + X_1 X_2) + (\overline{X_1}\overline{X_3} + X_1 X_2 + X_1 X_3) \, (\overline{\overline{X_1}\overline{X_3} + X_1 X_2}) = 1$$

After performing logic minimization using Boolean algebra, a single test vector emerges.

$$X_1 \overline{X_2} X_3 = 1$$

The result shows that the single test vector $X_1 X_2 X_3 = 101$ will test the specified condition. This agrees with the previous analysis.

The test vector for point B s-a-1 can be determined by taking the XOR of F and F_{A-1}. The solution is left for an exercise.

B.6   Controllability and Observability

Examples B.1 and B.2 illustrate the basic principles that are involved in fault detection techniques. It is necessary to submit a prescribed set of logic values to a component. The logic values, or test vector, is chosen to drive a logic gate to a prescribed condition. This concept is known as **controllability**. It is also necessary to observe system response to a test vector in order to compare it to a fault-free value. This concept is known as **observability**.

In relatively simple circuits using random logic components, all of the critical points that must be driven or sensed are readily accessible through probes or test points. In other situations, extensive amounts of circuitry may be internal to an integrated circuit and not directly accessible. Adding additional pins to the circuitry to increase the accessibility of test points is prohibitively expensive. In these situations the issues of controllability and observability are much more restrictive and must be solved by other methods.

Two general approaches are used. The first, **path sensitization**, attempts to create a input path from an access point on the circuit to a fault point for the purpose of driving the internal logic. This assures controllability. A sensitized output path is also created from a critical point to an access point for the purpose of sensing the internal condition. This provides observability.

If the internal circuitry is reasonably complicated, it may not be possible to assure controllability and observability without modifying the circuit or the way it is packaged. Additional I/O pins, for example, may be added to gain greater access to the circuitry.

A second, more methodical, approach involves **scan paths** to increase the controllability and observability of a complex circuit. A scan path is a long shift register which can be used to transport information in bit serial form to critical internal locations; this provides controllability. Similarly, the logic values of internal locations can be transferred via the shift register out the serial port for external examination. The shift register acts as a shuttle train for transporting information to and from internal locations. The serial data format minimizes the need for large numbers of extra data access I/O pins.

B.7   Scan Registers

The serial scan register, SSR, attempts to create a device that allows access to many internal test points by means of relatively few access points. This is accomplished by organizing test points, for which controllability and observability are required, into a long data stream using scan registers.

A serial scan register is a shift register with a single access point and single egress point from the DUT. Each of the component flip-flops, is associated with an internal data location in the DUT. Controllability is gained by shifting information into the device through the serial port and propagating it to the required internal location of the DUT. Observability is gained by loading internal logic values into the shift register and propagating them to the egress terminal where they can be examined.

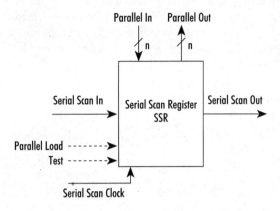

Figure B.3   A Serial Scan Register

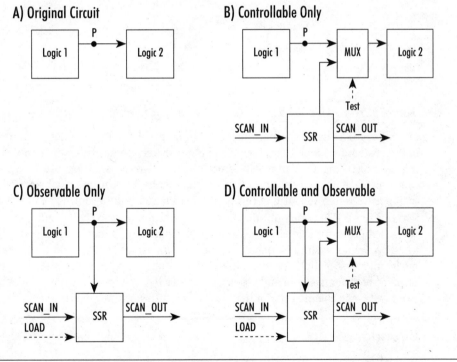

Figure B.4   Increasing Controllability and Observability with a Serial Scan Register

There are three possible internal conditions in which changes in the inherent controllability and observability of the device may need to be made. Figure B.4A shows point P, an internal point embedded in logic. P is not directly attached to an input/output pin and is neither controllable nor observable.

In Figure B.4B an SSR provides a known logic input into the block of logic labeled Logic 2. A multiplexer is used to switch between the normal mode, in which the logic value comes from Logic 1, and the test mode, in which the logic value comes from the SSR. In this circuit the input to Logic 2 has been made controllable but not observable.

In Figure B.4C, the parallel input to the SSR from point P allows P to be observable but not controllable. In Figure B.4D both features are combined to create a point which is both controllable and observable.

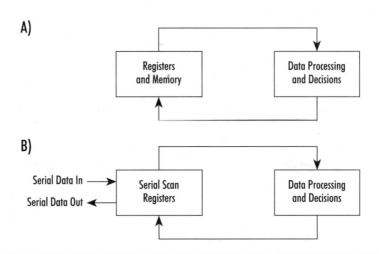

Figure B.5   Replacement of Register Flip-Flops with Serial Scan Register Elements

Figure B.5 shows another approach to the problem. Figure B.5A indicates the decomposition of the system into a block involving the register and memory elements and a second block including the processing elements. Data processing involves presenting data from the memory elements to the processing elements and storing the results back in memory elements. The memory elements are often in the form of registers. For simplicity, data routing elements are not included.

In Figure B.5B, each flip-flop in the register and memory elements is replaced by a modified serial scan register that can function in two modes. In the normal mode, the device functions as a system register with its actions controlled by its control inputs and the system clock. In the test mode, all register flip-flops are connected as a long serial scan register. This makes each memory element both controllable and observable. Since logic between memory elements is usually limited to a few layers, the majority of the internal circuitry can now be tested.

B.8   Boundary Scan

Scan techniques can also test circuit boards that mount and connect circuit components. If all circuit output ports are controllable and all circuit input ports are observable, the interconnecting wiring can be tested. In Figure B.6, serial scan register devices are placed at all input and output pins of the device. Prescribed patterns of 1s and 0s are placed at the output ports via the scan registers. Patterns of 1s and 0s can be sensed at the input ports of other devices. Analysis of the serial shift data can disclose the presence of open circuits and bridges. Boundary scan can also be used to test the logical correctness of "glue" logic on the circuit board.

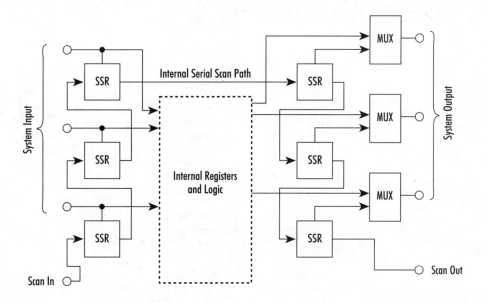

Figure B.6   Use of Boundary Scan Devices to Detect System Interconnection Faults

B.9   Built-in Self-test (BIST)

Proprietary specialized tests performed and analyzed by the device controller are an alternative to using external stimuli to test digital logic devices. For example, immediately following power-on, the controller may initiate a series of actions to test the viability of the major subsystems of the device. All test vectors are generated internally and hence are not subject to the constraints imposed by inadequate device-access points.

Periodically, the controller may again initiate testing procedures, even while the device is in operation. Inactive subsystems that are not performing task algorithms may be tested. This allows algorithm execution and parallel actions to test components. A skilled designer can specify algorithms that continuously perform self-testing operations with negligible loss of system performance.

When faults are detected during self-test, special algorithms may be initiated to either safely turn the system off or to notify a human operator via an externally observable operation. Fault detection may also be used to deactivate units and switch in redundant spares.

B.10   IEEE 1149.1 Testing Standard

Because of the importance of standardized testing procedures, IEEE has approved a standard based on the work of the Joint Test Action Group (JTAG). The standard is concerned with integrated use of a variety of testing procedures and a standard bus and protocols that can be used to implement them. IEEE 1149.1 provides a framework in which boundary scan can test board interconnections, scan registers can test controllability and observability of logic internal to chips, and BIST can test specialized test procedures.

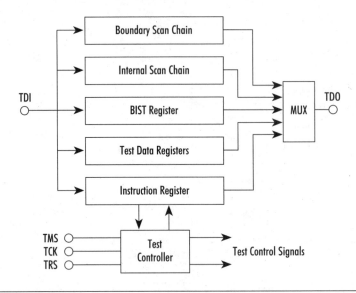

Figure B.7   Major Features of the IEEE 1149.1 Test Architecture

Figure B.7 shows a simplified view of a system designed according to IEEE 1149.1 standards. Only the major concepts will be covered in this brief presentation.

In IEEE 1149.1, five access pins are dedicated to a Test Access Port (TAP). Four of the pins are mandatory. TDI and TDO are the Test Data Input and Test Data Output serial access pins. TCK is the clock for the serial data. TMS is an input for selecting either the "normal" action mode or the "test" mode for the device. An optional pin TRS may be specified by the designer to initialize testing circuits.

Upon entering the test mode, an instruction is sent to the instruction register via the TDI pin. The instruction tells the controller to perform a specified test function. Additional data might be strobed into the boundary scan chain, internal scan chain, or the test data registers. A test algorithm is executed under the control of the test controller. The test may involve boundary scan activities, internal scan activities, or an internal self-test. Results of the test may be stored in the test data register. At the completion of the testing, results may be serially removed via the TDO output pin.

While the IEEE 1149 test procedure is controversial because of its complexity, several implementations have suggested that it is a workable standard and produces its intended results. The cost of the test circuitry is usually in the range of 5% to 25% of the available chip area. Data General Corporation recently reported the implementation of standard test-access architecture based on IEEE 1149 which detected 98% of stuck-at faults on devices with 10,000 to 50,000 gates with an overhead of less than 5% of the available chip area.

Key Terms

Boolean difference
boundary scan
built in self test
controllability
design for test
observability
path sensitization
scan paths
stuck-at-one
stuck-at-zero

Readings and References

Abramovici, M., M. Breuer, and A. Friedman. *Digital Systems Testing and Testable Design.* New York: Computer Science Press, 1990.

For additional information and references on the topics presented in this appendix, access the World Wide Web page provided at: http://www.ece.orst.edu/~herzog/docs.html.

Exercises

B.1 In the 2-level AND/OR network of Figure B.8, assume there is no more than one fault. Use a truth-table analysis to determine a test vector for each of the following possible faults:

(a) Point A stuck-at-0.

(b) Point B stuck-at-1.

(c) Point C stuck-at-0.

(d) The two inputs to the center AND gate are bridged with a low signal dominating.

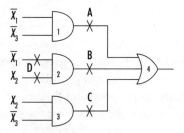

Figure B.8

B.2 Repeat Exercise B.1 using Boolean differences to determine a test for the four conditions.

B.3   In the 2-level NAND/NAND circuit in Figure B.9, assume there is no more than a single fault. Use a truth-table analysis to determine a test vector for each of the following possible faults:

(a) Point A stuck-at-1.

(b) Point B stuck-at-0.

(c) Point C stuck-at-1.

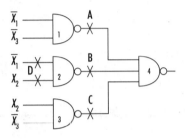

Figure B.9

B.4   In the logic diagram shown in Figure B.10, use a truth table to determine a test for a stuck-at-0 fault at point A. Find a test for a stuck-at-1 fault at point B.

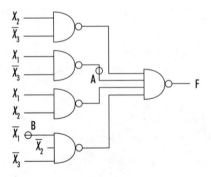

Figure B.10

B.5   In the circuit of Figure B.9, is it possible to distinguish the following faults:

(a) Point A stuck-at-one and point B stuck-at-one?

(b) Point A stuck-at-0 and point B stuck-at-0?

B.6   The circuit in Figure B.11 has been tested and the truth table determined experimentally. If there is a single fault at either point A, B, or C, what is it?

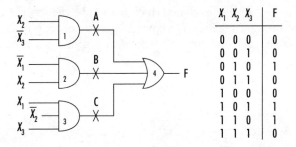

X_1 X_2 X_3	F
0 0 0	0
0 0 1	0
0 1 0	1
0 1 1	0
1 0 0	0
1 0 1	1
1 1 0	1
1 1 1	0

Figure B.11

B.7 Use the method of Boolean differences to determine if point B is stuck-at-1 in Figure B.2.

B.8 Determine a test for each of the following for the circuit in Figure B.12:
    (a) Point A stuck-at-1.
    (b) Point B stuck-at-0.
    (c) Point C stuck-at-1.

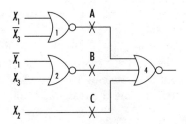

Figure B.12

B.9 Can the response of a network with two faults be determined if the response of a network to each of the faults individually is known? Verify by example.

B.10 Is it possible that a network with two faults might behave identically to the fault-free network? Give an example to support your conclusion.

B.11 An IC package contains 3 NAND gates, each with three inputs. One of the gate inputs on gate 3 is known to have a stuck-at-1 fault. Does the package have any further use as a logic circuit?

B.12 Two integrated circuit chips, A and B, are connected by copper traces on a printed circuit board. Five of the pins on each chip are interconnected to a scan register. Each chip has a separate scan register. Design a test to determine if the connections are correct for the printed circuit board under the following interconnection patterns:

(a) $A_1 \rightarrow B_1$
    $A_2 \rightarrow B_2$
    $A_3 \rightarrow B_3$
    $A_4 \rightarrow B_4$
    $A_5 \rightarrow B_5$

(b) $A_1 \rightarrow B_1, B_2$
    $A_2 \rightarrow B_3$
    $A_3 \rightarrow B_3$
    A_4, B_4 open circuit
    A_5 tied to + power supply
    B_5 tied to ground

B.13 There are many possible methods of implementing design-for-test procedures. What advantage is there in using IEEE 1149?

B.14 In the context of fault detection, what is controllability and observability? Why are they important concepts? Describe two ways to achieve controllability and two ways to achieve observability in a digital device.

B.15 In Figure B.2, assume that the circuitry lies deep inside an IC package and points X_1, X_2, X_3, and F are not available through pins. How might serial scan techniques be used to determine if gate 5 is stuck-at-1 (s-a-1)? Indicate the components that would need to be added to the circuitry and the test you would use. Assume that no other components need to be tested.

Glossary

Access time The period of time from the start of a read operation until the requested information is available.

Accumulator A register in a computer which is extensively used as a source or destination of information for an arithmetic and logic unit.

Address mode The method of translating the address field of an instruction to the actual memory location of instructions or data. *See also specific types of address modes.*

Address space The range of addresses that can be referenced by a digital computer. *See also specific types of address space.*

Amdahl's law An algebraic relationship which predicts the speedup possible in a parallel system based on the percentage of operations which must be performed sequentially.

Amplitude shift keying (ASK) A digital modulation scheme in which the two binary elements are represented by carrier frequency amplitude level.

Analytical Engine A mechanical device designed by Charles Babbage in 1856. Although never built, this was the first stored-program digital computer. Babbage's ideas came to fruition almost one hundred years later, when electronic components provided the speed and reliability necessary for full implementation.

Array multiplier A multiplier composed of an array of cells which combine summands to produce a product.

Arithmetic and logic unit (ALU) A logic device that can calculate arithmetic and logic operations using two data words.

Assembler A computer program that translates assembly language source code to executable machine code. The assembler also detects some programming errors and provides documentation for the code.

Assembly language A computer language with a one-to-one correspondence to the machine language of a computer. Assembly language allows symbolic addressing and the use of mnemonics to make programming easier and less subject to errors.

Associative cache A cache memory system in which cache lines may be stored in any storage location. A search of the cache requires a parallel search of all storage locations to determine if there is a hit. This is normally accomplished using a content addressable memory.

ASCII (American Standard Code for Information Interchange) An 8-bit binary code used extensively for encoding letters, numerals, and standard punctuation symbols.

Asynchronous signals Signals which start and stop at times not synchronous with the system clock.

Automatic repeat request (ARQ) A class of communication protocols used in digital communication. With ARQ, source transmission is halted until an acknowledgment of the previous packet is received.

Available address space The range of addresses which exist in the memory hierarchy of the computer.

Baseband encoding Modulation of a communication signal in which the two binary elements are mapped to two constant voltages.

Biased numbers A number system in which all of the magnitudes have a numeric offset from standard binary representation. The bias is usually a power of two. Biased number systems allow a smoother transition between positive and negative numbers. They are commonly used to represent exponents in floating-point number systems.

Binary-coded decimal The available bit positions are divided into groups of 4 bits. Each grouping is used to represent one decimal digit in a weighted decimal representation.

Binary switch A contact closure switching device with two input terminals and two output terminals. The connection between input and output is "straight through" or "crossed" depending on the control signal applied to the device.

Bit serial processor A processor in which the bits of the operands are introduced sequentially one bit at a time. These processors usually require few computational devices but process data much slower than parallel processing units.

Bit slice processor A vertical slice of a processor. Each bit slice unit contains a fixed number of bits (usually 4) of multiple registers, interconnection paths, and computational elements. The individual slices can be interconnected to service data words of arbitrary widths.

Bit stuffing A technique used in synchronous serial communication which modifies patterns of data bits so that they do not appear as special delimiter characters (flags). Bit stuffing modifies the message so that the pattern "01111110" does not appear.

Blocking A switching network is blocked if the use of one transmission path through the network between network nodes prevents a second transmission to otherwise free nodes.

Broadband encoding Modulation of a communication signal in which the two binary elements are used to control the amplitude or phase of a high frequency carrier signal.

Bus A set of wires used to transmit information between multiple components in a digital system. Connection to the bus is through a tri-state buffer.

Bus master A device which supervises the access and use of a shared data bus.

Butterfly An array of binary switches that allows connection between any source node and any destination node.

Cache memory A small high-speed memory that holds a portion of the main memory contents. When information is found in the cache, the acquisition time is much shorter than the time required to access the main memory.

Carry completion adder An adder that uses characteristics of the operands to determine the maximum number of carry ripples that will occur. On average, this shortens the processing time delay associated with ripple carries.

Carry look-ahead adder This device uses additional logic circuitry to determine the presence of a carry signal for each adder stage. Carries do not ripple from stage to stage. Addition time is significantly reduced.

Carry save adder This device is useful when a large number of operands are to be added. Carries generated during the addition of each operand are saved and applied as the next operand is introduced. This eliminates carry rippling for all except the last operand to be added.

Cascade processing Processing elements with no storage elements are arranged in sequence. Information enters the first processor element. Output from the first element is applied to the second without intermediate storage. The output from the last processing element is placed in a register. The total processing time may be longer than one clock cycle and require special attention from the controller.

Catenation The process of merging parts of several data elements to create a new element.

Central processing unit (CPU) A vague term originally used to indicate the central core of a digital computer, including registers, ALU, and the controller. It often refers to anything from a microprocessor to a mainframe computer.

Centralized protocol A single master unit makes all decisions in implementing communications protocol.

Circuit switching A method of establishing a communication path in a digital communication structure. In circuit switching, an end-to-end path between source and destination is established and maintained for the duration of the communication session.

CISC An acronym for complex instruction set computing. This design philosophy emphasizes a powerful and complex set of instructions that provides strong support for higher-level languages. CISC computers use multiple addressing modes, instructions of different lengths, and a microprogrammed controller.

Clock A timing signal which provides synchronization for all data processing and storage operations in a digital system.

Clock cycle The time period between clock pulses. The clock cycle starts after the active transition of the system clock and ends with the next active transition.

Clock skew A timing lag in a clock pulse caused by propagation delay or transmission through logic devices. Clock skew should be avoided since it can cause complications in the performance of sequential digital circuitry.

Closely coupled A parallel computing structure in which a shared memory is used to pass information between cooperating processors.

Collision A collision results when two or more nodes of a communication system simultaneously attempt to transmit information on a bus or other shared media. Data is corrupted and must be retransmitted.

Compiler A computer program which translates a high level computing language, such as C or FORTRAN, into machine language that can be run on the target computer. The compiler also provides additional services to the programmer such as checking for syntax errors.

Content addressable memory (CAM) A memory in which data is accessed using a portion of stored contents as a key.

Control point A terminal of a processing, storage, or data routing device that allows a control signal to affect the processing or routing action.

Control signal A signal generated in the control unit that directs and sequences processing, storage, and routing actions in the datapath.

Controller Portion of a computing structure that directs the processing algorithm of the datapath by monitoring status signals and generating control signals in the proper sequence.

Controller state diagram A graphical representation of the operation of the controller. It displays the generation of control signals and the response to status signals.

Crossbar switch A pattern of crossing horizontal and vertical bus elements. Each intersection point allows a coupling for data transmission between source and destination elements.

CSMA/CD An acronym for carrier sense media access/collision detection. This is an algorithm for a contention-based media access protocol. Transmission units wait for a free period on the media before transmitting their message. If a collision is detected, transmission is halted, the unit waits a period of time, and then tries again.

Cycle stealing A form of direct memory access used in digital computers. A device wishing to use the computer external memory and data buses waits until the computer provides a signal indicating that the buses are free for the next machine cycle. This allows the external device to effectively share the computer buses without reducing computer performance.

Cyclic redundancy check (CRC) A method of creating an error control field to protect a digital information packet. The field is created at the sending end and transmitted along with the message. The data is analyzed as it is received to assure that it agrees with the CRC field. Errors require retransmission of the data packet.

Data dependency A condition in a pipelined instruction processor which occurs when data produced by one instruction is required by a second instruction. If the instructions are close together in the pipeline, results from the first may not be available for the second. This can cause a pipeline stall.

Data flow computer A computational structure in which data processing activities are scheduled and performed based on the availability of data.

Data memory The portion of computer memory used for data storage.

Datapath The data processing portion of a computing structure. The datapath includes buses, data storage, and switching and data processing devices. The datapath is paired with a controller to make a complete computing structure.

Data register A group of flip-flops that can load, store, and output a data word.

Data routing devices Switches, multiplexers, demultiplexers, and other logic components that route information from sources to destinations.

Decision block Element of a microoperation diagram that specifies the effect of status variables on the processing action.

Decision elements Logic elements that process data to produce yes/no status information used by the controller to sequence the processing algorithm of a computing structure.

Delayed jump A jump instruction for a digital computer that is delayed in its execution to allow an instruction queue to be vacated and filled with instructions from the jump location. This allows higher performance by relieving the processor of flushing an instruction queue.

Demultiplexer A logic device with one input and multiple outputs. The variable at the input terminal is routed to an output terminal specified by a selection code.

Design The process of specifying the requirements for a device and then implementing a device fulfills those specifications. Design is a creative activity that combines aspects of technical knowledge with constraints imposed by time and economics.

Difference Engine A computational device designed in 1823 by Charles Babbage. It was designed to compute and print the values of polynomials using finite difference techniques. Its application was intended for tide tables, logarithms and other polynomial calculations.

Direct addressing A method of obtaining information for a computer instruction in which the address of the data location is presented as part of the instruction.

Discrete logic A term used to describe an assortment of individual gates that are assembled to perform a logic function.

Distributed protocol The nodes of the network share responsibility in implementing communications protocol.

Dual port memory A memory in which two independent access ports are provided. This device is useful for exchanging blocks of information between two independent devices.

Full duplex A communication channel in which information can flow in two directions simultaneously.

Half duplex A communication channel in which information can flow in two directions but not simultaneously.

Dynamic branch A branch in a computer program in which the branch location(s) cannot be predicted in advance. This usually occurs when the branch address is a computed result stored in a register. Instruction queues are difficult to manage when dynamic branching is used.

Dynamic random access memory (DRAM) A RAM that uses a charged capacitor as a memory element. Although the circuitry is simpler and less expensive than an SRAM, it requires periodic refreshing to maintain its contents.

Efficiency (memory hierarchy) Ratio of the time to access a data element from main memory to the average time to access the information if a multi-level memory hierarchy is used.

Erasable programmable read-only memory (EPROM) A ROM memory device that can be programmed by the user, erased, and reprogrammed as necessary. Erasure is usually performed by exposing the circuitry to an intense ultraviolet light source.

Ethernet A common name for a network which uses the CSMA/CD protocol.

Execute cycle In simple digital computers, instruction execution requires a fetch cycle to acquire the instruction and an execute cycle to perform the specified operation.

Fetch cycle *See execute cycle.*

Fixed point number A binary number representation in which the location of the binary point is in a fixed position in the word. Normally the binary point is assumed to follow the least significant bit. This creates an integer representation.

Flash memory A ROM memory device that may be erased and reprogrammed while in use in an application. Flash memories are well suited for upgrading programs within a working system. The reprogramming act is slower than a data write to a memory. Memory contents are non-volatile.

Floating-point numbers A binary number representation that includes fields for a mantissa to express a numeric magnitude and an exponent. Floating point numbers can represent a much wider range of values than possible with fixed point representation.

Flood routing A data routing technique used in computer networks. Data packets are dispatched from source nodes to all available destinations. This process is repeated at each node and creates a flood of duplicate packets. Delivery at the destination is assured if a pathway exists. While highly reliable, this technique wastes system resources by transmitting multiple duplicate packets.

Flow control A method to limit the rate of data transmission by a source node. This is necessary if network resources or the destination node is not capable of absorbing information at the rate it is produced by the source.

Flynn's classification A method of characterizing computing structures based on their ability to process multiple sets of independent data and their ability to perform multiple independent processing actions.

Fully connected network A communication network in which every node is connected to every other node by a dedicated communication path.

Generator polynomial A specially selected polynomial which is used as part of an error protection algorithm. The generator polynomial processes digital data to create a frame check sequence that is transmitted with the data packet.

Handshake An interlocked set of signals that are used to synchronize the movement of data between two nonsynchronized digital systems.

Hardwired control A method of implementing a control unit that utilizes digital hardware components such as gates and registers. These gates are faster than ROM components used in microprogrammed control. Hardwired control, because of its higher performance, is used in RISC computing structures.

Harvard architecture A computer architecture in which the data memory and program memory are physically distinct.

HDLC An acronym for high level data link control. It is a standardized packet format used in synchronous digital data communication systems.

Hit (cache hit) A memory access in which the desired information is found in the cache.

Hit ratio Ratio the number of successful memory access attempts to the total number of access attempts. A hit ratio can refer to acquisition from a cache memory or acquisition from the main memory.

Hop A node to node data transmission. Source to destination data transmission may require multiple hops through intermediate nodes.

Host A computer connected to a network.

Hypercube An interconnection structure for processors in a parallel computing structure. In a hypercube of degree n there are 2^n processors. Each has a communication link to n other processors.

IEEE 488 An IEEE standard for an instrumentation bus. The bus uses eight control lines and eight data lines. Data is asynchronously transmitted byte-by-byte using a three line handshake.

IEEE floating point An IEEE standard for representing floating point information. Short real format uses 32 bits of information. Long real format uses 64 bits of information. The standardized representation allows for the design and fabrication of special-purpose hardware devices to process floating point data.

Immediate addressing A computer addressing mode in which the data is presented as part of the instruction.

Indexed addressing A computer addressing mode in which the target address is obtained by adding an address field provided by the instruction to the contents of an index register.

In-place processing A processing structure in which processed data sets are returned to their original register locations after each processing action of an algorithm that requires multiple sequential processing actions.

Initial condition block A component of a microoperation diagram. This block specifies the initial conditions of registers following system initialization.

Input/output processor (IOP) A specialized processor which is optimized for performing input and output operations. IOPs commonly use direct memory access to acquire computer system buses, access information from computer memory, and perform input or output operations. IOPs off-load input/output operations from the main processor and improve system performance.

Instruction pipeline A method of processing instructions in which the total processing action is divided into several cycles. Each cycle is performed by one element of the instruction pipeline. This method is used in RISC processors to greatly increase the rate at which instructions are performed.

Interrupt An external signal applied to a computer which causes a variation in program execution. The original program is halted (interrupted) and the program counter is saved on the stack. The interrupting source is identified and a special program, the interrupt service routine, begins execution. At the conclusion of the service routine the program counter is restored and the original program resumes.

Iterative processor A processor that uses an algorithm that requires multiple successive processing activities to determine the final result.

Key The search criteria for an element in memory. In a content-addressable memory, the key may be the contents of any group of bits in the word.

Latency time The time required for the physical movement of magnetically-stored data to the read head. The latency is relatively short for high-speed disk memory. It may be quite long for digital tape storage.

Least recently used replacement algorithm The block to be replaced has the longest period of time since last use.

Line A set of several consecutive data words that are moved to and from main memory and managed as a group in a cache memory.

Linear memory A memory space that is managed without the use of segment pointers. Linear memory requires internal data registers with bit lengths adequate to address the logical memory space.

Linker A computer program that works with code modules produced by an assembler. It accepts multiple code modules, resolves any address conflicts, and creates a single machine language program.

Local area network (LAN) A communication network under single ownership which interconnects nodes within a 1 KM radius.

Logic delay The delay associated with signals as they pass through logic components.

Logical address space The range of addresses that can be addressed by a digital computer. This is usually related to the number of bits that can be specified for an address.

Loosely coupled A parallel computing system in which information is transferred between processing units using input/output or communication ports.

Machine language Computer code which may be directly loaded into a computer for execution. There are no comments or mnemonics. All information is in the form of {1,0}.

Machine cycle A portion of the operation set required to process a computer instruction. Simple computers may use a fetch cycle to acquire an instruction and an execute cycle to perform

the specified operation. RISC computers generally decompose an instruction into four or five processing actions that can be pipelined.

Macrocode The native instruction set of the computer. The term macrocode is used to distinguish between the code used by programmers (macrocode) and the code used to implement the control unit (microcode).

Manchester encoding An encoding system used in synchronous serial communication. The two binary data elements are encoded such that there is a low-to-high transition in the middle of the bit period for a "0" and a high-to-low transition in the middle of the bit period for a "1." The mid-period transitions allow clock and timing information to be extracted from the signal.

Memory The portion of a computational device used to store information, data, or instructions. *See also specific types of memory.*

Memory coherency When multiple memories are used in a memory hierarchy, for example a cache memory and a main memory, it is important to assure that the contents of the two are the same (or coherent). This action is normally performed by the cache controller.

Memory hierarchy A composite memory composed of a fast but relatively small main memory and a large but slow secondary memory. Through proper management and page swapping, the combination memory maintains an access time close to that of the main memory and a size equal to that of the large memory. A cache memory may also be included in the hierarchy.

Memory mapped input/output An input/output technique in which input and output ports are assigned addresses within the memory space of the computer. Data is accessed using the same instructions for memory read and write.

Mesh An interconnection topology for nodes in a digital communication network. There is no regular pattern of interconnection between network nodes.

Message switching A method of establishing communication between nodes in a digital communication system. With message switching, messages may be transmitted to intermediate nodes where they are stored and relayed toward the destination node.

Microcode Contents of a ROM in a microprogrammed control unit. The pattern of bits are divided into multiple fields. One field provides the control bits for the datapath. Other fields specify different aspects of the controller activity such as jump addresses and the selection of status variables.

Microcode compaction The process of compressing microcode so that it occupies the least possible amount of storage space.

Microcontroller A microprocessor designed to primarily perform control related activities. Special hardware support including counters, timers, enhanced interrupt capability, and communication ports are often included on the chip.

Microinstruction A group of register transfer operations performed concurrently.

Microoperation A register transfer operation performed in the datapath of a computing structure.

Microoperation diagram A graphical presentation of a processing algorithm in which each block specifies the operations to be performed in a specific clock cycle.

Microprogrammed control A control unit that uses microcode from a ROM to control the processing algorithm sequencing and control signals generation.

Modem A modulator-demodulator used to connect computers to the telephone system. Binary signal elements are encoded (modulated) as tones at the sending end. The tones are transferred back to logic signals (demodulated) at the receiving end.

Multibus An IEEE standard backplane interface. It is used to add special purpose cards for data acquisition, communication, or other purposes to an existing computer.

Multiplexer A logic device with several input terminals and one output terminal. A control signal specifies the output port that is to receive the contents from an input terminal.

Multiprogramming Running multiple independent computer programs on a single computer.

Multitasking Decomposing a single computational activity into tasks. Multiple tasks may then be run on the parallel structures of the computer. For example, integer activities may be run on an integer processor while a floating-point computation is performed on a floating-point unit.

Nanoprogramming Unique microcode words are identified and encoded using a binary code. The encoded information is stored in the microprogram ROM. A nanoprogram ROM is used to convert the nanocode to microcode. In some applications a net decrease in stored ROM space is achieved.

Network interface unit (NIU) A device that performs network services for a host computer connected to a network. Services might include error control, network access, and flow control.

Newton-Raphson technique An iterative technique for performing algebraic computations to determine the root of an equation.

Next state logic Logic used to generate the next state for a state machine controller.

Nibble One-half of a byte; four bits.

Node An element of a communication system formed by the combination of a host and a network interface unit.

Offered load A measure of the inter-network traffic in a communication system. The offered load includes messages which pass successfully from source to destination as well as those that are repeated due to errors.

One-hot controller A particularly simple form of control unit in which one flip-flop is associated with each block of the microoperation diagram. A single logic 1 is circulated among the flip-flops under the guidance of logic and the status signals.

Operand forwarding A technique used in RISC processors to avoid pipeline stalls. Operands produced during the execute cycle are forwarded immediately to the following instruction. The "write back" cycle is avoided.

Overlay Swapping of data between a main memory and a secondary memory (disk) that is performed under control of the programmer. High performance is possible, but the code becomes very tuned to particular memory configuration.

Output logic The logic in a controller that produces control signals as a function of the current state.

Packets A message element used in synchronous serial communication systems that usually contains hundreds of bits of information along with separate fields for packet addressing and error control.

Packet switching A form of message switching in which long messages are first broken into packets before transmission.

Page Pieces of memory swapped as a unit between primary and secondary memory.

Parallel efficiency The average percent of time in which processors are used in a parallel processing algorithm.

Parallel processing The simultaneous use of multiple processing units to perform a processing algorithm.

Phase shift keying A digital modulation technique in which the binary elements are represented by multiple phases of the carrier frequency.

Physical address space The range of addresses that exists in the main memory of a digital computer. The range can be accessed without memory swaps to the memory hierarchy.

Pipe segment A portion of a processing pipeline.

Pipeline processing An assembly line approach to performing processing actions on multiple data sets. Each pipe segment performs a processing action on data from a different data set. After the pipeline is filled, a new result is produced in each machine cycle.

Pipeline stall An event in which all processing actions in a pipeline are delayed to allow one stage of the pipeline to complete its action.

Polling An action associated with an interrupt in a digital computer. Following assertion of an interrupt, the computer polls possible sources to determine the source of the signal. Once identified, the proper service routing can be initiated.

Pop The process of retrieving information from a data queue or stack. Only the last item stored can be retrieved. A pop is usually accompanied by the decrement of a pointer register.

Porting The translation of code from one computing device to another. If a compiler language is used for source code, porting can usually be accomplished by a compiler written for the second processor.

Prefetch queue A queue of instructions that are fetched in advance of their execution. By prefetching, delays associated with acquiring the instructions from memory can be avoided and performance improved.

Priority A measure of importance assigned to interrupt requests. By using priority measures, the most important interrupt can be serviced first.

Programmable logic devices (PLDs) Chip level arrays of logic devices, including gates and registers that allow final customization by the user. Software is commonly used to aid design. Programming hardware performs the final customization.

Programmable read-only memory (PROM) A ROM memory device that can be programmed by the user and thereafter may only be read.

Program counter A register in a computer datapath that holds the address of the next instruction to be fetched.

Program memory A memory used to store computer instructions.

Programmable system A system that has characteristics determined by the contents of its program memory. A digital computer is one example of a programmable system.

Programmed input/output A method of performing input/output activity completely under program control. Performance is greatly improved by using interrupts and direct memory access techniques.

Propagation delay The delay associated with the propagation of electromagnetic energy in the medium. This energy propagates at a velocity close to the speed of light. Signals originating in one part of a system must propagate to other parts before they can be sensed and used.

Protocol A set of rules for allocating resources and assuring error free data transmission in digital communication systems.

Pseudocode A method of representing programming activities that is a generic form of assembly code. Pseudocode can easily be changed to actual assembly code for a specified digital computer.

Push The process of putting information onto a data queue or computer stack. A push is usually accompanied by an increment of a pointer register.

Queue A digital storage element that can accept data at one end and output data at the other. A first-in, first-out (FIFO) storage device.

Random access memory (RAM) A read/write memory in which data is referenced by its address. The time for accessing data from the memory is constant and independent of its location.

Random replacement algorithm The block to be replaced is chosen at random.

Register addressing A computer addressing mode in which the address field specifies a register which contains one of the operands.

Register file A modular collection of registers accessed for storage and retrieval of information by means of addresses.

Register indirect addressing A computer addressing mode in which the address field specifies a register. The register contains the address of the operand.

Register transfer language (RTL) A mathematical language that describes the movement and processing of information.

Register window A method of referencing a set of registers. Changing the location of the window changes the relative location of the registers referenced by software. Register windows allow a clean set of registers to be quickly accessed for servicing a subroutine or interrupt.

Relative addressing A computer addressing mode similar to indexed addressing. The target address is created by adding the address field provided by the instruction to the program counter. Relative addressing allows the program to be relocated in memory without affecting program functionality.

Replacement algorithm Used to decide which block in memory must be discarded to provide room for a new memory block from a memory unit higher in the memory hierarchy.

Read-only memory (ROM) A memory device that is programmed during the manufacturing process and thereafter may only be read.

Residue number system A number system in which an integer value is represented by its remainder with respect to a set of bases. The bases are usually a set of prime numbers. Arithmetic and multiplication of residues numbers is particularly easy. Conversion between residue representation and a weighted representation, such as binary or decimal, is difficult.

Reverse Polish notation (RPN) A technique for sequencing a set of algebraic operations. This technique is especially useful when operands are stored on an internal stack.

Ring An interconnection structure used in a digital communication network. All nodes are interconnected to a single source node and a single destination node. This forms a single circular interconnection path which connects all nodes.

Ripple processing *See cascade processing.*

RISC An acronym for reduced instruction set computing. RISC processors use simple instruction formats, few memory references, many registers, pipelined instruction processing, and fast control units to provide high performance computation.

Round robin replacement algorithm The block to be replaced is chosen based on a predetermined ordering of blocks. The ordering is normally sequential.

Routing algorithm An algorithm used by a node in a communication network to route messages to the destination node.

RS-232-C Acronym for recommended standard 232. A standard byte-oriented protocol for transmitting asynchronous serial data on communication channels.

Scheduling The assignment of processing tasks to processors.

Segmented memory A method of decomposing memory into blocks that follows the natural boundaries of the code and data. Since it is rarely possible to make all segments the same size, segmented memory is more difficult to use in virtual memory and paged memory systems.

Semi-synchronous signals Signals that are synchronous with the system clock but do not appear in a predictable clock cycle. Handshaking is required.

Serial data channels These channels propagate data in a sequence of single bits.

Set-up time The minimum period of time data must be provided to a memory element to assure reliable acceptance.

Sequencer A simple form of controller in which there are no status or decision variables to control state transitions.

Shift register sequencer A sequencer comprised of a set of flip-flops in the form of a shift register. A single bit is circulated among the flip-flops to generate control signals.

Sign-magnitude A method of representing signed integers in which the most significant bit is designated as a sign bit. If the sign bit is 0, the integer is positive; if it is 1, the integer is negative.

Simulator A computer program that accepts instruction code for a target device. It simulates the actions of the code and displays register contents and other significant aspects of the target computer. The simulation is used for test and debug purposes. Simulation action is much slower than real time.

SIMD structure An acronym for single instruction, multiple data processing. Multiple processors perform the same processing action on several data streams.

Simplex A communication channel that can provide information flow in only one direction.

Single bus structure A processing structure in which a single bus is used to provide two operands from a register file to an ALU and to return processing data to the register file. Two temporary registers are needed.

Slave A participant in an asynchronous data transfer operation. The slave responds to handshake signals initiated by the master.

Sliding window protocol A protocol used in digital communication networks that allows a limited number of message blocks to be transmitted without receiving an acknowledgment. Sliding window protocols improve efficiency when satellite communication, with long transit delays, is used.

Spanning diameter If the minimum number of hops required to pass a message between all network node pairs is determined, the spanning diameter is the largest of these numbers. It represents the maximum number of hops that must be allowed to permit a message to traverse between any two nodes of the network.

Spatial division multiplexing A switching network in which independent messages use different network resources.

Spatial locality References made to an area of data or code memory are commonly followed by additional references to the same area.

Speedup The ratio of the time required to perform a processing activity with a single processor to the time required with multiple processors.

Stack A region of memory addressed by a stack pointer and used to provide last-in first-out storage. Stacks are commonly used to store return address during subroutine calls. They are also used for temporary data storage.

Stack pointer A register in a computer used to maintain the address of the first available storage location on the stack.

State machine A sequential device composed of flip-flops and logic components. The state is the content of the flip-flops. The state changes in a predictable manner with each clock pulse. The next state is a function of the present state and the inputs to the device. The output of the state machine is a function of the current state (Moore state machine) or the current state and input (Mealy state machine). State machines are used as controllers for digital processing structures.

Static branch A branch activity in a computer program in which possible destinations are predictable in advance. This allows possible branch targets to be fetched in advance.

Static random access memory (SRAM) A RAM that uses flip-flop technology to store individual bits. In contrast to a DRAM, an SRAM does not require refreshing to maintain its contents.

Status register A register in a computing structure that stores information relevant to program execution. Common contents are bits indicating that the last processing action produced a value of zero, an overflow carry, or a negative number.

Status signal A signal that originates from a logic decision in the datapath and is used by the controller to sequence the processing algorithm.

Summand Result of the 1-bit multiplication of a bit of the multiplier with a bit of the multiplicand.

Superscalar computer A computer that uses multiple processing units to simultaneously execute multiple instructions.

Supervisory control signals Signals originating outside a computing structure that supervise processing actions. Examples include signals which start, pause, resume, and abort processing action.

Synchronous signals Signals in which transitions occur synchronously with the system clock.

Systolic array A repetitive array of arithmetic and logic components. Systolic arrays can be created using integrated circuit design techniques. The interconnection of array components can be chosen to provide a variety of computational actions.

Tag A portion of the logical address produced by a digital computer when cache and/or virtual memory is used. A comparison of the tag field of the address and the tag field of stored information determines if the desired information is available.

Talker An information source element in a digital information exchange.

Temporal access memory A memory device, such as a tape or disk, in which data elements are located by physical position on the storage media.

Temporal locality Memory fetches that occur closely in time often occur in close proximity in the memory space. When there is a "miss" in a cache or paged memory, a block of information is accessed, since the next memory address is likely to be close to the first one and therefore in the acquired block.

Thrashing This is an undesirable action that occurs when a cache or virtual memory spends a large amount of time and effort ineffectively swapping memory blocks.

Three bus structure A datapath structure in which two buses are allocated to provide data from source registers to an ALU. A third bus returns data to the destination register for storage.

Throughput A measure of the total amount of information that is transferred between hosts of a communication network. Repeated messages, due to errors, are not considered part of the throughput.

Time-out The predetermined time limit allowed for an asynchronous event. If the event does not occur within the time-out period an alternative processing action is initiated.

Token A small control message which allocates permission to use a communication structure to the recipient.

Top of the stack The last data item entered onto the stack.

Topology The interconnection mechanism of network components. Common topologies include buses, rings, trees, and meshes.

Trial and error processing A processing technique in which a sequence of yes/no decisions is used to determine the value of a computed result.

Tri-state buffers An electronic device for connecting logic signals to a bus structure. When the control signal to the buffer is not active, the buffer presents a high impedance output to the bus. This prevents the buffer from interfering with other signals on the bus.

Tuning Optimizing parallel processing activities by modifying scheduling rules.

Turing machine A conceptual computational machine created by Alan Turing. The Turing machine has an endless tape and a finite state machine. It has the ability to read the tape, move the tape forward, move the tape backward, and to write a symbol on the tape. Turing proved that the device could perform any computable task in a finite length of time.

Two bus structure A structure for a computing device in which operands are provided to an ALU via one bus and a temporary storage register. Results are returned to the registers using a second bus.

Two's complement A signed number system in which positive integers have a 0 in the most significant bit position. Negative integers are formed by taking the logical complement of all bits and adding 1. Negative numbers have a 1 in the most significant bit position. The addition of a positive number with its negative equivalent results in a value of 0.

Vectoring In the context of microprogramming, vectoring is the process of locating the section in microcode associated with the execute cycle for an instruction.

Vector processor A parallel processing device specially structured to perform vector operations such as addition.

von Neumann architecture An architecture characterized by a single memory unit that stores both instructions and data.

Wait state An extra clock cycle added to a computer machine cycle to accommodate a memory device with a long access time.

Wallace tree adder A very high-performance arithmetic adder. It uses conventional binary adders but combines operands as rapidly as possible in a multi-level structure. Carries do not ripple but are saved for insertion at lower levels of the structure.

Word A grouping of bits moved and processed as a unit in a computing structure.

Index